Magnetochemistry

Materials and Applications

Edited by

**Inamuddin[1,2,3], Rajender Boddula[4]
and Abdullah M. Asiri[1,2]**

[1]Chemistry Department, Faculty of Science, King Abdulaziz University, Jeddah 21589, Saudi Arabia

[2]Centre of Excellence for Advanced Materials Research, King Abdulaziz University, Jeddah 21589, Saudi Arabia

[3]Department of Applied Chemistry, Faculty of Engineering and Technology, Aligarh Muslim University, Aligarh-202 002, India

[4]CAS Key Laboratory of Nanosystem and Hierarchical Fabrication, National Center for Nanoscience and Technology, Beijing 100190, PR China

Published by **Materials Research Forum LLC**
Millersville, PA 17551, USA

Published as part of the book series
Materials Research Foundations
Volume 66 (2020)
ISSN 2471-8890 (Print)
ISSN 2471-8904 (Online)

Print ISBN 978-1-64490-060-4
eBook ISBN 978-1-64490-061-1

This book contains information obtained from authentic and highly regarded sources. Reasonable efforts have been made to publish reliable data and information, but the author and publisher cannot assume responsibility for the validity of all materials or the consequences of their use. The authors and publishers have attempted to trace the copyright holders of all material reproduced in this publication and apologize to copyright holders if permission to publish in this form has not been obtained. If any copyright material has not been acknowledged please write and let us know so we may rectify this in any future reprints.

Distributed worldwide by

Materials Research Forum LLC
105 Springdale Lane
Millersville, PA 17551
USA
https://www.mrforum.com

Manufactured in the United States of America
10 9 8 7 6 5 4 3 2 1

Table of Contents

Preface

Magnetochemistry has gained considerable attention in the last few decades due to the exceptional physical, chemical, and magnetic properties of these materials. Magnetic materials generate a great deal of attention for researchers working in the area of material science because these magnetic materials are used in various applications starting from daily life to the industry.

The book "Magnetochemistry: Materials and Applications" aims to provide in-depth literature review on magnetic nanomaterials and their applications in allied fields. This book explores magnetic properties, fundamental concepts, and applications of magnetic nanomaterials. It also discusses the engineering strategies and current applications of magnetic nanomaterials in analytical chemistry, biomedical science, spintronic, electrochemistry, energy storage and conversion, membranes, and fuel cells. The chapters are contributed by various international experts from industry and academia to ensure the highest quality information from multiple perspectives.

This book covers the entire spectrum of magnetochemistry and its recent applications from a basic scientific background to the latest advances in materials, to challenges and opportunities for practical applications. Therefore, all sections of this book emphasize recent advances, current challenges, and state-of-the-art studies through detailed reviews. Finally, academics, researchers, scientists, engineers, and students will be able to resolve current industry issues by referring to this book.

Inamuddin[1,2,3], Rajender Boddula[4] and Abdullah M. Asiri[1,2]

[1]Chemistry Department, Faculty of Science, King Abdulaziz University, Jeddah 21589, Saudi Arabia

[2]Centre of Excellence for Advanced Materials Research, King Abdulaziz University, Jeddah 21589, Saudi Arabia

[3]Department of Applied Chemistry, Faculty of Engineering and Technology, Aligarh Muslim University, Aligarh-202 002, India

[4]CAS Key Laboratory of Nanosystem and Hierarchical Fabrication, National Center for Nanoscience and Technology, Beijing 100190, PR China

Chapter 1

Magnetic Nanomaterials for Bio-Sensors based on SERS Effect

Cixue Xu[a], Wenxian Wei[b]*, Chengyin Wang[a,b]*

College of Chemistry and Chemical Engineering, Jiangsu Key Laboratory of Environmental Engineering and Monitoring, Yangzhou University, 180 Si-Wang-Ting Road, Yangzhou, 225002, P.R. China

Testing center, Yangzhou University, Yangzhou, Jiangsu 225002, P.R. China

wangcy@yzu.edu.cn

Abstract

The discovery of surface-enhanced Raman spectroscopy technology especially the magnetic nanoparticle-based SERS technology has promoted the detection of trace amounts of biomolecules. The magnetic action of the nanoparticles causes the analytes to aggregate together, which promotes the detection of small amounts of biomass. Compared to other detection systems, analytical results will be more easily interfered in the complex biomass systems. Fortunately, magnetic nanoparticles can separate the detectors by external magnetic fields to avoid interference from impurities. Magnetic nanoparticles are often used in noble metal recombination to greatly enhance SERS by synergy. Magnetic nanoparticle/precious metal composites combine the advantages of both to improve the SERS effect. Magnetic nanoparticle SERS substrates can detect specific biomass (such as bacteria, antibodies, cells, RNA, etc.) by modifying definite aptamers. In this review, the research progress of SERS and magnetic nanoparticle-based SERS detection mechanism are described in detail. At the same time, different structure magnetic nanoparticle including Fe, Ni, $CoFe_2O_4$ used as SERS substrates are introduced and compared in detail. Moreover, the magnetic nanoparticle-based SERS detection types of detected biomass are also discussed.

Keywords

Magnetic Nanoparticles, SERS, Biosensors, Nanocomposites, Applications

Contents

1.   Introduction

The detections of biomolecule, bacteria or cells play an important role in food safety, clinical diagnosis, environmental monitoring, etc. [1, 2]. Due to the urgent need to detect biomass and the advancement of science and technology, the improvement of detection technology, such as fluorescence, electrochemical detection and surface-enhanced Raman scattering (SERS), has been promoted [3, 4]. Advantages of the SERS technology such as simple pre-processing, non-invasive detection, high sensitivity, etc. [5-9]. are conducive to its application in biosensors.

Among the SERS sensors, the magnetic interaction facilitates separation [10, 11], concentration [12], and immobilization [13, 14] of the analyte, which promotes the application of magnetic nanoparticle-based SERS as a biosensor. Of course, this magnetic interaction also has its disadvantages, such as poor dispersion in solution and easy agglomeration. The method of complexing noble metals (such as gold, silver) with magnetic nanoparticles can solve the above problem [15]. Magnetic nanoparticle/precious metal composites combine the advantages of both to improve SERS activity. The loading effect of noble metals is affected by the structure of magnetic nanoparticles. For example, the three-dimensional network Fe_3O_4 exhibits better SERS enhancement than the Fe_3O_4 microspheres in favor of the noble metal loading [16, 17]. The enhancement effect of SERS depends on the morphology of the noble metal and its gap [18-22]. Sharp precious metals and small particle gaps promote the increase in SERS.

Magnetic nanoparticles/special precious metal complexes facilitate the application of SERS in biosensors because magnetic nanoparticle-based SERS sensors can improve the sensitivity, selectivity and stability of biomarker detection [6, 23, 24]. Biomass can be detected by modifying an aptamer that strongly interacts with the analyte and specific binding by SERS substrate, which shows an excellent detection effect. For example, the nickel-chelated N-nitrilo-triacetic acid (Ni-NTA) was modified on a SERS substrate to detect the surviving protein dimer, which showed excellent detection performance [25]. For example, $Fe_3O_4@Ag$ is composited with Raman tags modified DNA to detect RNA in cancer cells, and the duplex-specific nuclease (DSN) is used to repeatedly increase the detection signal, which shows a low detection limit (0.3 fM) [26].

In this review, the progress of SERS research is elaborated. Different types and structures of magnetic nanoparticle-based SERS substrates have been described and compared in detail, and the magnetic nanoparticle-based SERS detection mechanism has been described in detail from different types of detected biomass.

2.   Surface enhanced Raman spectroscopy

Surface-enhanced Raman spectroscopy is an ultra-sensitive fingerprint spectroscopy technology that enables trace non-destructive testing and development of portable devices [27]. The detection uses a suitable frequency of excitation light to illuminate a metal surface which has a certain nanostructure to excite its surface plasmon resonance, and the metal surface is amplified by local electric field amplification of the resonance interaction to realize detection of a low concentration analyte. This technology not only has the characteristics of high resolution and rapid detection of traditional Raman spectroscopy, but also improves the sensitivity of Raman spectroscopy, thus achieving wider application in the fields of electrochemistry, biomedicine, environmental science and other fields [28].

2.1   Mechanism of surface enhanced Raman scattering

Raman scattering was first discovered by the Indian physicist Raman in the 20th century. The principle of explaining the Raman scattering phenomenon from quantum theory is that when the light wave is irradiated onto the material, the photons in the incident light collide with the material molecules, mainly divided into elastic and inelastic collisions. When an elastic collision occurs, there is no energy exchange process between the photon and the molecule, and the photon frequency is fixed but the direction of motion changes. However, when an inelastic collision occurs, photons will interchange energy with molecules, along with former's change of direction and frequency. This inelastic scattering is defined as Raman scattering [29]. In 1974, Fleischmann studied the Raman spectrum of pyridine molecules on a rough silver electrode substrate, founding that the Raman signal of the pyridine molecule is greatly enhanced, and the conclusion that the Raman signal of the analyte can be enhanced by some means is obtained. However, the enhancement mechanism was not further explored. Until 1977, Jeanmaire et al. found that the rough silver electrode was the main reason for enhancing the Raman signal of pyridine molecule, making a breakthrough in Raman spectroscopy. The phenomenon is called surface-enhanced Raman scattering (SERS). The acquisition of SERS requires the adsorption of molecules on a special metal surface, and the development of SERS can greatly increase the Raman signal of the probe by adsorbing the probe molecules to the surface of a specially prepared substrate, sol or blend. [30].

Since SERS has been found, the enhancement mechanism of SERS is being discussed. Although the application of SERS is widely used in modern detection, its complex enhancement mechanism has not yet been fully understood. At present, there are two types of enhancement mechanisms which are recognized by physical enhancement and chemical enhancement respectively while the acceptance of general physical enhancement mechanisms is higher.

The physical enhancement, which is the electromagnetic field enhancement mechanism, is mainly affected by the local electric field enhancement caused by the plasmon resonance of the substrate surface. When a certain frequency of incident light is irradiated onto the surface of the nano-scale rough substrate, the plasma of the substrate surface and the incident light will have a coupling effect to generate a near-field electromagnetic wave, if it has oscillation frequencies, resonance phenomenon occurs. The resonance phenomenon greatly enhances the local electric field enhancement ability of the substrate surface, causing the enhancement of Raman scattering [31]. As the metal nanoparticles of the SERS substrate, different particle sizes and their respective special properties have a great influence on the enhancement effect of SERS. The resonance intensity produced by the coupling effect is also related to the distance between the nanoparticles and the size of the nanoparticles themselves [32]. Different from physical enhancement, chemical enhancement which is charge transfer mechanism, is mainly affected by the change of polarizability between the substrate and the adsorbed molecules, and the distance between the adsorbed molecules and the substrate is required to be close enough [28]. Under the influence of the optical electric field, the SERS substrate on the rough surface and its adsorbed probe molecules are chemically bonded to change the polarizability of the system to enhance the Raman signal. The necessary condition for chemical enhancement is that a chemical reaction occurs between the SERS substrate and the adsorbed molecule to form a chemical bond, and the reinforcing effect mainly depends on factors such as the bonding mode of the adsorbed molecule on the surface of the SERS base metal. The currently accepted explanation is that the SERS effect is the result of a combination of the two, in which the plasmon resonance of metal nanostructures can be theoretically enhanced 10^{14} times [33].

Compared with ordinary Raman spectroscopy, surface-enhanced Raman spectroscopy has higher sensitivity, mainly depending on the properties of SERS substrate. Common SERS-reinforced substrates are noble metals, transition metals, semiconductor quantum dots and graphene. Among them, noble metals such as Au, Ag, and Pt have strong electromagnetic enhancement, while semiconductors, quantum dots, and graphene have only chemical enhancement. Although the SERS enhancement effect of some magnetic transition metals such as Fe, Co, Ni, etc. is weaker than that of noble metals, in order to achieve better enhancement effect, the surface roughness of the nano-substrate can be synthesized by synthesizing composite nanoparticles with noble metals. Therefore, control of morphology and distribution of SERS substrates reaches the enhanced effect of SERS [30].

2.2   Development of SERS active substrates

The use of SERS technology to detect analytes is affected by many factors, and the SERS substrate itself plays a key role in enhancing the activity. The SERS active substrate needs to have a strong reinforcing effect, and should also have the characteristics of easy preparation and easy storage. The preparation process of the active substrate basically determines the size, shape and spatial morphology of the surface microstructure, which basically influence the substrate enhancement effect of the adsorbate. In recent years, a series of research work on nanostructured SERS active substrates have been carried out, and various high-sensitivity SERS active substrates have been developed using noble metal nanostructures in combination or template-assisted approach [34]. Hu et al. designed a peroxidase-simulating nanozyme that oxidizes Raman inactive reporter white blood green oxide to active malachite green (MG) with hydrogen peroxide.

Simultaneously as a SERS substrate to enhance the Raman signal of the generated MG [35]. They assembled glucose oxidase (GOx) and lactate oxidase (LOx) onto AuNP@MIL-101 to form an integrated nanozyme for in fast detection of lactic acid and glucose by SERS. Research work on nanostructures as SERS active substrates will be well applied to environmental monitoring and clinical samples.

Hu et al. realized the high adsorption capacity of gold nanoparticle (AuNP)-embedded metal-organic framework and found that SERS substrate has high sensitivity to several different target analytes [36]. The method in situ grows AuNP by solution impregnation strategy and encapsulates it in the main matrix of MIL-101 (Fig.1a). It can be found from TEM and SEM images that AuNP is inside MIL101 crystal and the composite has good stability. It can be seen from Fig.1b that the octahedral crystal structure is not destroyed, and AuNP is uniformly distributed and embedded in MIL-101. The synthesized AuNPs / MIL-101 nanocomposites combine the local surface plasmon resonance properties of gold nanoparticles with the high adsorption capacity of metal-organic frameworks. The analyte near the electromagnetic field is effectively pre-concentrated, making it a highly sensitive SERS matrix. Fig.1c shows that the method exhibits a linear relationship of 0.9950 between p-phenylenediamine 1.0 and 100.0 $ng \cdot mL^{-1}$, and Fig.1d shows the use of p-phenylenediamine for the detection of sewage. The SERS substrate can also be used for bioassay. It can be seen that there is a good linear relationship between alpha-fetoprotein 1.0-130.0 $ng \cdot mL^{-1}$ in Fig.1e with a correlation coefficient of 0.9938. Fig.1f shows the SERS of AFP applied to serum samples. ELISA. The composites they synthesized retained nanovoids, and AuNPs/MIL-101 readily enriched the analyte and tested for excellent SERS activity by Raman signal of the analyte in aqueous solution. A highly sensitive SERS active matrix can be designed using a metal-organic framework.

Considering the practical application, the SERS substrate should have good stability and reusable properties while enhancing the SERS effect. With the development of nanotechnology, the performance of magnetic nanomaterials has received more and more attention. Many researchers have introduced magnetic nanomaterials into the construction of SERS active substrates. On one hand the composite magnetic nanomaterials can be quickly separated and aggregated under the action of external magnetic field, which effectively simplifies the detection process of complex analytes, on other hand, the magnetic materials also exhibit excellent SERS activity because of the introducing of noble metals.

Therefore, the prepared magnetic nanoparticles combined with noble metals are magnetic as well as have excellent SERS performance, which caused widely attention by researchers. For example, Baniukevic et al. used surface-enhanced Raman scattering (SERS) as a detection method for detecting bovine leukemia virus antigen gp51. The antibody modified magnetic gold nanoparticles (MNP-Au) in a targeted or random manner to detect bovine leukemia virus antigen gp51. Antibody immobilization and antigen capture efficiency were recorded by high performance liquid chromatography HPLC [37]. This immunoassay has been successfully applied to the rapid, reliable and selective detection of gp51 in milk actual samples. The LOD and LOQ of this immunoassay were 0.95 mg·mL^{-1} and 3.14 mg·mL^{-1}.

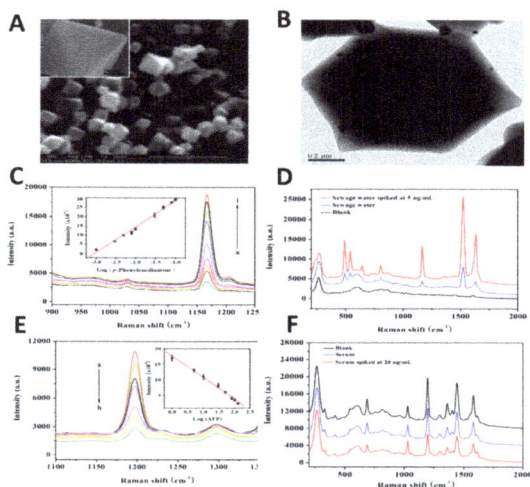

Figure 1.   (A) A SEM image of composite AuNPs/MIL-101;(B) A TEM images of smooth surface of the composite;(C) Relationship of the SERS intensity and lgC (p-diphenylamine);(D) Raman detection of p-diphenylamine in actual samples;(E) SERS intensity varies with AFP concentration;(F) AFP detection application in actual samples[36].

3.   Progress of magnetic SERS substrate research

Magnetic nanomaterials are magnetic materials with a particle size up to the nanometer scale. They not only have magnetic guiding properties, superparamagnetic properties, but also have the advantages of nanomaterials. Modified magnetic nanomaterials will have better water solubility, biocompatibility and dispersion stability [38, 39]. Precious metal nanoparticles, especially gold and silver nanoparticles, are often used in conjunction with magnetic nanoparticles for sensor development, and composites as magnetic SERS substrates. Magnetic separation can be used to achieve rapid separation and enrichment of analytes, while noble metal nanoparticles can improve the sensitivity of biomarker detection.

3.1   Iron oxide based SERS substrate

The use of magnetic nanomaterials for sample separation and preconcentration, i.e. magnetically assisted surface-enhanced Raman spectroscopy (MA-SERS), can further enhance the application potential of SERS. Due to its superparamagnetic and simple preparation, iron oxide is most widely used in composite magnetic SERS substrates for surface adsorption of targets and enhanced Raman signals [40]. Balzerova et al. synthesized ferromagnetic $Fe_3O_4@Ag@streptavidin@anti-IgG$, which effectively separates the target from the complex matrix by applying external magnetic force and then directly measures it using SERS. The use of anti-IgG on the surface of silver nanoparticles and inactivation by ethylamine to achieve high selectivity enables determination of IgG in whole blood samples at 600 fg·mL^{-1} [41]. Yang et al. developed a magnetic assisted SERS scheme to detect trace adenosine in urine samples from lung cancer patients and healthy people by synthesizing stable $Fe_3O_4/Au/Ag$ nanocomposites. The sample is tested within 20 min with sensitivity as low as 1×10^{-10} M [42].

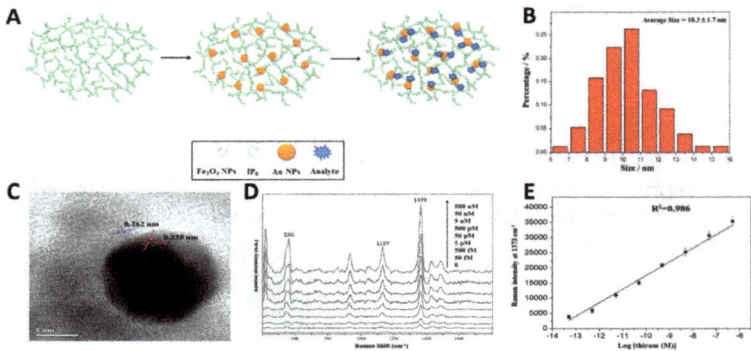

Figure 2.   (A) Process of preparing a magnetic network structure; (B) Distribution of different sizes of Au NPs; (C) A HR-TEM image of composite magnetic nanomaterials; (D) Raman detection with different concentrations of thiram; (E) Relationship between SERS intensity and lgC (thiram) [17].

For example, Yang et al. developed a nano-structure of AuNPs dot-like magnetic network in aqueous solution. The synthesis process is shown in Fig.2a. Fe_3O_4NP forms a magnetic network nanostructure under the action of a stabilizer and a crosslinking agent. Utilizing the specific surface area and sensitivity of Au-MNN, we can achieve the analysis of fly-molar levels to identify vegetable pesticide residues [17]. Fig.2b shows that the average size of the Au NP synthesized in this work was 10.3 ± 1.7 nm. The HR-TEM characterization of the dot-like magnetic network nanostructures is shown in Fig.2c, indicating that the structure is not simple under the action of the cross-linking agent. It can be seen from Fig.2d that the Raman intensity (1373 cm^{-1}) of thiram concentration is 50 fM to 500 nM, and the logarithm of the concentration of thiram has a good linear relationship with the Raman intensity, and R^2=0.986 (Fig.2e) presents a good linear relationship. They use IP6 as a stabilizer and bridging agent to capture Fe_3O_4 NP into magnetic network nanostructures and easily embellish Au NP, which uses the high sensitivity of Au-MNN to amplify Raman signals. Successful combination with portable Raman spectroscopy provides an idea for intelligent nanosensor research.

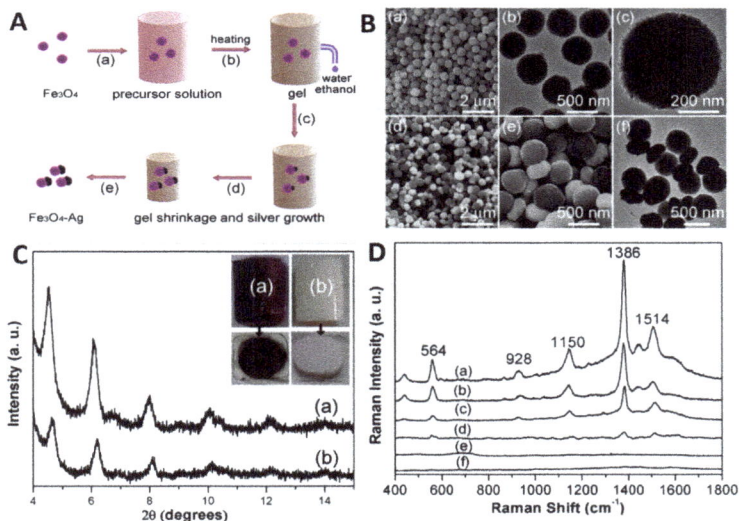

Figure 3.   (A) Schematic diagram of composite microsphere growth process;(B)SEM images and TEM images of Fe_3O_4 composite microspheres;(C)XRD comparison of composite microsphere gel products;(D) SERS detection of different concentrations of pesticides [43].

Zhang et al. synthesized dumbbell-shaped Fe_3O_4-Ag composite microspheres in a gel system by simple solvothermal method. This schematic diagram of composite microsphere growth process can immobilize Fe_3O_4 microspheres in the gel to avoid their aggregation, and the silver ions in the gel can be gradually released. And the process of diffusing the closest Fe_3O_4 microspheres to produce a dumbbell-like Fe_3O_4-Ag structure is shown in Fig.3a. Fig.3b shows the morphology change of the composite microspheres during the synthesis. As shown in Fig.3b(a) and (b), the microspheres are composed of a large number of tiny Fe_3O_4 nanoparticles, as shown in Fig.3b(c). After the completion of the gel reaction, the morphology of the sample characterized by SEM, as shown in Fig.3b(d), the solvent-formed dumbbell-like structure can be clearly presented in Fig.3b(e) and (f), and this structure is used as selective detection of surface-terminated oleic acid chain thiram with rapid magnetic separation of SERS substrate [45]. In the study of the precursor solution, the structure was studied by low-angle X-ray diffraction (XRD), and the presence or absence of Fe_3O_4 gel slice comparison in Fig. 3c showed that the presence of Fe_3O_4 magnetic microspheres had little effect on the gel system. The gel limits the limitation of magnetic microspheres in the process of synthesizing dumbbell-shaped Fe_3O_4-Ag composite microspheres, and also limits the release and reduction of Ag^+ ions. In this work, synthetic dumbbell-like Fe_3O_4-Ag is used. The composite microspheres were used as SERS substrates for the detection of thiram in water. As shown in Fig.3d (d), the low level of 10^{-7} M also clearly observed all the characteristic peaks of the Fumex dual band [43].

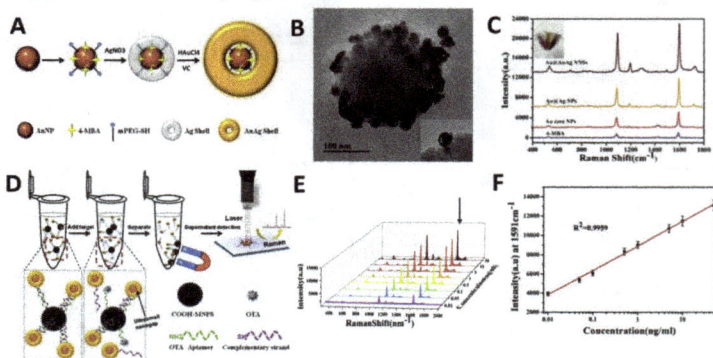

Figure 4.   (A) Synthesis of Au@AuAg NNSs (B)A TEM image of Au@AuAg NNSs;(C) Comparison of Raman Strength of Different Nanoparticles Synthesized;(D) Schematic diagram of OTA SERS detection principle (E) Raman intensity with different concentrations of OTA (F) Curve of SERS intensity with different concentrations of OTA [46].

Compared with the above simple synthetic magnetic nanostructures, the core-shell nanocarrier containers can meet the accuracy of detection of complex substrates, requiring clean substrates and low background signals. Shao et al designed an ultra-sensitive SERS aptamer sensor for ochratoxin A (OTA) detection [46]. Fig.4a clearly shows the construction principle of the sensor. The TEM morphology of the synthesized Au@AuAgNNS nanostructure is shown in Fig.4b. SERS activity differs among 3 different nanoparticles. Fig.4c shows that Au @ Au-Ag NNS has the most obvious SERS enhancement. Au(nuclear)@Au-Ag(shell) nanocarrier nanostructures (Au@AuAgNNS) were coupled to Fe_3O_4MNP as shown in Fig.4d. Au@AuAgNNSs can exhibit better Raman enhancement on the basis of pure metal nanoparticles (4-MBA) through ultra-small nano-gap, improve the sensitivity of the sensor and Fe_3O_4MNP can be green, economical and simple f captured the target rom solution (Fig.4e). It can be clearly seen that the logarithm of the OTA concentration at 1591 cm^{-1} has a good correlation coefficient with the Raman intensity (R^2=0.9959). The linear range is 0.01-50 ng mL^{-1} (Fig.4F) Detection limit (LOD) reaches 0.004 ng·mL^{-1}.

Cai et al. designed a core/shell structure Fe_3O_4/C/Au synthesized with different sizes of AuNP for in situ SERS monitoring of p-nitrothiophenol (pNTP) and p-aminothiophenol (p-ATP). Catalytic reaction and catalytic efficiency [47]. Zhang et al. successfully synthesized a novel multifunctional Fe_3O_4@TiO_2@Au triple-core magnetic microsphere with unlike sizes of gold nanoparticles [48]. The magnetic microspheres they prepared have excellent catalytic activity and are self-cleaning, exhibiting excellent SERS activity, prominent stability and high repeatability In order to maximize the SERS activity, different particle sizes and distributions of Au nanoparticles were selected to achieve optimal conditions. Under the optimal conditions, the detection limit of R6G can reach 10^{-11} M, and Fe_3O_4@TiO_2@Au has obvious enhancement effect on SERS. Core-shell magnetic microspheres prepared by composite materials are also widely used in other fields, and their advantages are well developed in the field of electromagnetism and biomedical research.

SERS detection needs to achieve fast and stable conditions; the development of new substrates based on magnetically assisted SERS has become a hot research direction. The sensitivity and reproducibility of SERS spectra are largely determined by specific nanostructures, and the preparation of special metal nanostructures is critical [49, 50]. The morphology of the nanomaterials used in the SERS substrate, size and particle size will have an impact on its properties. Iron oxide as a widely used magnetic nanoparticle has mature synthetic techniques to adjust its properties [44]. The biocompatibility of the platform is enhanced by controlling the aggregation and dispersion of Fe_3O_4. Functionalized magnetic composite nanoparticles can also be used as a concentration tool for SERS substrates, improving the stability of biosensors for recycling [29, 44, 51]. The network structure synthesized by the aqueous solution has the advantages of microstructure, the synthesis method is simple and easy to control, and the product particle size is uniform. The core-shell composite nanoparticle uses the iron core as a reducing agent, and can be synthesized in situ without additional reducing agent. An iron-based magnetic SERS substrate with a low background signal.

3.2   Nickel-based SERS substrate

Nickel nanomaterials have unique electrical and magnetic physical properties and are hotspots in magnetic storage and nanoelectronic devices. Nickel nanomaterials have large specific surface area and high activity. They can process their processes into magnetic nanomaterials with special morphological properties, which have great application and research value [52, 53]. In order to expand the limitations of SERS substrates, it is found that the SERS active substrate can extend the application of Raman spectroscopy in the field of nanomaterials by using the charge transfer mechanism. Ni ions can generate defect level due to their special properties and doping with noble metals or semiconductor materials. Thereby promoting charge transfer, to some extent enhance the SERS signal [54]. More and more research has been done on the application of nickel nanomaterials to precious metals or semiconductor materials to prepare nanocomposites for SERS substrates [55].

In recent years, the rapidly developing SERS detection method has been widely used in the fields of life sciences, environmental testing, etc., and this application requires stability while high sensitivity detection is also important. The synthetic active substrate is the basis for a sensitive and stable SERS signal.

For example, Liu et al. combine magnetic induction with recrystallization to synthesize a Ni@Ag special structure which has ultra-tip structure with a certain internal clearance using a simple solution-based method (Fig. 5a) [43]. SEM is characterized as Fig.5b. The nanostructured tips they synthesized are in the size of 5 nm. Due to the magnetic induction assembly, there are huge amounts of contact points between particles of a special urchin-like morphology. These advantages can be used for ultra-sensitive SERS probe for biochemical analysis. XRD Characterization (Fig.5c) shows that the core-shell structure formed by Ag and Ni is not alloyed. Using a large proportion of Ni can reduce expensive precious metal as shown in Fig.5d, which is less costly. Nanostructure based on Ni @ Ag can be applied to SERS detection of R6G.Ni@Ag special nanostructure of the present invention can achieve significantly higher magnetic field enhancement as shown in Fig.5e. In addition, Ni@Ag special nanostructure are able to collect magnetic fields after use and are recycled due to the magnetization of the nickel core.

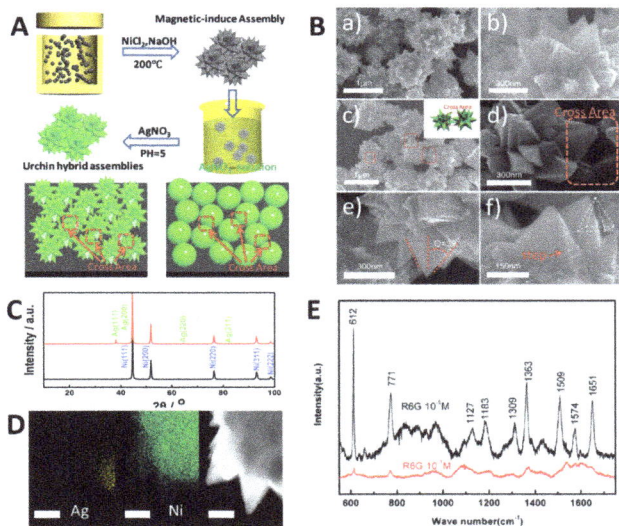

Figure 5.   (A) Schematic diagram of Ni @Ag nanostructure manufacturing principle;(B)A SEM image of urchin-like nanostructure;(C) XRD pattern comparison of composites (D) Elemental mapping;(E) Based on composite Ni @ Ag for R6G Raman detection [43].

Ondrej Petrus et al. successfully constructed Ni/Ag nanocavities by combining colloidal lithography with electrodeposition [56]. The thickness of the Ni/Ag nanocavity and the amount of Ag nanoparticles on the Ni nanocavity substrate were varied by controlling the two electrochemical deposition times. The best SERS signal enhancement was observed at 500 nm Ag nanocavity. The detection limit of Rhodamine 6G aqueous solution was 1×10^{-12} mol·dm^{-3}, and the enhancement factor was 1.078×10^{10}.

At the current stage in the manufacture of new SERS substrates, porous metal structures are well suited for improving the sensitivity of analytes with good reproducibility and stability. Nickel foam is a sponge-like porous metal nickel material. Although it does not have SERS activity itself, the unique three-dimensional network of foamed nickel can be used as a base for synthesizing SERS. The work of Zhao et al. demonstrated a three-dimensional superhydrophobic SERS substrate for detection of polycyclic aromatic hydrocarbons (PAH) [57]. They deposit Au nanoparticles on the surface of nickel foam, the prepared nanostructures are superhydrophobic, and hydrophobic interactions can be used to detect PAHs. In the SERS detection, the detection limit can reach 10^{-8}M, and the enhancement factor can reach 1.2×10^4.

3.3   Cobalt- ferrite based SERS substrate

$CoFe_2O_4$ is a common magnetic nanomaterial with spinel type. $CoFe_2O_4$ is a kind of magnetic material with better magnetic properties in nanomaterials. It is more cost-effective than other magnetic nanomaterials [58, 59]. At present, $CoFe_2O_4$ is used as a composite nanomaterial for SERS substrate. In the fields of drug separation, environmental testing, etc. [60, 61]. He et al. have designed a SERS sensor for NT-proBNP ultrasensitive detection. The CoFe2O4 @ AuNPs composite nanospheres prepared by them have good uniformity and can be used for immobilized antibody, and Raman probe [58]. In order to improve the uniformity of the SERS substrate, functionalized magnetic nanospheres ($CoFe_2O_4$@AuNPs) were prepared to assemble the primary antibody, and the ultra-sensitive immunoassay of NT-proBNP was carried out by the principle of double-anti-sandwich method. This design produces a strong SERS signal with a detection range of $1fg \cdot mL^{-1}$ to $1ng \cdot mL^{-1}$ with a detection limit of $0.75fg \ mL^{-1}$. Guo et al. developed a dual colorimetric and SERS detection of Hg^{2+}. They use the microwave method to assist in the synthesis of composite nanomaterials, and they change color after oxidation with TMB. Their affinity with Hg^{2+} can be applied to heavy metal detection. This method can detect Hg^{2+} as low as 0.67 nM and is suitable for sensitive assays under real environmental conditions.

4.   Application of SERS in biosensors

Biosensors are very helpful for environmental testing, drug testing and food safety [4,5, 62]. SERS applications are feasible in biosensors because of its high molecular specificity and high sensitivity [15,27], albeit a complex biomass system. Magnetic nanoparticle-based SERS as a biosensor can detect trace amounts of biomass because of its magnetic field focusing, enrichment of targeted analytes and abundant interparticle hotspots created [63]. Even in complex biological systems, the presence of magnetic nanoparticles can help separate the analytes and ensure proper operation of the assay [63]. Biosensors can be divided into immunosensors, microbial sensors, nucleic acid sensors, cell sensor, and other biomolecular sensors due to differences in analytes and mechanisms.

4.1   Immunosensors

Immunosensors rely on the specific interaction between antigen and antibody to achieve specific recognition. Immunosensors generally use antibody sandwich methods to detect antigens or specific substances. The first antibody acts as a "detection substrate", the second antibody acts as a "marker carrier", and the antigen is sandwiched between the two antibodies, and the content of the antigen directly affects the content of the second antibody and reflects the content of the label. The amount of the marker reflects the amount of the antigen (analyte) [64] by a special detection method. Common methods for detecting immune proteins, such as colorimetric and fluorescent methods, can be greatly affected, such as complex biomass interference in actual samples. Magnetic-based SERS effectively solves this problem [3].

The actual sample contains a lot of biomass, which has a great interference effect on sensor detection. The antigen to be detected is isolated by immunospecific binding prior to detection and further tested [7]. For example, Wang et al. specifically binds to the antigen to be detected by Fe_3O_4/polyethyleneimine (PEI)/Ag-antibody, and the purpose of separation is achieved by the action of an external magnetic field, which can eliminate the influence of other interference. These proteins do not exhibit strong signals in the Raman spectrum, and the exposed antigen on the Fe_3O_4/polyethyleneimine (PEI)/Ag-antibody surface continues to bind to the antibody. When AuNRs-5,5'-Dithiobis (2-nitrobenzoic acid) (DNTB)-antibody binds to specific surface antigens, magnetic Fe_3O_4 can help aggregate, separate and immobilize the complex (Fig.6a) [7]. Among them, DTNB is a marker of SERS, which can lead to an evident increase in Raman intensity (Fig.6b). The detection limit of human Ig G by specific immunization of Fe_3O_4/PEI/Ag complex reached 10^{-14} $g \cdot mL^{-1}$ (Fig.6c) [7]. Biomass can also be purified by other methods, such as Zou et al. using microfluidics to pre-treat samples [65].

The immune protein has a weak signal in the Raman spectrum. SERS is used to improve the detection, especially the magnetically assisted SERS [66-69]. The magnetically assisted SERS-based immunosensor utilizes the double antibody sandwich method. The magnetic nanoparticles are all carriers of the first antibody, and due to their magnetic properties, it is possible to achieve separation, purification, and aggregation of the analyte. Moreover, the commonly used Fe_3O_4 has good biocompatibility and is easy to combine with biomass [70]. The second antibody is mainly composed of noble metal Ag or Au, which can improve detection sensitivity. The SERS marker is bound by the second antibody, further increasing the Raman signal. A common magnetically assisted SERS-based immunosensor is the simultaneous binding of two antibodies to an antigen-containing solution, followed by magnetic separation to detect Raman intensity. For example, He et al. used Antigen to link $MOF@Au@TB@Ab_2$ and

15

CoFe$_2$O$_4$@Au@Ab1@BSA and detect their Raman signals (Fig.6d). As can be seen from Fig.6e, at 1420 cm^{-1}, as the antibody (NT-proBNP) concentration (C) increases, the Raman signal increases. Moreover, the logarithm of antibody (NT-proBNP) concentration showed a good linear relationship with Raman intensity, and R^2=0.996 (Fig.6f) [71].

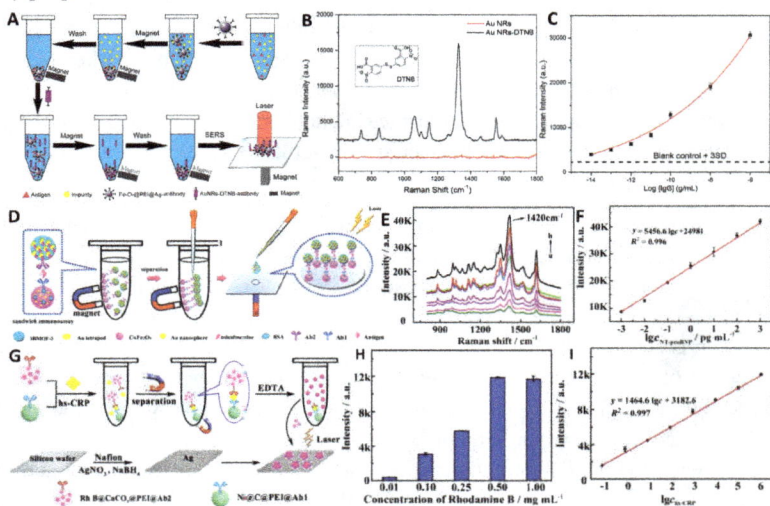

Figure 6.   (A) The SERS immune mechanism of Fe$_3$O$_4$@PEI@Ag (B) Raman intensity of Au NRs/DTNB (C) Dose–response curve at different human IgG concentrations [65]. (D) Schematic Diagram of the SERS-Based Immunosensor (E) Raman intensity with the increasing concentration of NT-proBNP (pg mL^{-1}) [71]. (G) SERS detection mechanism for hs-CR. (H) Raman intensity with different rhodamine B concentrations. (I) Relationship of the Raman intensity and lg c(hs-CRP) [72].

In addition to detecting the Raman signal of the specific structure of antibody-antigen-antibody, some researchers have also measured the antigen content by detecting the Raman label on the second antibody. For example, Wang et al. coated rhodamine B with CaCO$_3$ and linked the antibody to form a special "second antibody" with magnetic Ni@C@PEI@Ab$_1$ as a special "primary antibody". The above two antibodies were pulled by the hypersensitive C-reactive protein (hs-CRP) and magnetically separated to obtain a sandwich structure complex containing rhodamine B. The ethylene diamine tetraacetic acid (EDTA) was used for the dissolution of CaCO$_3$, then rhodamine B is released into the solution and the solution is dropped onto the SERS substrate to reflect the antigen content in the Raman spectrum (Fig.6g). It can be seen from Fig.6h that at a certain antigen content, the concentration of rhodamine B increases, and the Raman intensity also increases, and the concentration is 0.5 mg mL^{-1} to achieve the best effect. As can be seen in Fig.6i, the logarithm of the hs-CRP concentration of the antigen showed a good linearity with the Raman intensity, and R^2=0.997 [72].

In summary, in a magnetically assisted SERS-based immunosensor, two antibodies can bind to the antigen simultaneously in one solution. Alternatively, the antigen can be bound by the first antibody, then separated, and then ligated by the second antibody. In one solution, there is a competition between the two antibodies, and multi-step binding can inhibit its action. Label coating re-separation detection, antibodies, antigens or other complexes are excluded from the detection system, avoiding the interference of these substances on the Raman signal, of course, only the signal of the marker, and there is no other effect of improving SERS which will affect the Raman intensity.

4.2   Microbial sensors

Bacteria are ubiquitous and pose a great threat to the safety of human drinking water and food [73]. Bacterial testing in food and water is safe for human safety. The surface of the bacteria contains a charge, and the opposite charge material can combine with the bacteria through electrostatic interaction to achieve the separation detection effect [74]. The bacterial surface also contains a variety of functional groups, and some of the detection substrates can aid in the detection of bacteria by aptamer modification. If the detection substrate contains magnetic nanomaterials, under the action of an external magnetic field, the purpose of bacterial separation, aggregation and fixation can be achieved [75].

Bacterial surfaces generally contain negative charges, and aminated modified iron oxide nanoparticles contribute to bacterial adsorption [74]. Such as, Yang et al. reported that this special particle can capture a variety of Gram-positive and Gram-negative bacteria [76]. Wang et al. modified the negatively charged bacteria by the modification of $Fe_3O_4@Au$ by PEI, and used the magnetic action of Fe_3O_4 to achieve the separation and purification of the detected substances (Fig.7a). The Raman signal is increased by Au@Ag NPs. Fig.7b shows that the $Fe_3O_4@Au@$polydopamine (PEI) complex successfully adsorbs Escherichia coli (1) by electrostatic matching. Fig.7c shows the detection of E. coli with different concentrations using SERS spectroscopy. $Fe_3O_4@Au$ shows a good heat transfer effect [74]. Moreover, PEI also contributes to near-infrared-absorbing [77, 78], which greatly helps the elimination of bacteria. For example, Escherichia coli can be killed by $Fe_3O_4@Au@PEI$ [79].

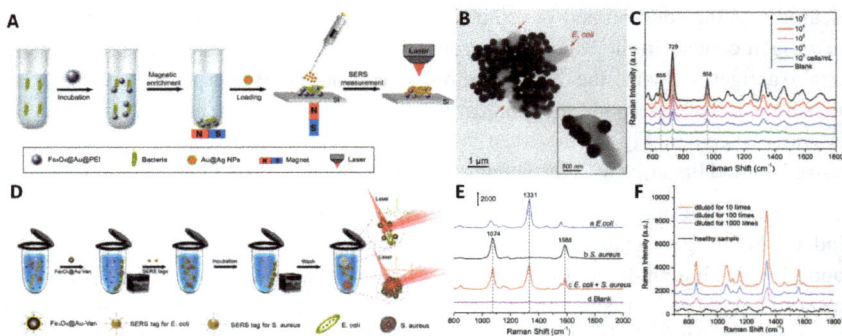

Figure 7.   (A)The process of electrostatic interaction to detect bacteria. (B)The TEM image of the $Fe_3O_4@Au@PEI$/bacterial. (C) SERS spectra of different concentrations of E. coli [74] (D)The process of detecting bacteria by Fe_3O_4/Au-Van. (E) SERS spectra of different bacteria. (F) SERS spectra of E. coli with different dilution ratios [75].

In immunosensors, specific detection is achieved by antigen-antibody binding. Bacterial sensors can also detect bacteria by means of antibodies, and achieve the detection effect [80]. But antibodies are relatively expensive, which limits their use. Then choose some suitable aptamers as the capture agent of bacteria, which is very helpful for the application of bacterial sensors. For example, vancomycin-modified $Fe_3O_4@Au$ ($Fe_3O_4@Au$-Van), combined with the SERS marker, successfully detected a variety of bacteria (Fig.7d). As can be seen in Fig.7e, the Raman characteristic peaks of each of the bacteria do not overlap each other, indicating that the mixed bacteria can also be detected by SERS Fig.7f shows that the dilution of bacteria with different multiples also changes the Raman intensity, which confirms the feasibility of quantitative detection of bacteria [75]. For example, Zhang et al. use Fe_3O_4 to provide magnetic separation, aggregation and fixation effects, use Au to improve SERS, and use mercaptobenzoic acid (MBA) and DNTB to label S. typhimurium and S. aureus, respectively. The content of mixed bacteria was measured, which shows a very low detection limit [81].

Bacterial sensors mainly use electrostatic interaction and specific binding of aptamers to achieve the combination of SERS substrate and bacteria, showing the detection effect. The materials required for electrostatic interaction and aptamer action are relatively inexpensive and can be used to promote their application. Bacteria can be killed by heat or light. In addition to detecting bacteria, this special sensor can achieve sterilization by a certain heat or light treatment.

4.3   Nucleic acid sensors

Nucleic acid sensors are devices used to detect DNA or RNA content. In magnetic nanoparticle-based SERS nucleic acid sensors, the interaction between DNA/RNA is used to achieve specific recognition, or the aptamer is combined with nucleic acid to obtain its content, reflecting the condition of the organism. The magnetic nanoparticle-based nucleic acid sensor can detect trace amounts of nucleic acid by magnetic field aggregation and separation of magnetic nanoparticles.

A precious metal (Ag, Au, etc.) containing many empty orbitals promotes the immobilization of nucleic acids. For example, $Fe_3O_4@Au@Ag$ is used as a SERS substrate to bind small molecule adenosine in nucleic acids. Because of the magnetic action of the substrate, adenosine is separated from the actual sample and its content is obtained by means of a Raman signal [82]. The nucleic acid extraction process is relatively complex, which promotes the need for trace nucleic acid detection. Yin et al. use duplex-specific nuclease (DSN) to achieve signal amplification. After the probe DNA binds to the miRNA, the DSN cuts off this particular structure and disrupts the structure of the DNA probe, while the miRNA can continue to bind to other DNA probes, so that it is cycled multiple times for signal amplification purposes [83]. Such methods are used by Pang et al. to detect RNA in cancer cells. $Fe_3O_4@Ag$ NPs acts as the substrate for SERS, and the Raman tag Cy3 labeled DNA binds to Ag to form the $Fe_3O_4@Ag$-DNA-Cy3 probe, which binds to the miRNA extracted from cancer cells. The DNA is then separated by DSN and destroyed by the structure, and the cut miRNA can continue to bind to the DNA on the surface of $Fe_3O_4@Ag$. This is repeated to achieve the effect of miRNA detection signal amplification (Fig.8a). As can be seen from Fig.8b, when the $Fe_3O_4@Ag$-DNA-Cy3, miRNA let-7b, and DNA are simultaneously present, the SERS signal intensity is reduced, thereby ensuring the stability of the Raman intensity in the absence of DNS. DNS can repeatedly cut the DNA/RNA structure, but it is not wirelessly cycled. As shown in Figure 8C, it can be seen that the best effect is achieved when the DNS acts for 60 minutes. Controlling the DNS effect for 60 min, the logarithm of miRNA concentration showed a good linear relationship with Raman intensity (Fig.8c). Due to the synergistic effect of magnetic nanoparticles and DNS signal amplification, the detection limit reaches 0.3 FM [26].

In addition to utilizing DNS signal amplification techniques, aptamers are often used to bind nucleic acids, such as doxorubicin and chlorambucil, or DNA/RNA interactions are relied upon for detection purposes. However, the detection limits and linear relationships achieved by these techniques are weaker than the DNS signal amplification technique [85-87]. Of course, aptamers are less expensive than nucleic acids. It is necessary for the aptamer development class DNS signal amplification technology.

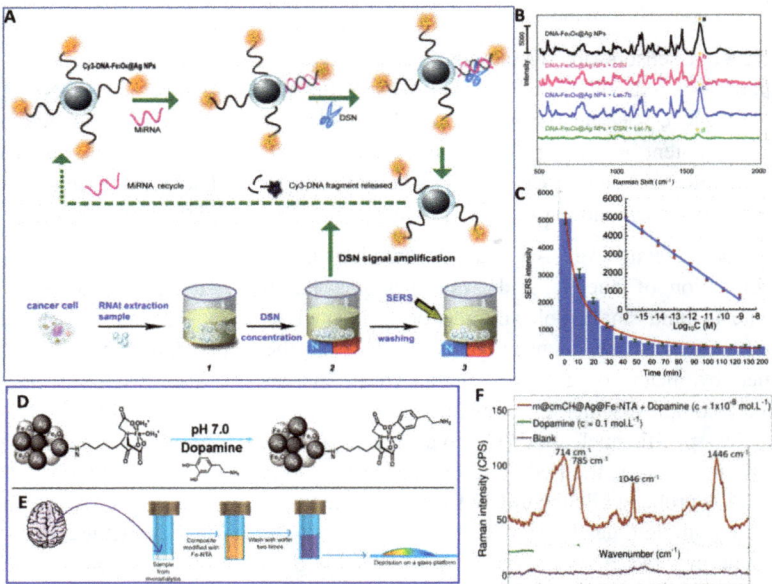

Figure 8.   (A) Schematic diagram of miRNA detection strategy. (B) SERS intensity of DNA–Fe₃O₄@Ag NPs (C)SERS intensity with different DSN processing ties. Inset: relationship between SERS intensity and lg C(miRNA) [26]. (D) The binding mechanism of dopamine by aptamer. (E) Analysis scheme of dopamine. (F) SERS spectra of different detection [84].

4.4   Cell sensor

Cancer cells are mutants of normal human cells. Cancer cells, like antigens, can be combined with specific antibodies for detection purposes [88]. There is a special tumor antigen on the surface of cancer cells, mucin 1 (MUC1) [89, 90]. The MUC1 on the surface of cancer cells is several hundred times larger than that of normal cells, and is evenly distributed on the surface of cancer cells. This special antigen provides great help for the detection of cancer cells [88, 91, 92]. The antibody (25-base oligonucleotide) that has captured this specific protein has been found [93]. Yarbakht et al. used this special antibody-modified Fe_3O_4@Au to capture MUC1 mucin, and then bind the antibody-modified Ag@SERS tag, magnetic nanoparticles can be used for the separation and purification of cancer cells [88]. Precious metals provide sites that bind to antigens and also increase hot spots. These are strong enhancement effects for Raman detection of cancer cells [63].

Programmed cell death receptor ligand 1 (PD-L1) is an immunodetection protein, and PD-L1 on the surface of tumor cells can be bound by anti-PD-L1[94].Due to the specificity of PD-L1 on cancer cells, Cupc-Fe_3O_4@GO@TiO_2 nanocomposites were used to bind anti-PD-L1 due to immunospecificity, anti-PD-L1 modified Cupc-Fe_3O_4@GO@TiO_2 nanocomposites bind to PD-L1, and the special action of PD-L1 on cancer cells causes the nanocomplex to adsorb on the surface of cancer cells (Fig.8d). Due to the magnetic aggregation of Fe_3O_4 and the enhanced charge transfer of GO, labeled cancer cells perform better in Raman spectroscopy. Fig.8f shows the spectrum of the Raman intensity of different cancer cell contents, the inset is its linear graph, and $R^2 =$ 0.993[95]

Cancer cells are a major threat to human health. It is necessary to develop better methods for detecting cancer cells. Nano-magnetic nanoparticle-based SERS is very helpful for the detection of cancer cells. By modifying the magnetic nanoparticle-based substrate to achieve specific binding to cancer cells, the content can be well detected, and even in situ detection is possible. If the substrate is modified by some materials that can absorb heat, and the magnetic aggregation between the magnetic nanoparticles locally increases the temperature around the material, it is expected to achieve the effect of locally destroying the cancer cells.

4.5   Other biomolecular sensors

In addition to the sensors mentioned above, magnetic nanoparticle-based SERS is also used for the detection of other biomolecules due to different bioassay requirements, such as Ochratoxin A [96, 97], dopamine [84], and Survivin Protein Dimer [25]. Ochratoxin A is a mycotoxin that can be detected in some foods such as rice, wine, coffee, etc. Therefore, detection of Ochratoxin A is critical for food safety [98]. Compared with many instruments, SERS detection technology is inexpensive and easy to operate, which promotes the detection of biomolecules by SERS [97]. For instance, Shao et al. found that magnetic nanoparticles are capable to Ochratoxin A (OTA) aptamers for capturing OTA while using noble metals to increase sensitivity, and the modified magnetic nanoparticles can aggregate under the action of a magnetic field. Eventually a very low detection limit is obtained. Dopamine is a neurotransmitter and an important part of the brain. Ranc et al. used the aptamer iron nitriloacetic acid (Fe-NTA) to bind dopamine, magnetic Fe_3O_4 as an important particle for aggregation and separation, and the noble metal silver enhanced the hot spot and could bind to the amino group in Fe-NTA. The binding mechanism of the ligand to dopamine is shown in Fig. 8d. Fig. 8e is a diagram showing the mechanism of physical analysis of dopamine, Pure dopamine has a relatively weak signal in the Raman spectrum, but its Raman intensity is greatly improved by the action of magnetic Fe_3O_4 and Ag (Fig. 8f) [84].

Conclusions and Outlook

In recent years, the application of magnetic nanoparticle-based SERS technology has been developed, and SERS sensors have become a powerful and sensitive analytical tool. More and more magnetic composite materials and special nanostructures have become the research direction. In addition to precious metal nanoparticles, semiconductor materials and two-dimensional materials have gradually become research hotspots. For example, the combination of graphene and SERS for biosensing and other research not only helps to obtain higher SERS activity, but also provides more possibilities for practical applications. The sensor technology developed by the combination of SERS and magnetic particles has the advantages of high resolution and no radioactivity, which makes it have a good development prospect in the medical field. But it has become a routine diagnostic tool to enter people's daily lives and still face many challenges. While studying the theoretical basis, we must explore fast, safe and stable solutions. Facing complex biological sample analysis, improving accuracy and low-cost automated testing may become a new direction for future research. Real-time testing will lay the foundation for life science and food safety progress. The improved SERS technology with continuous improvement of functions will inevitably become an important research direction in the field of biosensing.

The use of electric and magnetic fields to form a controllable metal nanoparticle active substrate can be designed into a special morphology to optimize the SERS active substrate by using different material properties such as gel, thereby realizing an efficient and intelligent biosensor system. With the continuous advancement of science, the superiority of Raman's enhancement is further enhanced. The exploration of SERS research may provide more effective practical means in the future, making SERS technology a more valuable analytical tool.

Magnetic nanoparticle-based SERS as a biosensor exhibits excellent performance. Complex material systems exist in biomass, which interfere with the Raman signal of the analyte and greatly limit the detection of biomass. Magnetic nanoparticle-based SERS can capture specific analytes by modifying aptamers (e.g., antibodies, DNA, etc.) at the substrate. Moreover, the magnetic action of the magnetic nanoparticles helps the separation of the analytes. Moreover, this magnetic property concentrates the analyte and facilitates the immobilization of the analyte on the SERS sample stage, which results in magnetic nanoparticle-based SERS as a biosensor exhibiting outstanding performance, which is of great significance in the areas of food safety, environmental testing and clinical diagnosis. The SERS substrate of magnetic nanoparticles can specifically bind and remove harmful substances such as bacteria. Moreover, the fluctuating magnetic field can heat the magnetic nanoparticles, which indicates that the specifically bound biomass may be heated, which will lead to the development of magnetic nanoparticle-based SERS to biological treatment, environmental optimization and the like.

References

[1] R. Das, C.D. Vecitis, A. Schulze, B. Cao, A.F. Ismail, X. Lu, J. Chen, S. Ramakrishna, Recent advances in nanomaterials for water protection and monitoring, Chem. Soc. Rev. 46 (2017) 6946-7020. https://doi.org/10.1039/C6CS00921B

[2] L. Wu, X. Qu, Cancer biomarker detection: recent achievements and challenges, Chem. Soc. Rev. 44 (2015) 2963-2997. https://doi.org/10.1039/C4CS00370E

[3] A. Balzerova, A. Fargasova, Z. Markova, V. Ranc, R. Zboril, Magnetically-assisted surface enhanced raman spectroscopy (MA-SERS) for label-free determination of human immunoglobulin G (IgG) in blood using $Fe_3O_4@Ag$ nanocomposite, Anal. Chem. 86 (2014) 11107-11114. https://doi.org/10.1021/ac503347h

[4] H. Zhou, J. Liu, J.J. Xu, S.S. Zhang, H.Y. Chen, Optical nano-biosensing interface via nucleic acid amplification strategy: construction and application, Chem. Soc. Rev. 47 (2018) 1996-2019. https://doi.org/10.1039/C7CS00573C

[5] T. Lee, T.H. Kim, J. Yoon, Y.H. Chung, J.Y. Lee, J.W. Choi, Investigation of Hemoglobin/Gold Nanoparticle Heterolayer on Micro-Gap for Electrochemical Biosensor Application, Sensors16 (2016), 660-671. https://doi.org/10.3390/s16050660

[6] Y. Liu, Y. Liu, Y. Xing, X. Guo, Y. Ying, Y. Wu, Y. Wen, H. Yang, Magnetically three-dimensional Au nanoparticles/reduced graphene/ nickel foams for Raman trace detection, Sensor. Actuat. B-Chem.273 (2018) 884-890. https://doi.org/10.1016/j.snb.2018.07.003

[7] C. Wang, J. Xu, J. Wang, Z. Rong, P. Li, R. Xiao, S. Wang, Polyethylenimine-interlayered silver-shell magnetic-core microspheres as multifunctional SERS substrates, J. Mater. Chem. C 3 (2015) 8684-8693. https://doi.org/10.1039/C5TC01839K

[8] T.H. Seefeld, A.R. Halpern, R.M. Corn, On-chip synthesis of protein microarrays from DNA microarrays via coupled in vitro transcription and translation for surface

plasmon resonance imaging biosensor applications, J. Am. Chem. Soc. 134 (2012) 12358-12361. https://doi.org/10.1021/ja304187r

[9] Y. Zhao, F. Chen, Q. Li, L. Wang, C. Fan, Isothermal Amplification of Nucleic Acids, Chem. Rev. 115 (2015) 12491-12545. https://doi.org/10.1021/acs.chemrev.5b00428

[10] L. Gao, J. Zhuang, L. Nie, J. Zhang, Y. Zhang, N. Gu, T. Wang, J. Feng, D. Yang, S. Perrett, X. Yan, Intrinsic peroxidase-like activity of ferromagnetic nanoparticles, Nat. Nanotechnol. 2 (2007) 577-583. https://doi.org/10.1038/nnano.2007.260

[11] P.C. Lin, P.H. Chou, S.H. Chen, H.K. Liao, K.Y. Wang, Y.J. Chen, C.C. Lin, Ethylene glycol-protected magnetic nanoparticles for a multiplexed immunoassay in human plasma, Small 2 (2006) 485-489. https://doi.org/10.1002/smll.200500387

[12] A.H. Latham, R.S. Freitas, P. Schiffer, M.E. Williams, Capillary magnetic field flow fractionation and analysis of magnetic nanoparticles, Anal. Chem. 77 (2005) 5055-62. https://doi.org/10.1021/ac050611f

[13] J.C. Liu, P.J. Tsai, Y.C. Lee, Y.C. Chen, Affinity capture of uropathogenic Escherichia coli using pigeon ovalbumin-bound Fe3O4@Al2O3 magnetic nanoparticles, Anal. Chem. 80 (2008) 5425-5432. https://doi.org/10.1021/ac800487v

[14] P.H. Chou, S.H. Chen, H.K. Liao, P.C. Lin, G.R. Her, A.C. Lai, J.H. Chen, C.C. Lin, Y.J. Chen, Nanoprobe-based affinity mass spectrometry for selected protein profiling in human plasma, Anal. Chem. 77 (2005) 5990-5997. https://doi.org/10.1021/ac050655o

[15] S. Moraes Silva, R. Tavallaie, L. Sandiford, R.D. Tilley, J.J. Gooding, Gold coated magnetic nanoparticles: from preparation to surface modification for analytical and biomedical applications, Chem. Commun. 52 (2016) 7528-7540. https://doi.org/10.1039/C6CC03225G

[16] C. Wang, P. Li, J. Wang, Z. Rong, Y. Pang, J. Xu, P. Dong, R. Xiao, S. Wang, Polyethylenimine- interlayered core-shell-satellite 3D magnetic microspheres as versatile SERS substrates, Nanoscale 7(44) (2015) 18694-18707. https://doi.org/10.1039/C5NR04977F

[17] T. Yang, X. Guo, H. Wang, S. Fu, J. Yu, Y. Wen, H. Yang, Au Dotted Magnetic Network Nanostructure and Its Application for On- Site Monitoring Femtomolar Level Pesticide, Small 10 (2014) 1325-1331. https://doi.org/10.1002/smll.201302604

[18] D. Liu, X. Wang, D. He, T.D. Dao, T. Nagao, Q. Weng, D. Tang, X. Wang, W. Tian, D. Golberg, Y. Bando, Magnetically assembled Ni@Ag urchin-like ensembles

with ultra-sharp tips and numerous gaps for SERS applications, Small 10 (2014) 2564-2569. https://doi.org/10.1002/smll.201303857

[19] J.F. Li, Y.F. Huang, Y. Ding, Z.L. Yang, S.B. Li, X.S. Zhou, F.R. Fan, W. Zhang, Z.Y. Zhou, D.Y. Wu, B. Ren, Z.L. Wang, Z.Q. Tian, Shell-isolated nanoparticle-enhanced Raman spectroscopy, Nature 464 (2010) 392-395. https://doi.org/10.1038/nature08907

[20] J. Zheng, Y. Ding, B. Tian, Z.L. Wang, X. Zhuang, Luminescent and Raman active silver nanoparticles with polycrystalline structure, J. Am. Chem. Soc. 130 (2008) 10472-10473. https://doi.org/10.1021/ja803302p

[21] O. Lioubashevski, V.I. Chegel, F. Patolsky, E. Katz, I. Willner, Enzyme-catalyzed bio-pumping of electrons into au-nanoparticles: a surface plasmon resonance and electrochemical study, J. Am. Chem. Soc. 126 (2004) 7133-7143. https://doi.org/10.1021/ja049275v

[22] J. Xie, Q. Zhang, J.Y. Lee, D.I. Wang, The synthesis of SERS-active gold nanoflower tags for in vivo applications, ACS Nano 2 (2008) 2473-2480. https://doi.org/10.1021/nn800442q

[23] Y. Xia, Y. Xiong, B. Lim, S.E. Skrabalak, Shape-controlled synthesis of metal nanocrystals: simple chemistry meets complex physics?, Angew. Chem. Int. Ed. Engl. 48 (2009) 60-103. https://doi.org/10.1002/anie.200802248

[24] S. Song, Y. Qin, Y. He, Q. Huang, C. Fan, H.Y. Chen, Functional nanoprobes for ultrasensitive detection of biomolecules, Chem. Soc. Rev. 39 (2010) 4234-4243. https://doi.org/10.1039/c000682n

[25] J. Wissler, S. Backer, A. Feis, S.K. Knauer, S. Schlucker, Site-Specific SERS Assay for Survivin Protein Dimer: From Ensemble Experiments to Correlative Single-Particle Imaging, Small 13 (2017), 1700802-1700812. https://doi.org/10.1002/smll.201700802

[26] Y. Pang, C. Wang, J. Wang, Z. Sun, R. Xiao, S. Wang, Fe(3)O(4)@Ag magnetic nanoparticles for microRNA capture and duplex-specific nuclease signal amplification based SERS detection in cancer cells, Biosens. Bioelectron. 79 (2016) 574-580. https://doi.org/10.1016/j.bios.2015.12.052

[27] D. Cialla-May, X.S. Zheng, K. Weber, J. Popp, Recent progress in surface-enhanced Raman spectroscopy for biological and biomedical applications: from cells to clinics, Chem. Soc. Rev. 46 (2017) 3945-3961. https://doi.org/10.1039/C7CS00172J

[28] H. Lai, F. Xu, Y. Zhang, L. Wang, Recent progress on graphene-based substrates for surface-enhanced Raman scattering applications, J. Mater. Chem. B 6 (2018) 4008-4028. https://doi.org/10.1039/C8TB00902C

[29] X. Zhu, J. Li, H. He, M. Huang, X. Zhang, S. Wang, Application of nanomaterials in the bioanalytical detection of disease-related genes, Biosens. Bioelectron. 74 (2015) 113-133. https://doi.org/10.1016/j.bios.2015.04.069

[30] X. Chi, D. Huang, Z. Zhao, Z. Zhou, Z. Yin, J. Gao, Nanoprobes for in vitro diagnostics of cancer and infectious diseases, Biomater. 33 (2012) 189-206. https://doi.org/10.1016/j.biomaterials.2011.09.032

[31] A.R. Ferhan, J.A. Jackman, J.H. Park, N.-J. Cho, D.-H. Kim, Nanoplasmonic sensors for detecting circulating cancer biomarkers, Adv. Drug Deliver. Rev. 125 (2018) 48-77. https://doi.org/10.1016/j.addr.2017.12.004

[32] J.E. Kim, J.H. Choi, M. Colas, D.H. Kim, H. Lee, Gold-based hybrid nanomaterials for biosensing and molecular diagnostic applications, Biosens. Bioelectron. 80 (2016) 543-559. https://doi.org/10.1016/j.bios.2016.02.015

[33] J. Reguera, D. Jimenez de Aberasturi, M. Henriksen-Lacey, J. Langer, A. Espinosa, B. Szczupak, C. Wilhelm, L.M. Liz-Marzan, Janus plasmonic-magnetic gold-iron oxide nanoparticles as contrast agents for multimodal imaging, Nanoscale 9 (2017) 9467-9480. https://doi.org/10.1039/C7NR01406F

[34] S.-C. Luo, K. Sivashanmugan, J.-D. Liao, C.-K. Yao, H.-C. Peng, Nanofabricated SERS-active substrates for single-molecule to virus detection in vitro: A review, Biosens. Bioelectron. 61 (2014) 232-240. https://doi.org/10.1016/j.bios.2014.05.013

[35] Y. Hu, H. Cheng, X. Zhao, J. Wu, F. Muhammad, S. Lin, J. He, L. Zhou, C. Zhang, Y. Deng, P. Wang, Z. Zhou, S. Nie, H. Wei, Surface-Enhanced Raman Scattering Active Gold Nanoparticles with Enzyme-Mimicking Activities for Measuring Glucose and Lactate in Living Tissues, Acs Nano 11 (2017) 5558-5566. https://doi.org/10.1021/acsnano.7b00905

[36] Y. Hu, J. Liao, D. Wang, G. Li, Fabrication of Gold Nanoparticle-Embedded Metal-Organic Framework for Highly Sensitive Surface-Enhanced Raman Scattering Detection, Anal. Chem. 86 (2014) 3955-3963. https://doi.org/10.1021/ac5002355

[37] J. Baniukevic, I.H. Boyaci, A.G. Bozkurt, U. Tamer, A. Ramanavicius, A. Ramanaviciene, Magnetic gold nanoparticles in SERS-based sandwich immunoassay for antigen detection by well oriented antibodies, Biosens. Bioelectron. 43 (2013) 281-288. https://doi.org/10.1016/j.bios.2012.12.014

[38] J. Wackerlig, P.A. Lieberzeit, Molecularly imprinted polymer nanoparticles in chemical sensing - Synthesis, characterisation and application, Sensors and Actuators B-Chemical 207 (2015) 144-157. https://doi.org/10.1016/j.snb.2014.09.094

[39] X. Huang, J. Song, B.C. Yung, X. Huang, Y. Xiong, X. Chen, Ratiometric optical nanoprobes enable accurate molecular detection and imaging, Chem. Soc. Rev. 47 (2018) 2873-2920. https://doi.org/10.1039/C7CS00612H

[40] N. Lee, P.J. Schuck, P.S. Nico, B. Gilbert, Surface Enhanced Raman Spectroscopy of Organic Molecules on Magnetite (Fe3O4) Nanoparticles, J. Phys. Chem. Lett. 6 (2015) 970-974. https://doi.org/10.1021/acs.jpclett.5b00036

[41] A. Balzerova, A. Fargasova, Z. Markova, V. Ranc, R. Zboril, Magnetically-Assisted Surface Enhanced Raman Spectroscopy (MA-SERS) for Label-Free Determination of Human Immunoglobulin G (IgG) in Blood Using Fe3O4@Ag Nanocomposite, Anal. Chem. 86 (2014) 11107-11114. https://doi.org/10.1021/ac503347h

[42] T. Yang, X. Guo, Y. Wu, H. Wang, S. Fu, Y. Wen, H. Yang, Facile and Label-Free Detection of Lung Cancer Biomarker in Urine by Magnetically Assisted Surface-Enhanced Raman Scattering, Acs Appl. Mater. Interf. 6 (2014) 20985-20993. https://doi.org/10.1021/am5057536

[43] Y. Wu, Y. He, X. Yang, R. Yuan, Y. Chai, A novel recyclable surface-enhanced Raman spectroscopy platform with duplex-specific nuclease signal amplification for ultrasensitive analysis of microRNA 155, Sensors and Actuators B-Chemical 275 (2018) 260-266. https://doi.org/10.1016/j.snb.2018.08.057

[44] X. Zhang, C. Niu, Y. Wang, S. Zhou, J. Liu, Gel-limited synthesis of dumbbell-like Fe3O4-Ag composite microspheres and their SERS applications, Nanoscale 6 (2014) 12618-12625. https://doi.org/10.1039/C4NR03301A

[45] B. Shao, X. Ma, S. Zhao, Y. Lv, X. Hun, H. Wang, Z. Wang, Nanogapped Au-(core) @ Au-Ag-(shell) structures coupled with Fe3O4 magnetic nanoparticles for the detection of Ochratoxin A, Anal. Chim. Acta 1033 (2018) 165-172. https://doi.org/10.1016/j.aca.2018.05.058

[46] W. Cai, X. Tang, B. Sun, L. Yang, Highly sensitive in situ monitoring of catalytic reactions by surface enhancement Raman spectroscopy on multifunctional Fe3O4/C/Au NPs, Nanoscale 6 (2014) 7954-7958. https://doi.org/10.1039/C4NR01147C

[47] X. Zhang, Y. Zhu, X. Yang, Y. Zhou, Y. Yao, C. Li, Multifunctional Fe3O4@TiO2@Au magnetic microspheres as recyclable substrates for surface-

enhanced Raman scattering, Nanoscale 6 (2014) 5971-5979.
https://doi.org/10.1039/C4NR00975D

[48] M. Ye, Z. Wei, F. Hu, J. Wang, G. Ge, Z. Hu, M. Shao, S.-T. Lee, J. Liu, Fast assembling microarrays of superparamagnetic Fe3O4@Au nanoparticle clusters as reproducible substrates for surface-enhanced Raman scattering, Nanoscale 7 (2015) 13427-13437. https://doi.org/10.1039/C5NR02491A

[49] H. Ilkhani, T. Hughes, J. Li, C.J. Zhong, M. Hepel, Nanostructured SERS-electrochemical biosensors for testing of anticancer drug interactions with DNA, Biosens. Bioelectron. 80 (2016) 257-264. https://doi.org/10.1016/j.bios.2016.01.068

[50] Y. Pang, C. Wang, J. Wang, Z. Sun, R. Xiao, S. Wang, Fe3O4@Ag magnetic nanoparticles for microRNA capture and duplex-specific nuclease signal amplification based SERS detection in cancer cells, Biosens. Bioelectron. 79 (2016) 574-580. https://doi.org/10.1016/j.bios.2015.12.052

[51] J. Docherty, S. Mabbott, E. Smith, K. Faulds, C. Davidson, J. Reglinski, D. Graham, Detection of potentially toxic metals by SERS using salen complexes, Anal. 141 (2016) 5857-5863. https://doi.org/10.1039/C6AN01584K

[52] Z. Pirzadeh, T. Pakizeh, V. Miljkovic, C. Langhammer, A. Dmitriev, Plasmon-Interband Coupling in Nickel Nanoantennas, Acs Photonics 1 (2014) 158-162. https://doi.org/10.1021/ph4000339

[53] Q. Fu, K.M. Wong, Y. Zhou, M. Wu, Y. Lei, Ni/Au hybrid nanoparticle arrays as a highly efficient, cost-effective and stable SERS substrate, RSC Adv. 5 (2015) 6172-6180. https://doi.org/10.1039/C4RA09312G

[54] Z. Cai, Y. Yan, L. Liu, S. Lin, X. Hu, Controllable fabrication of metallic photonic crystals for ultra-sensitive SERS and photodetectors, RSC Adv. 7 (2017) 55851-55858. https://doi.org/10.1039/C7RA11721C

[55] D. Liu, X. Wang, D. He, D. Thang Duy, T. Nagao, Q. Weng, D. Tang, X. Wang, W. Tian, D. Golberg, Y. Bando, Magnetically Assembled Ni@Ag Urchin-Like Ensembles with Ultra-Sharp Tips and Numerous Gaps for SERS Applications, Small 10 (2014) 2564-2569. https://doi.org/10.1002/smll.201303857

[56] O. Petrus, A. Orinak, R. Orinakova, Z.O. Kralova, E. Mudra, M. Kupkova, K. Koval, Colloidal lithography with electrochemical nickel deposition as a unique method for improved silver decorated nanocavities in SERS applications, Appl. Surf. Sci.423 (2017) 322-330. https://doi.org/10.1016/j.apsusc.2017.06.149

[57] H. Zhao, J. Jin, W. Tian, R. Li, Z. Yu, W. Song, Q. Cong, B. Zhao, Y. Ozaki, Three-dimensional superhydrophobic surface-enhanced Raman spectroscopy substrate for sensitive detection of pollutants in real environments, J. Mater. Chem. A 3 (2015) 4330-4337. https://doi.org/10.1039/C4TA06590E

[58] Y. He, Y. Wang, X. Yang, S. Xie, R. Yuan, Y. Chai, Metal Organic Frameworks Combining CoFe2O4 Magnetic Nanoparticles as Highly Efficient SERS Sensing Platform for Ultrasensitive Detection of N-Terminal Pro-Brain Natriuretic Peptide, Acs Appl. Mater. Interf. 8 (2016) 7683-7690. https://doi.org/10.1021/acsami.6b01112

[59] L. Xie, X. Yang, Y. He, R. Yuan, Y. Chai, Polyacrylamide Gel-Contained Zinc Finger Peptide as the "Lock" and Zinc Ions as the "Key" for Construction of Ultrasensitive Prostate-Specific Antigen SERS lmmunosensor, Acs Appl. Mater. Interf. 10 (2018) 15200-15206. https://doi.org/10.1021/acsami.7b19717

[60] R. Berenguer, A. La Rosa-Toro, C. Quijada, E. Morallon, Electrocatalytic oxidation of cyanide on copper-doped cobalt oxide electrodes, Applied Catalysis B-Environmental 207 (2017) 286-296. https://doi.org/10.1016/j.apcatb.2017.01.078

[61] R. Liu, J.-f. Sun, D. Cao, L.-q. Zhang, J.-f. Liu, G.-b. Jiang, Fabrication of highly-specific SERS substrates by co-precipitation of functional nanomaterials during the self-sedimentation of silver nanowires into a nanoporous film, Chem.l Commun. 51 (2015) 1309-1312. https://doi.org/10.1039/C4CC08016E

[62] C. Krafft, J. Popp, The many facets of Raman spectroscopy for biomedical analysis, Anal BioAnal. Chem. 407 (2015) 699-717. https://doi.org/10.1007/s00216-014-8311-9

[63] H.M. Kim, D.M. Kim, C. Jeong, S.Y. Park, M.G. Cha, Y. Ha, D. Jang, S. Kyeong, X.H. Pham, E. Hahm, S.H. Lee, D.H. Jeong, Y.S. Lee, D.E. Kim, B.H. Jun, Assembly of Plasmonic and Magnetic Nanoparticles with Fluorescent Silica Shell Layer for Tri-functional SERS-Magnetic-Fluorescence Probes and Its Bioapplications, Sci. Rep. 8 (2018) 13938. https://doi.org/10.1038/s41598-018-32044-7

[64] W. Mao, J. He, Z. Tang, C. Zhang, J. Chen, J. Li, C. Yu, A sensitive sandwich-type immunosensor for the detection of MCP-1 based on a rGO-TEPA-Thi-Au nanocomposite and novel RuPdPt trimetallic nanoalloy particles, Biosens. Bioelectron. 131 (2019) 67-73. https://doi.org/10.1016/j.bios.2019.02.021

[65] K. Zou, Z. Gao, Q. Deng, Y. Luo, L. Zou, Y. Lu, W. Zhao, B. Lin, Picomolar detection of carcinoembryonic antigen in whole blood using microfluidics and surface-enhanced Raman spectroscopy, Electrophoresis 37 (2016) 786-789. https://doi.org/10.1002/elps.201500535

[66] M.H. Shin, W. Hong, Y. Sa, L. Chen, Y.-J. Jung, X. Wang, B. Zhao, Y.M. Jung, Multiple detection of proteins by SERS-based immunoassay with core shell magnetic gold nanoparticles, Vib. Spectrosc.72 (2014) 44-49. https://doi.org/10.1016/j.vibspec.2014.02.007

[67] A. Ali, E.Y. Hwang, J. Choo, D.W. Lim, PEGylated nanographene-mediated metallic nanoparticle clusters for surface enhanced Raman scattering-based biosensing, Anal. 143 (2018) 2604-2615. https://doi.org/10.1039/C8AN00329G

[68] A. Ali, E.Y. Hwang, J. Choo, D.W. Lim, Nanoscale graphene oxide-induced metallic nanoparticle clustering for surface-enhanced Raman scattering-based IgG detection, Sensor. Actuat. B-Chem.255 (2018) 183-192. https://doi.org/10.1016/j.snb.2017.07.140

[69] V. Ranc, R. Zizka, Z. Chaloupkova, J. Sevcik, R. Zboril, Imaging of growth factors on a human tooth root canal by surface-enhanced Raman spectroscopy, Anal BioAnal. Chem. 410 (2018) 7113-7120. https://doi.org/10.1007/s00216-018-1311-4

[70] F.Q. Hu, L. Wei, Z. Zhou, Y.L. Ran, Z. Li, M.Y. Gao, Preparation of Biocompatible Magnetite Nanocrystals for In Vivo Magnetic Resonance Detection of Cancer, Adv. Mater. 18 (2006) 2553-2556. https://doi.org/10.1002/adma.200600385

[71] Y. He, Y. Wang, X. Yang, S. Xie, R. Yuan, Y. Chai, Metal Organic Frameworks Combining CoFe2O4 Magnetic Nanoparticles as Highly Efficient SERS Sensing Platform for Ultrasensitive Detection of N-Terminal Pro-Brain Natriuretic Peptide, ACS Appl. Mater. Interf. 8 (2016) 7683-7690. https://doi.org/10.1021/acsami.6b01112

[72] S. Wang, J. Luo, Y. He, Y. Chai, R. Yuan, X. Yang, Combining Porous Magnetic Ni@C Nanospheres and CaCO3 Microcapsule as Surface-Enhanced Raman Spectroscopy Sensing Platform for Hypersensitive C-Reactive Protein Detection, ACS Appl. Mater. Interf. 10 (2018) 33707-33712. https://doi.org/10.1021/acsami.8b13061

[73] Z. Teng, N. Yang, H. Lv, S. Wang, M. Hu, C. Wang, D. Wang, G. Wang, Edge-Functionalized g-C3N4 Nanosheets as a Highly Efficient Metal-free Photocatalyst for Safe Drinking Water, Chem 5 (2019) 664-680. https://doi.org/10.1016/j.chempr.2018.12.009

[74] C. Wang, J. Wang, M. Li, X. Qu, K. Zhang, Z. Rong, R. Xiao, S. Wang, A rapid SERS method for label-free bacteria detection using polyethylenimine-modified Au-coated magnetic microspheres and Au@Ag nanoparticles, Anal. 141 (2016) 6226-6238. https://doi.org/10.1039/C6AN01105E

[75] C. Zhang, C. Wang, R. Xiao, L. Tang, J. Huang, D. Wu, S. Liu, Y. Wang, D. Zhang, S. Wang, X. Chen, Sensitive and specific detection of clinical bacteria via vancomycin-modified Fe3O4@Au nanoparticles and aptamer-functionalized SERS tags, J. Mater. Chem. B 6 (2018) 3751-3761. https://doi.org/10.1039/C8TB00504D

[76] Y.F. Huang, Y.F. Wang, X.P. Yan, Amine-functionalized magnetic nanoparticles for rapid capture and removal of bacterial pathogens, Environ. Sci. Technol. 44 (2010) 7908-7913. https://doi.org/10.1021/es102285n

[77] S.H. Kim, I. In, S.Y. Park, pH-Responsive NIR-Absorbing Fluorescent Polydopamine with Hyaluronic Acid for Dual Targeting and Synergistic Effects of Photothermal and Chemotherapy, Biomacromolecules 18 (2017) 1825-1835. https://doi.org/10.1021/acs.biomac.7b00267

[78] R. Zheng, S. Wang, Y. Tian, X. Jiang, D. Fu, S. Shen, W. Yang, Polydopamine-Coated Magnetic Composite Particles with an Enhanced Photothermal Effect, ACS Appl. Mater. Interf. 7 (2015) 15876-15884. https://doi.org/10.1021/acsami.5b03201

[79] W.-E. Hong, I.L. Hsu, S.-Y. Huang, C.-W. Lee, H. Ko, P.-J. Tsai, D.-B. Shieh, C.-C. Huang, Assembled growth of 3D Fe3O4@Au nanoparticles for efficient photothermal ablation and SERS detection of microorganisms, J. Mater. Chem. B 6 (2018) 5689-5697. https://doi.org/10.1039/C8TB00599K

[80] R. Najafi, S. Mukherjee, J. Hudson, Jr., A. Sharma, P. Banerjee, Development of a rapid capture-cum-detection method for Escherichia coli O157 from apple juice comprising nano-immunomagnetic separation in tandem with surface enhanced Raman scattering, Int. J. Food Microbiol. 189 (2014) 89-97. https://doi.org/10.1016/j.ijfoodmicro.2014.07.036

[81] H. Zhang, X. Ma, Y. Liu, N. Duan, S. Wu, Z. Wang, B. Xu, Gold nanoparticles enhanced SERS aptasensor for the simultaneous detection of Salmonella typhimurium and Staphylococcus aureus, Biosens. Bioelectron. 74 (2015) 872-877. https://doi.org/10.1016/j.bios.2015.07.033

[82] T. Yang, X. Guo, Y. Wu, H. Wang, S. Fu, Y. Wen, H. Yang, Facile and label-free detection of lung cancer biomarker in urine by magnetically assisted surface-enhanced Raman scattering, ACS Appl. Mater. Interf. 6 (2014) 20985-20993. https://doi.org/10.1021/am5057536

[83] B.C. Yin, Y.Q. Liu, B.C. Ye, One-step, multiplexed fluorescence detection of microRNAs based on duplex-specific nuclease signal amplification, J. Am. Chem. Soc. 134 (2012) 5064-5067. https://doi.org/10.1021/ja300721s

[84] V. Ranc, Z. Markova, M. Hajduch, R. Prucek, L. Kvitek, J. Kaslik, K. Safarova, R. Zboril, Magnetically assisted surface-enhanced raman scattering selective determination of dopamine in an artificial cerebrospinal fluid and a mouse striatum using Fe(3)O(4)/Ag nanocomposite, Anal. Chem. 86 (2014) 2939-2946. https://doi.org/10.1021/ac500394g

[85] Z. Chaloupková, A. Balzerová, Z. Medříková, J. Srovnal, M. Hajdúch, K. Čépe, V. Ranc, R. Zbořil, Label-free determination and multiplex analysis of DNA and RNA in tumor tissues, Appl. Mater. Today 12 (2018) 85-91. https://doi.org/10.1016/j.apmt.2017.12.012

[86] H. Ilkhani, T. Hughes, J. Li, C.J. Zhong, M. Hepel, Nanostructured SERS-electrochemical biosensors for testing of anticancer drug interactions with DNA, Biosens. Bioelectron. 80 (2016) 257-264. https://doi.org/10.1016/j.bios.2016.01.068

[87] Y. Wu, Y. He, X. Yang, R. Yuan, Y. Chai, A novel recyclable surface-enhanced Raman spectroscopy platform with duplex-specific nuclease signal amplification for ultrasensitive analysis of microRNA 155, Sensor. Actuat. B-Chem.275 (2018) 260-266. https://doi.org/10.1016/j.snb.2018.08.057

[88] M. Yarbakht, M. Nikkhah, A. Moshaii, K. Weber, C. Matthaus, D. Cialla-May, J. Popp, Simultaneous isolation and detection of single breast cancer cells using surface-enhanced Raman spectroscopy, Talanta 186 (2018) 44-52. https://doi.org/10.1016/j.talanta.2018.04.009

[89] J. Zhao, L. Zhang, C. Chen, J. Jiang, R. Yu, A novel sensing platform using aptamer and RNA polymerase-based amplification for detection of cancer cells, Anal. Chim. Acta 745 (2012) 106-111. https://doi.org/10.1016/j.aca.2012.07.030

[90] T. Li, Q. Fan, T. Liu, X. Zhu, J. Zhao, G. Li, Detection of breast cancer cells specially and accurately by an electrochemical method, Biosens. Bioelectron. 25 (2010) 2686-2689. https://doi.org/10.1016/j.bios.2010.05.004

[91] W. Wei, D.F. Li, X.H. Pan, S.Q. Liu, Electrochemiluminescent detection of mucin 1 protein and MCF-7 cancer cells based on the resonance energy transfer, Anal. 137 (2012) 2101-2106. https://doi.org/10.1039/c2an35059a

[92] Y. Li, X. Qi, C. Lei, Q. Yue, S. Zhang, Simultaneous SERS detection and imaging of two biomarkers on the cancer cell surface by self-assembly of branched DNA-gold nanoaggregates, Chem. Commun. 50 (2014) 9907-9909. https://doi.org/10.1039/C4CC05226A

[93] C.S. Ferreira, C.S. Matthews, S. Missailidis, DNA aptamers that bind to MUC1 tumour marker: design and characterization of MUC1-binding single-stranded DNA aptamers, Tumour Biol. 27 (2006) 289-301. https://doi.org/10.1159/000096085

[94] S.F. Chiang, C.Y. Huang, T.W. Ke, T.W. Chen, Y.C. Lan, Y.S. You, W.T. Chen, K.S.C. Chao, Upregulation of tumor PD-L1 by neoadjuvant chemoradiotherapy (neoCRT) confers improved survival in patients with lymph node metastasis of locally advanced rectal cancers, Cancer Immunol Immunother 68 (2019) 283-296. https://doi.org/10.1007/s00262-018-2275-0

[95] E. Feng, T. Zheng, X. He, J. Chen, Y. Tian, A novel ternary heterostructure with dramatic SERS activity for evaluation of PD-L1 expression at the single-cell level, Sci. Adv. 4 (2018) eaau3494-eaau3506. https://doi.org/10.1126/sciadv.aau3494

[96] B. Shao, X. Ma, S. Zhao, Y. Lv, X. Hun, H. Wang, Z. Wang, Nanogapped Au(core) @ Au-Ag(shell) structures coupled with Fe3O4 magnetic nanoparticles for the detection of Ochratoxin A, Anal. Chim. Acta 1033 (2018) 165-172. https://doi.org/10.1016/j.aca.2018.05.058

[97] D. Song, R. Yang, S. Fang, Y. Liu, F. Long, A. Zhu, SERS based aptasensor for ochratoxin A by combining Fe3O4@Au magnetic nanoparticles and Au-DTNB@Ag nanoprobes with multiple signal enhancement, Mikrochim Acta 185 (2018) 491-501. https://doi.org/10.1007/s00604-018-3020-2

[98] V. Sorrenti, C. Di Giacomo, R. Acquaviva, I. Barbagallo, M. Bognanno, F. Galvano, Toxicity of ochratoxin a and its modulation by antioxidants: a review, Toxins 5 (2013) 1742-1766. https://doi.org/10.3390/toxins5101742

Magnetochemistry - Materials and Applications   Materials Research Forum LLC
Materials Research Foundations **66** (2020) 34-86   https://doi.org/10.21741/9781644900611-2

Chapter 2

Magnetic Nanomaterials for Electrocatalysis

V.N. Nikolić

Laboratory for Theoretical Physics and Condensed Matter Physics, Institute of Nuclear Sciences Vinča, University of Belgrade, Serbia

violeta@vin.bg.ac.rs

Abstract

Application of nanomaterials for electrocatalysts requires usage of low-cost and eco-friendly materials, characterized by high surface activity and stability. Due to low prices, high electrocatalytic activity, and eco-friendly behavior, transition metal-based nanocomposite materials are recognized as a potential replacement of precious metal electrocatalysts. The most straightforward way to improve properties of applied magnetic nanomaterials in electrocatalysts is to alter the synthesis approach, since it is known that variation of synthesis conditions enables tailoring of materials magnetic and catalytic properties. In this chapter is discussed the application of magnetic nanomaterials in hydrogen production and in biomedicine, since these areas nowadays attract the highest scientific attention and offer the most attractive future returns.

Keywords

Magnetic Nanomaterials, Electrocatalysts, Synthesis, Structure

Contents

1. Introduction

Nanomaterials attract a lot of attention from the scientific community due to the presence of numerous new phenomena that could be of significance for different applications. Nanomaterials behavior shows various peculiarities and improved properties (electrical, magnetic, mechanic, or catalytic) in comparison with bulk materials. Investigation of nm-sized materials in recent few decades resulted in the progress in different industry areas, enabling lowering costs and improving the quality of the final product which is offered to the market. It is important to notice that one of the industry areas where the usage of nanomaterials could result in the significant lowering of the costs is electrocatalysis.

Electrocatalysis has been established by Bowden and Rideal in the begin of the 20th century [1], although more significant achievements in this area have been reported after the 1950s [2]. Electrocatalysts represent catalysts that are involved in the electrochemical reactions, which take place in the interface between the electrode and electrolyte [3]. Their most important role is facilitation of electron transfer between an electrode and reactants. Accordingly, electrocatalysts are usually deposited on the surface of the electrode; moreover, the electrode surface itself could present electrocatalyst [3].

Applied electrocatalysis requires improvement of the electrodes activity for the desired reaction and prevention of side reactions. Choice of electrode and catalyst materials should meet certain requirements, such as stability of the used materials (absence of corrosion), high surface activity, as well as usage of low-cost and eco-friendly materials [2]. Since the noble metal electrocatalysts are expensive, the research for appropriate replacement of this type of catalyst is intensively performed.

One of the replacements for precious electrocatalysts is found to be magnetic nanoparticles. Noteworthy, observation of first mixed-metal oxide coatings in 1965 [2] opened the way for the development of the electrocatalysts based on the mixed metal oxides (mixed-metal oxide catalysts), which usage in electrocatalysis is significantly wider since the discovery of nanomaterials.

In comparison with bulk magnetic materials, nanomagnetics show superior behavior: alteration of point of magnetic phase transition, increased coercivity, or appearance of new magnetic states that opened its new applications. To get deeper insight in the origin of the changed magnetic behavior of nanomaterials, let us remember that only few of the elements (Fe^0, Co^0, Ni^0, Gd^0) are characterized with the presence of magnetic ordering near room temperature. Noteworthy, magnetic ground states of various ions (for example, 3d cations: Ti^{3+}, V^{4+}, V^{3+}, V^{2+}, Mn^{3+}, Mn^{2+}, Cr^{3+}, Cr^{2+}, Fe^{3+}, Fe^{2+}, Co^{2+}, Ni^{2+}, Cu^{2+}, Zn^{2+}) could be determined by Hund's rules, considering hyperfine structure and spin-orbit interaction [4]. Bulk materials based on the elements that are magnetically ordered near room temperature represented ordered systems consisting of multidomain particles, that shows various magnetic configurations (ferromagnetic, antiferromagnetic, ferrimagnetic, spin glasses, spiral and helical magnetic structures), which is dependent on the type of interactions between particles [4].

Lowering dimensions brings to the appearance of single domain particles, resulting in new magnetic states, such as superparamagnetism (SPM) or super spin glass (SSG).

The radius of a single-domain particle can be found considering the magnitude of the exchange energy and anisotropy energy (eq. 1):

$$r_c \approx 9[(J_{ab} \cdot K_a)^{1/2} / \mu_0 \cdot M_s^2]$$ (1)

where is J_{ab} - exchange integral, K_a - anisotropy constant, M_s - saturation magnetization. First rough estimation of a radius of spherical single-domain ferromagnetic nanoparticle is given by Kittel, who estimates the value of r_c between 10-1000 nm [5]. Literature review revealed that the spherical nanoparticles of the most popular magnetic element, metallic iron, experiences ferromagnetism to superparamagnetism transition when its size was close to 14 nm, while single-domain r_c for spherical metallic nickel and cobalt estimates 55 nm and 70 nm, respectively [6]. Single-domain sizes of magnetic oxides are shifted to higher values (for example, critical radius of spherical magnetite and maghemite nanoparticles are: r_c (Fe_3O_4) = 128 nm, r_c (γ-Fe_2O_3) = 166 nm [6]). Also, it is important to notice that the main feature of SPM nanoparticles presents absence of intrinsic coercivity at room temperature. Since a nanoparticle is characterized by a much higher magnitude of magnetic moment (compared to bulk particles), SPM nanoparticle system behave as a system of paramagnetic nanoparticles with large magnetic susceptibility, what expands its potential applications.

In catalysis, the transition from bulk to lower dimensions enabled increased surface activity, which is of importance for improvement of different industrial processes. Important factor in determination of the catalytic activity presents catalysts mass activity. The mass activity is measured at specific electrode potential value, and could be described as the ratio of the normalized current and loaded catalyst mass deposited on the electrode [7]. The increase of the catalysts specific surface area enable alteration of the mass activity, that is of importance for its application [8]. One of the synthesis strategies which ensure desired materials properties is to modify size, morphology and structure of the nanoparticles. Possibility to vary catalyst mass activity by alteration of the synthesis parameters is attributed to the catalyst structural sensitivity: the metal-support interaction, the electronic state and the dependence of surface geometry [9].

On the other hand, although variation of synthesis parameters enables the improvement of the nanomaterials properties, this scientific field is still not sufficiently investigated and even nowadays presents many open questions. What is known is that surface and structure of nanoparticles has a major role in determining the properties of the nanomaterials. The impact of the surface brings to the change of materials electrical conductivity as well as enhancement of catalytic activity, resulting in widening of various magnetic nanomaterials applications [3]. Consequently, alteration of synthesis conditions enables tailoring of the magnetic and catalytic properties of nanoparticles.

As mentioned before, in spite of high catalytic activity of noble metal electrocatalysts (ruthenium (Ru-), platinum (Pt-) and iridium (Ir-) based), potential useful alternatives are intensively investigated due to the limited natural resources and high price of precious metals [10]. Since magnetic transition metal elements satisfied desired requirements for usage in electrocatalysis (low price, high electrocatalytic activity, eco-friendly behavior), in recent time is intensified search for magnetic nanocomposite materials that could be used as effective replacement of precious metal-based electrocatalysts.

Comparison of the prices of three mostly used noble metal electrocatalysts (Ru-, Pt-, and Ir-based) with the prices of low-cost transition metals that could be used as a replacement of noble metal-based catalysts (Fe-, Ni-, and Co-based), explains intensive search for appropriate precious metal-based electrocatalyst replacement.

Figure 1 gives information about the market prices of the elements used for electrocatalysts preparation (statistical data are collected at 01.02.2019).

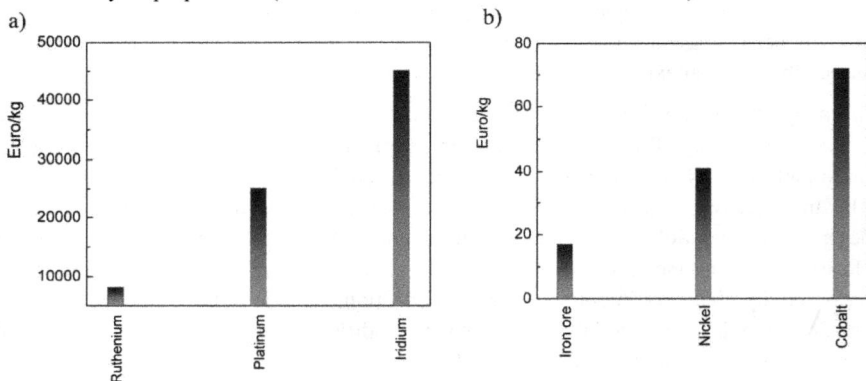

Figure 1. Prices, in Euro/kg for: a) noble elements (Ru, Pt, Ir); b) transition metal elements (Fe, Ni, Co) (accessed 01 February 2019).

Due to the fact that nanocomposite materials possess better properties than any single phase, in this chapter will be discussed applications of different magnetic nanocomposite materials in electrocatalysis. Having in mind the presence of magnetic properties, standard applications of magnetic nanoelectrocatalysts could be found in a wide range [11]. Conventional applications of this type of electrocatalyst are based on the possibility of easy separation of the catalysts from the electrode-electrolyte system, as well as on the potential reusage of magnetic electrocatalyst, characterized by the same catalytic activity [11]. Nevertheless, in this chapter comments on the usage of magnetic nanoparticles as electrocatalysts mostly in H_2 production processes and in biomedicine, because these areas of scientific research appears to offer the most attractive future returns. It is important to note that some of the magnetic nanocomposites possesses even superior properties (such as electrochemical stability), compared with noble metals [12], which encourages further deeper scientific search in this area.

1.1 Industrial needs for energy and electrocatalysis

Today, the efforts of the scientific community which is investigating potential usage of magnetic nanoparticles as electrocatalysts are mostly focused on successful replacement of electrocatalysts based on precious metals in the energy production industry.

The industrial needs for energy are mainly oriented toward direction of hydrogen energy production. One of the reactions that is often used for industrial production of hydrogen, is water-gas-shift (WGS) reaction. WGS considers reaction between carbon monoxide and water vapor, resulting in the release of hydrogen. Accordingly, WGS reaction presents a less expensive and efficient method for production of hydrogen [13]. Noteworthy, as a co-product of the WGS reaction, carbon dioxide is also developed during reaction (eq. 2):

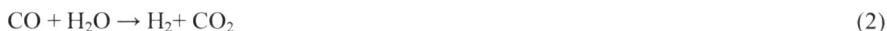

$$CO + H_2O \rightarrow H_2 + CO_2 \tag{2}$$

Having in mind the increased demand for energy, renewable energy has attracted more interests in the scientific community in the XXI century [14-20]. Renewable energy sources considered usage of wind, solar and geothermal energy. In order to meet energy demand, conversion of energy obtained from renewable energy resources into chemical fuels is intensively investigated, and has been recognized as a potential satisfactorily solution [21]. In the light of this, it could be noticed that hydrogen is characterized by high gravimetric energy density, the relatively high prevalence in nature, and zero emission during consumption, which consequently resulted in the fact that hydrogen presents the most important fuel candidate [22,23].

Regardless, at the moment, more than 90% of the H_2 supply is obtained from sources such as coal reforming, crude oil, natural gas, biomass, wood and organic wastes [24]. For this reason, WGS reaction is of high importance to obtaining clean H_2 for different industrial applications [24]. Lack of this process presents appearance of carbon dioxide, which is lowering the purity of hydrogen and increases the costs of the hydrogen production [25]. With the aim to increase purity of the reaction, as well as having in mind energy demand, the search for an alternative technique of hydrogen production revealed that potential replacement could be found in water splitting induced by electricity or sunlight [26-29]. Electrochemical energy storage that consists of H_2 as an energy carrier obtained avia completely water splitting is based on four elemental reactions: oxygen evolution reaction (in literature abbreviated as OER), hydrogen evolution reaction (abbreviated as HER), oxygen reduction reaction (abbreviated ORR) and hydrogen oxidation reaction (abbreviated HOR) [30].

These processes could be explained in the example of water splitting under alkaline conditions [31]. Water splitting could be represented as consisting of two half-reactions: OER (eq.3) and HER (eq.3). OER and HER processes require a four-proton-coupled electron-transfer (eq. 3 and eq. 4) [31]:

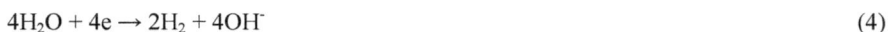

$$4OH^- \rightarrow O_2 + H_2O + 4e \tag{3}$$

$$4H_2O + 4e \rightarrow 2H_2 + 4OH^- \tag{4}$$

In these reactions, electrocatalysts are used in order to decrease the overpotential, and, what is expected for catalysts, with the aim to accelerate the reaction rate. Since the catalysts based on precious-metal materials expresses the best performing properties, this type of catalyst is used in electrocatalysis [32-34]. Let us remember that the main problem regarding precious metal-based catalysts presents limited availability and the high prices of precious metal, and for this reason, the scientific community performs intensive search for cheap and abundantly widespread materials, characterized by eco-friendly behavior, that could be a replacement for noble metal-based electrocatalysts.

Noteworthy, having in mind an industrial point of view, it is interesting to note that beside H_2 production, water electrolysis produces O_2 in pure form as well (eq.3). One of the area where production of both of hydrogen and oxygen in pure form is of importance presents coal-to-oil (liquids) industry, which is today intensively developed in China [35,36]. Integration of electrolysis and coal-to-liquids technology enables usage of oxygen in the gasification process, with simultaneous usage of hydrogen as additive to the synthesis gas. When the gas is obtained from the coal, then it can be characterized by inherently low H_2/CO ratio [35], which minimizes the process of CO_2 production in the water gas shift process. This is one of the examples how the combination of WGS reaction and coal industry technology enables the reduction of pollutant environmental influence in industrial areas which are recognized as a high pollutants, such as the coal industry [37].

Furthermore, the new, still not sufficiently investigated aspect in water splitting considered investigation of nanoelectrocatalysts magnetic properties in oxygen evolution, with the emphasis on the impact of magnetic behavior of the electrocatalysts onto the formation of O_2 [38]. Influence of the electrocatalysts magnetism on the oxygen formation is enabled by the fact that the oxygen molecule is characterized by the magnetically ordered quantum triplet state, which is obtained when ionic precursors are oriented in the appropriate direction [38]. In combination with kinetic studies conducted with the aim to reveal the mechanistic pathway of the reaction under realistic conditions, as well as by getting deeper insight in the variation of synthesis parameters and consequently occurrence of structural changes, performed studies could significantly improve efficiency of investigated nanoelectrocatalyst. This presents one of the areas of electrification of the chemical industry which will be improved in the near future by increased scientific search focused toward this direction.

1.2   Iron-based nanocomposite materials as nanoelectrocatalyst in biomedicine

Application of electrocatalysts based on composites containing magnetic nanoparticles in biomedicine has been thoroughly investigated in recent years. Electrocatalysts employed in biomedicine usually required preparation of nanocomposites based on magnetic nanoparticles in combination with precious metal nanoparticles (platinum, palladium, or gold) [39,40]. One of potential nanoelectrocatalysts that could be used for biomedical applications are metal-coated SPM magnetic iron oxide nanospheres. Increased catalytic activity of metallic nanoparticles brings to the acceleration of the overall rate of electron transfer and provides an opportunity to enlarge the electrochemical signal; in other words, excellent catalytic properties ensured usage of this class of nanomaterials as electrocatalysts [41].

Xu et al. [42] pointed to the advantages of the usage of nanocomposites containing conjugated magnetic and noble nanoparticles, such as nanocomposite containing gold and iron spinel nanoparticles. Such multifunctional system is characterized by biocompatibility, which is of key importance for biomedicine. Then, synergy of Au^0 and Fe_2O_3 nanoparticles enabled both a magnetic and an optically active plasmonic unit, which enable simultaneous optical and magnetic detection, which could enhance nanocomposites usage. Also, the presences of iron spinel and gold surfaces facilitate the attachment of various chemical functionalities. Finally, the size of either of these two kinds of nanoparticles could be varied in order to achieve optimization of its magnetic and optical properties, and to tailor its catalytic activity [42].

One of the benefits of using metallic nanomaterials as electrocatalysts in biomedicine is the possibility to avoid the shortcoming of enzymes such as thermal and environmental instability [43]. Good examples are Fe_3O_4 nanoparticles, characterized by presence of an intrinsic horseradish peroxidase-mimicking activity, that enable the application of this material in catalysis of the chemical or electrochemical reduction of H_2O_2 [43], which is of importance in the preparation of biosensors.

Nevertheless, although usage of nanomagnetic electrocatalysts in biomedicine enables the improvement of the properties of electrochemical sensing devices, the disadvantage is that the usage of magnetic nanoparticles as electrocatalysts often requires the application of additional substrates (usually glucose and H_2O_2) and/or requires redox mediators to increase the speed of the electron transfer. Having in mind that electrochemical sensors mostly use water as reaction medium, this presents one of the obstacles that should be overcome in the future.

In the next part is presented an overview of some chosen applications of nanoelectrocatalysts based on nanocomposites containing low-cost magnetic elements (iron, cobalt, nickel) which could be used in H_2 production processes or in biomedicine.

2.   Fe-, Co-, Ni-based nanocomposite materials as electrocatalysts

2.1   Iron-based nanocomposite materials as electrocatalyst

Since iron represents the cheapest and the most widespread magnetic element, in the text below are listed some applications of iron-based electrocatalyst, in water splitting and in biomedicine.

2.1.1   Iron-based nanocomposite materials as electrocatalysts in water splitting

Lee et al. found that low-cost Fe-based nanocomposites, such as Fe_3O_4-Fe nanohybrids, could be used as a nanoelectrocatalyst for oxygen electrode, utilized in lithium-oxygen batteries [44]. The investigated material was prepared by the process of an electrical wire pulse [44]. Fe_3O_4-Fe nanomaterials–containing electrodes were characterized with significantly improved properties: high discharge capacity, long cycle stability, and low overpotential. Observed electrode features appeared as a consequence of the good electrical conductivity of the Fe metal nanoparticles, which improve ORR and OER activities [44].

Zhang et al. found that FeP nanoparticles on candle soot (new type of carbon-based nanomaterials) shows excellent performance and stability [45] toward the HER process, and could be used as an promising replacement for the platinum-based catalysts, that are usually employed in this reaction [45]. Nanocomposite material is prepared via the synthesis of nanoparticles on candle soot supports; magnetite nanoparticles were grown on mentioned supports by solvothermal treatment [45].

Also, iron-based nanocomposites showed promising properties for electrocatalysis employed in microbial fuel cells (MFC). MFC presents a technology that enables synergy of the processes of wastewater cleaning and electricity generating [46]. Lu et al. found that material based on nitrogen-doped carbon, containing iron nanoparticles (Fe-N_x/C), could be used as efficient electrocatalyst in ORR reaction, which is important for practical applications of MFCs [47]. Prepared material is obtained by thermal decomposition of an organic polymer-based on iron(II) porphyrins. Cyclic voltammetry investigation reveals that prepared material showed the same or better performance in MFCs, compared with material based on carbon loaded with 20% of platinum (Pt/C), confirming increased tolerance and durability of investigated electrocatalyst [47].

Seyyedi recently performed an investigation of the usage of the nanoelectrocatalyst based on the material consisting of metallic iron and carbon black (FeN_4-S-C), that could be used for the ORR reaction in alkaline medium, which is of importance for alkaline fuel cells [48]. Obtained nanomaterial is prepared by using chlorinated iron phthalocyanine nanoparticles as iron nitride template [48]. Prepared material is recognized as a good replacement for the platinum/carbon electrocatalyst.

In 2018 Saruyama et al. found that NiP_x@FeP_yO_z core@shell nanoparticles could be used as nanocatalysts during OER in electro- and photo-catalytic water oxidation processes [49]. Nanoparticles around 11 nm in-size are obtained via few step thermal decomposition, starting from transition metal precursors $Fe(CO)_5$ and $Ni(acac)_2$. Appropriate nanoelectrocatalysts could be prepared by altering the structure of $NiFe(OH)_x$ [49]. Obtained materials proved that intuitive literature review could result in wrong conclusions: although investigated nanomaterial, $NiPx$@$FePyOz$, in combination with foreign metal showed very good electro/photocatalyst properties, this hybrid system combinations are reported very rarely in the literature, because in a micrometer scale, NiFe layered double hydroxides [50,51], as well as composites with carbon, are characterized by small interfacial contact area, so it is difficult to directly hybridize such electrocatalysts with semiconductor photocatalysts by simple mixing. This example depicted to the importance of the size effect in nanoparticle dimensions, and pointing to how much deceive could be caused if scientist proposes the same behavior of micro and nano compartments.

Magnetochemistry - Materials and Applications
Materials Research Foundations **66** (2020) 34-86

Materials Research Forum LLC
https://doi.org/10.21741/9781644900611-2

2.1.2 Iron-based nanocomposite materials as electrocatalyst in biomedicine

Few examples that lighting the role of iron-based nanocomposites used as nanoelectrocatalysts in biomedicine will be given in the text below.

Sun et al. proposed the preparation of an assay that could be used in order to detect tumor-associated plasma (and serum) p53 autoantibody, by using magnetic nanoelectrocatalysts consisted of porous ferric spinel nanoparticles of the cubic shape, in which pores are loaded gold nanoparticles ($Au@Fe_2O_3$) [52,39]. This assay is prepared in the next manner: carbon electrode is modified with a neuravidin and functionalized with biotinylated p53. With the aim to quantify autoantibody by using the colorimetric and electroanalytical measurement, authors used HRP-conjugated antihuman IgG [42]. Nanocatalyst $IgG/Au@Fe_2O_3$ is involved in the process of the incubation of plasma and serum samples which already contained p53-specific autoantibody placed on the carbon electrode surface. Oxidation of 3,3′,5,5′-tertamethylbenzidine, which has been occurred in the presence of hydrogen peroxide, is catalyzed by the surface-attached $Au@Fe_2O_3$ nanoparticles. Finally, the level of p53 autoantibody could be monitored by UV/VIS and amperometric current measurements [42,52].

Ji et al. reported immunoassay that could be performed for quantitative detection of alpha fetoprotein (AFP) [53]. Immunoassay is facilitated by usage of nanoelectrocatalysts contained precious metal palladium nanoparticles and secondary antibodies (Ab2) loaded on carbon, decorated with magnetite microspheres ($Fe_3O_4@C$) - which are characterized by high surface area and satisfied adsorption behavior [53]. The observed high electrocatalytic activity for H_2O_2 reduction and the ease of magnetic separation provide preparation of highly sensitive immunoassay.

Zheng et al. revealed that Fe_3O_4 nanoparticles show an electrocatalytic activity towards the reduction of small dye molecules. Enhance of magnetite catalytic electrochemical signal could be achieved by attaching metallic nanocages (bimetallic Ag–Pd) to the magnetite surface. Performed investigation revealed that prepared nonenzymatic nanoelectrocatalyst could be employed as signal label for highly sensitive electrochemical cytosensing [54].

Furthermore, in electrochemical bioassays, not only magnetic nanoparticles, but redox-doped magnetic beads also show attractive properties [55]. Zhang et al. used magnetic zirconium hexacyanoferrate nanoparticles (ZrHCF) as signal nanoprobes, in order to prepare electrochemical biosensor that could be used for highly sensitive DNA assay. Nanocomposite material used as an electrocatalyst is obtained by preparation of core-shell nanoparticles, consisted of magnetic core coated with ZrHCF shell [55]. Obtained nanoelectrocatalyst showed enhanced electrocatalytic behavior, which is desirable for the reaction of hydrogen peroxide reduction, and resulting in the increase of the electron transfer rate which occurs on the modified electrode.

Another example presents electrochemical immunosensor that could be used for detection of cancer biomarker squamous cell carcinoma antigen (SCC-Ag), fabricated by Wu et al. Immunosensor was prepared by using material based on graphene nanocomposite consisted of palladium nanotubes and magnetite/graphene sheets (Pd@Fe_3O_4/graphene). Pd@Fe_3O_4/graphene nanocomposite presented matrix, employed for immobilization of primary anti-SCC antibody (Ab1), as well as mark, used for conjugation with secondary anti-SCC antibody Ab2 [56]. Prepared immunosensor showed very good reproducibility and stability. Moreover, the immunosensor was probed in order to analyze of clinical serum samples, confirming its applicability in this area [56].

2.2   Cobalt-based nanocomposite materials as electrocatalyst

A literature review showed that nanocomposites based on cobalt-containing compounds could be an appropriate replacement for precious-metal based electrocatalysts in ORR, OER and HER reactions.

Behret et al. performed investigation of electrocatalytic properties of transition metal chalcogenides containing the cationic (Fe, Co, Ni) and anionic (S, Se, Te) substitution [57]. Results revealed the next trend for ORR electrochemical performance: in the case of chalcogenide anions: S > Se >Te, and in the case of metal cations: Co > Ni > Fe. The authors found that the best ORR replacement for noble metal electrocatalysts in ORR reaction in alkaline media is CoS compound, due to similar energies of the d orbital of sulfides and 2p orbital of oxygen [58]. In acidic media, Co_9S_8 is found to be the best replacement [59]. Dai et al. noticed that the ORR reaction is dependent on the size and structure of nanoelectrocatalysts [60]. Different particle sizes influences the electrocatalytic active surface [61]. Also, a literature review revealed that influence of crystal structure of the materials characterized by the same chemical composition resulted in various ORR catalytic activity [62-65]. The particle size and structure can be approximately adjusted in the desired interval by controlling the concentrations of the precursors involved in the synthesis, which presents a parameter that strongly affects the processes of the nucleation and Ostwald ripening [66]. Specifically, DFT studies revealed that in ORR [66], the process of oxygen adsorption influences the nanoparticles size of which the catalyst is consisted, what appeared as a consequence of the presence of catalysts active sites [67].

On the other hand, Co_3O_4 is recognized as an appropriate nanoelectrocatalyst for OER [68,69]. Modification in the synthesis method brings to the variations in morphology of Co_3O_4, which increases its electrochemical activity, favoring the OER reaction. Zheng et al. showed that, compared with Co_3O_4 nanoparticles, mesoporous Co_3O_4 nanowires are characterized by seven-fold increased electrochemical activity of OER reaction [68]. After the chemical reduction process, experimental results depicted to the increased number of active sites, appeared as a result of the presence of the oxygen vacancies [68].

Xia et al. synthesized 3D ordered mesoporous cobalt oxide particles (Co_3O_4), cubic in shape, by employing the hard templating method [70]. X-ray photoelectron spectroscopy measurements showed that the molars surface ratio Co^{3+}/Co^{2+} of the prepared Co_3O_4 is lowered in comparison with bulk cobalt, pointing to the fact that nano Co_3O_4 is characterized by more surface oxygen vacancies, which facile the oxygen adsorption and increased activity of molecular oxygen.

As a potential candidate for HER, Li et al. [71] performed research with the aim to examine the combined effect of morphology and structure of $CoSe_2$ [71]. Authors synthesized $CoSe_2$ nanotubes in different structures, orthorhombic (o-$CoSe_2$) and cubic (c-$CoSe_2$) by the simple precursor transformation method. Experimental investigation revealed that orthorhombic phase possessed the highest HER performance [71].

Also, a literature review revealed that cobalt-containing compounds in combination with conducting carbonaceous materials could be an appropriate replacement for precious-metal based electrocatalysts in ORR, OER and HER reactions. Nanocomposite materials based on cobalt-containing compounds in combination with conducting carbonaceous materials [72-80] are characterized by increased surface area and electrical conductivity, as well as by increased rate of the electron transfer, which are properties that ensure increased catalytic activities of these materials.

Furthermore, bimetallic cobalt based nanocomposites could also be used in ORR, OER and HER processes. Kuang et al. [81] synthesized an N-doped carbon framework with a bimetal nanoparticles of copper and cobalt embedded in this material (CuCo@NC), by using the preparation method based on the in situ growth of ZIF-67 polyhedrons on copper hydroxide nanowires, which is followed by the treatment of thermal decomposition. Investigation revealed that the existence of Cu cations ensure additional active sites, which resulted in the increased amount of nitrogen within the prepared carbon framework, and in that way enabled alteration of the porous structure and increase of the surface area, which is of significance for the rise of the overall ORR electrocatalytic activity. According to literature data, the prepared CuCo@NC material is characterized by very high ORR activity (E_{onset} = 0.96 V and $E_{1/2}$ = 0.88 V), in comparison with commercial 30% Pt/C (E_{onset} = 1.04 V and $E_{1/2}$ = 0.84 V) [81].

2.3   Nickel-based nanocomposite materials as electrocatalyst

Similar as different bimetallic Co-based nanoelectrocatalysts, bimetallic Ni-based compounds could be used toward the OER process, due to their low conductivity [82]. NiFe-based compounds could be used in water splitting reactions by loading these compounds on the appropriate electrodes. Zhu et al. found that material based on composite consisted of nickel hydroxide and iron hydroxide particles deposited on microarrays containing coral-shape copper oxide and copper ($Fe(OH)_3/Ni(OH)_2/c$-CuO/Cu), could be employed as electrode and OER electrocatalyst which shows very good properties [82]; it is expected that prepared material can be used for the purpose of construction of semiconductor electrocatalysts, which could be used in water splitting reactions [82].

Further, the literature reported possibility to use nickel phosphide (Ni_2P) hollow nanoparticles as nanoelectrocatalysts in HER processes [83]. Investigation performed by Popczun et al. revealed that nanoelectrocatalyst, based on hollow Ni_2P nanoparticles prepared in a way which enables an increased surface density of (001) plane, is characterized by very high activity for HER reaction. Comparison of the HER activity of this electrocatalyst and non-precious metal electrocatalysts confirmed that electrocatalyst based on Ni_2P nanomaterial exerted the highest electrocatalytic activity for HER processes [83].

Noteworthy, hydrothermally prepared Co-Ni-P ternary nanowires could also be used as nanoelectrocatalysts in HER reaction [84]. Li et al. found that electrodes comprising nanocomposite based on cobalt, nickel and phosphorus (Co-Ni-P) wire arrays, deposited on metal (Ni) foam as a collector, ensure electrocatalytically active sites, efficient conductivity, and facilitate mass transport. Observed features are explained in the term of the monolithic electrodes structure, and unusual porous nanowire morphology [84].

Represented examples confirmed the fact that doping process and variations in synthesis parameters enables improvement and tailoring of nanoelectrocatalysts. For this reason, preparation of different alloys based on nickel and other metal is intensively investigated in the literature, and experimental results showed that alloying enable facilitation HER processes by prevention of Ni-hydride forming and rise the electrodes durability toward the HER process [85,86]. What is interesting, from all nickel alloy catalysts, Ni-Mo-based alloys are recognized as the most suitable for HER electrocatalysis, since it shows the best HER activity in alkaline media, although the mechanism which brings increased electrocatalytic activity of this type of catalysts has not yet been clarified [21]. In acidic media, nickel sulfide and nickel phosphide (NiS and Ni_2P) showed good electrocatalytic activity for HER reaction [21].

Magnetochemistry - Materials and Applications   Materials Research Forum LLC
Materials Research Foundations 66 (2020) 34-86   https://doi.org/10.21741/9781644900611-2

Ni-based compounds are also represented as catalysts in hydrogen oxidation reaction (HOR) [87,88]. Among 3d transition metals, nickel shows the best HOR activity [81]. In spite of this, undoped Ni nanoelectrocatalyst still possess nearly two order of magnitude lower activity than platinum or palladium [87]. It is important to notice that nowadays there is a high discrepancy in the level of understanding the role of dopant in the synthesis of nanoelectrocatalysts, as well as regarding knowledge of the influence of dopant on the electrocatalytic activity of HOR catalysts; in this scientific area, various experimental results were reported. Nevertheless, Davydova et al. investigated electrocatalytic properties of material based on metal nickel nanoparticles deposited on carbon, doped with different transition metal dopants (Fe, Co and Cu), and obtained results revealed that Ni-Fe/C compound is the most promising catalyst for HOR in alkaline media [89].

3.   Structure and morphology of magnetic nanoparticles used in electrocatalysis

Let us recall the fact that ordered and symmetrical arrangement of atoms in a crystalline material presents its crystal structure. In order to better imagine a crystal structure, the term "crystal lattice" has been introduced. It represents mathematical concept describing crystal structure as a set of infinite, arranged atoms related to each other by transitional symmetry. According to symmetry and space rotations, crystal lattices are classified into seven crystal systems, characterized by different symmetries. Hume-Rothery postulated correlation between crystal structure of transition metal elements and electron concentration [90]. Also, the literature pointed to the correlation between electrical conductivity and catalytic properties. The activation energies for some catalytic reactions are comparable with activation energies for conductivity; studies performed on the noble metal-based catalysts revealed that the investigated systems behave as semiconducting systems [91].

On the other hand, magnetic behavior of materials could be explained by different theories that include the symmetry and principal of the quantum mechanics. In the crystal field theory, the shape of the atomic orbitals and filling the orbitals with electrons are of key importance for the formation of the atomic structure, which determines the other properties, such as structural, magnetic, electric, catalytic, etc. In this process it is of high significance to pair the energy, appearing in the system as a result of influence of different energies, such as energy of crystal field and the Coulomb energy required for the positioning of two different spins in the same orbital. Dependent on the type of local environments (for ex. octahedral/tetrahedral), it could be predicted the precise order in adding electrons [4], due to the fact that the symmetry of the local environment dominantly influences the size and nature of crystal field effects [4]. In the most simply

Magnetochemistry - Materials and Applications   Materials Research Forum LLC
Materials Research Foundations **66** (2020) 34-86   https://doi.org/10.21741/9781644900611-2

case, symmetry determines the local crystal environments, what enables deduction of the electronic structure and, dependent on the number of electrons which fill up the energy levels, magnetic properties [4]. In the huge number of transition metal compounds, a transition metal cation is positioned in the centre of an octahedron, while the anion is placed on each corner. In such compounds, crystal field appears as a consequence of electrostatic repulsion from the negatively charged electrons in the oxygen orbitals [4], and usually it can be relatively easy to predict how the electrons will be added [4]. The magnetic ordering of the compound is maintained by the presence of magnetic interactions (forming so-called "magnetic lattice"), and magnetic interactions define final magnetic state of the crystalline material.

Contrary, in certain number of cases, the magnetic properties could impact the symmetry of the local environment. This phenomenon presents Jahn-Teller effect, and occurs when it is more energetically favorable for octahedron/tetrahedron symmetry to spontaneously distort in order to keep balance between energy cost of summary electronic energy (which is saved by a distortion) and increased elastic energy [4].

Obviously, compounds containing transition metal elements and characterized by the same chemical composition, could show different magnetic properties (for example iron oxide species: γ-Fe_2O_3, ε-Fe_2O_3, α-Fe_2O_3[92, 93]), due to the fact that every phase and/or structural transformation induces changes in the crystal field of the atomic structure, consequently leading to the alteration of magnetic lattice. Vice versa also could be applied; compounds characterized by various chemical compositions could belongs to the same magnetic state showing characteristic magnetic behavior.

Having in mind correlation between catalytic and electrical properties, as well as presence of correlation between electron concentration and electrical properties (which are additionally coupled with magnetic properties through spin-orbit coupling), it is clear that there is connection between materials magnetic and catalytic behavior, although this relation is more sophisticated and still not sufficiently investigated.

It is important to notice that transition from bulk to nano dimensions induces change in the ratio of the number of surface atoms and the number of the atoms represented in the residue of the nanoparticle, what result in the appearance of surface effects [94]. Due to the presence of surface effects, nanoparticles characterized by the same structure, but different size and morphology, could show various properties. Regardless, the nanomaterials behavior is still mainly dependent on the symmetrical arrangement of atoms, what determined its structure. Since synthesis parameters enable tailoring of the structure, nanomaterials structure has a special importance for determination of materials behavior, because controlling the synthesis of the various structures enables control of materials properties, such as magnetic, electric, catalytic, etc.

The crystal structure of different magnetic nanomaterials which electrocatalytic properties are intensively investigated usually belongs to spinel, perovskite or garnet structure. A literature review revealed that electrocatalytic behavior of nano-sized spinels and perovskites is highly and intensively investigated. Figure 2. presents the number of articles devoted to this topic (statistical data are taken from Google Scholar).

In Figure 2. is shown that nearly same number of the articles is devoted to the examination of electrocatalytic properties of both perovskites and spinels, although statistical treatment of the literature data showed a gentle difference in favor spinels. Studies about garnets as an electrocatalysts are also represented in literature. Noteworthy, in recent years, electrocatalytic properties of perovskite-spinel nanocomposites have started to be examined although it has not yet been sufficiently probed.

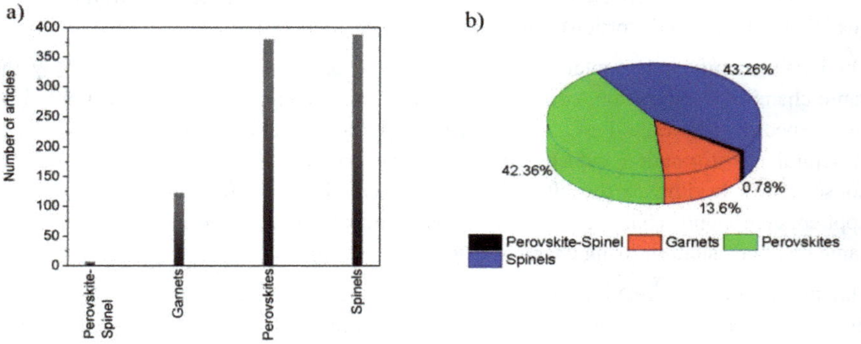

Figure 2. a) The number of articles devoted to electrocatalytic activity research of nanostructured perovskite-spinels, garnets, perovskites and spinels. b) 3D Color Pie Chart. Statistical data are obtained by search of the first 40 pages of Google Scholar, 2019 (accessed 01 February 2019).

In the text below is listed some basic properties of spinel, perovskite, and garnet structures. It is important to notice that although bulk ferrites and garnets belongs to a family of ferrimagnets [4], while perovskites could show various magnetic ordering in bulk form [95-97], magnetic behavior of mentioned structures in the nano dimensions could be significantly altered. The presence of new magnetic states characteristic for nanomaterials, such as SPM or SSG [98-101], expands their usage.

Spinel structure could be explained by formula XY_2Z_4; letter "X" refers to tetrahedrally, and "Y" to octahedrally coordinated cations, while the letter "Z" presents anion. Till today there are known three types of spinels: normal, inverse and mixed. Structure of some compound belongs to the normal or inverse spinel structure is dependent, primarily, on the crystal field stabilization energies of ions positioned in the A and B crystallographic sites [102,103]. Structure of normal spinels proposed close-packed array of anions at A- and B- crystallographic sites. At the A-sites, cations are placed in 1/8 tetrahedral holes, while at B-sites, cations are positioned in 1/2 of the octahedral holes. Structure of inverse spinels is described similar (unit cell is characterized by the same large as in the case of normal spinels), with the difference that A-site cations and B-site cations switch places [102]. Accordingly, inverse spinels could be described with the formula $Y(XY)Z_4$. The letter "Y" denotes cations placed on tetrahedral sites, while "(XY)" cations occupy cation positions in octahedral sites. Mixed spinels are characterized by the structure of intermediate (normal/inverse) spinels.

In the Fig. 3 is presented structure of Fe_3O_4, that could be used as a nanoelectrocatalyst for oxygen reduction reaction [101]. All structures are obtained by using Mercury Program [104], and Inorganic Crystal Structure Database [105].

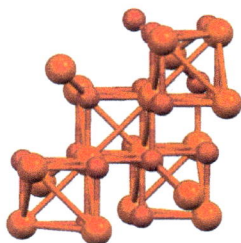

Figure 3. Fe_3O_4 structural fragment. Light red colored balls presented iron ions, red balls - oxygen ions.

Perovskites, as well as spinels structure, belongs to the close-packed crystal lattices systems [102]. Perovskites presents ternary oxides that could be described by the general formula XYZ_3. The letter "X" presents the larger cation (such as Rb^+, Sr^{2+}, Ba^{2+} or lanthanide 3+ cations), "Y" is the smaller transition metal cation (such as Ti^{4+}, Nb^{5+}, Ru^{4+}, etc.), while the letter "Z" is ascribed to the anion (such as oxygen, nitrogen or halogen). Perovskite structure proposed simple cubic symmetry, although the A-site cations and the "Z" anions comprise a fcc lattice. In other words, the structure is flexible; the coordination of "X" cations and the arrangement of "YZ_3" octahedra enable rotation and cooperatively tilt of the octahedra. Octahedra surrounded the site of the X cations presenting cuboctahedral cavity [106,107].

Due to the flexibility of the structure, some perovskites experienced the distortion of the structure resulting in the appearance of an electrical dipole and ferroelectricity [108]. Multiferroics presents a special class of perovskites showing ferroelectric and ferromagnetic behavior simultaneously, and for that reason are very interesting for examination.

The example of the material characterized by perovskite structure that could be used as nanoelectrocatalyst in different processes (for example, oxygen reduction reaction), is lanthanum manganite, $LaMnO_3$ [109,110], Fig. 4.

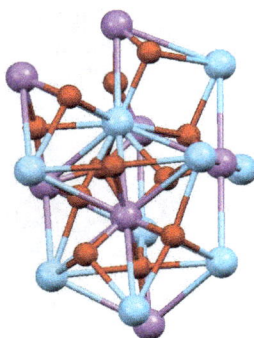

Figure 4. LaMnO₃ structural fragment. Blue balls presents - lanthanum; violet - manganese; and red - oxygen.

Perovskite-spinel structure presents combination of both structures: perovskite and spinel, and belongs to the group of multiferroics. Let us recall the fact that ferromagnetic behavior is characterized by a stable and switchable magnetization, while ferroelectric is characterized by a stable and switchable electrical polarization [111]. As a result, perovskite-spinel structures show novel properties and large potential from an applications point of view. A literature review revealed that this type of material could be used as an electrocatalyst [112], although the number of the investigation of electrocatalytic activity of the materials based on perovskite-spinel nanocomposites is still small (Fig. 2).

The garnet structure accommodates various cations characterized by significantly different sizes and valence states [113]. Crystal structure of garnets could be described by the general chemical formula: $X_3Y_2Z_3O_{12}$, where the letter "X" presents cations Mg^{2+}, Fe^{2+}, Ca^{2+}, letter "Y" is ascribed to Al^{3+}, Cr^{3+}, Fe^{3+}, while letter "Z" presents silicon. The crystal structure is cubic, although the unit cell is complex and consists of the X-site being eight coordinated (dodecahedral symmetry), the Y-site being six-coordinated (octahedral symmetry), and the Z-site being four-coordinated (tetrahedral symmetry) [114]. The complex cubic crystal lattice implies identical oxygens, related to one octahedron, one tetrahedron, and to two divalent dodecahedral sites (circles), all together building 96 per unit cell [114]. Two subgroups of garnets are ugrandite, usually consisting chromium and/or vanadium, and pyralspite, consisting ferrous iron [115].

Magnetic properties of nanogarnets are strongly dependent on its phase composition and the size of the particles [100]; although nanogarnets could be characterized by SPM behavior, the magnetism of the garnets is originated from the presence of higher concentrations of magnetically ordered elements, such as iron (up to 35% iron oxide by wt.) [115].

As an example of a garnet structure, Figure 5. shows a structure of lithium lanthanum zirconate, $Li_7La_3Zr_2O_{12}$, that could be used as nanoelectrocatalyst for Li-ion batteries [116].

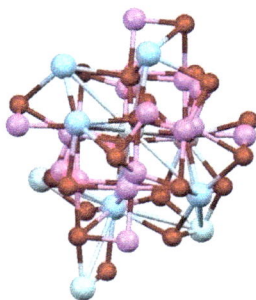

Figure 5. $Li_7La_3Zr_2O_{12}$ structural fragment derived. Blue colored balls presents lanthanum ions; light blue - zirconium; violet - lithium; and red - oxygen.

3.1 Spinel ferrites in nanoelectrocatalysis

Former is shown that nano-sized spinels are thoroughly examined as potential replacement for noble metal-based catalysts in industrial processes for electrochemical energy conversion (Figure 2). Crystallography classified compounds characterized by spinel structure based on its chemical composition into next spinel groups: iron spinels, aluminum spinels, chromium spinels, and spinels with the structure of ringwoodite $((Mg,Fe)_2SiO_4)$, coulsonite (FeV_2O_4) or magnesiocoulsonite $(MgCr_2O_4)$ [117]. From the crystallographic point of view, the most important class of spinels used in electrocatalysis are iron spinels, ferrites - mixed oxides with general formula MFe_2O_4 [118]. Although ferrites could be characterized by different structures (spinel, garnet, hexaferrite and orthoferrite [119]), accurately spinel ferrites presents the most investigated class of magnetic spinels in term of potential application in electrocatalysis.

Among them, species based on mixed transition metal oxide spinels are intensively investigated due to their prominent electrocatalytic activity and low costs, as well as due to variety of interesting catalytic, magnetic, electronic and optical properties [120]. The most interesting potential nanoelectrocatalysts could be found within the iron cubic spinel ferrites described by formula - MFe_2O_4, where M refers to 2+ cations of Cu, Co, Ni, Zn, Mn, Mg, Cd, Ba. In the next part will be more discussed ferrite spinel nanomaterials as candidates for electrocatalysis [121-131].

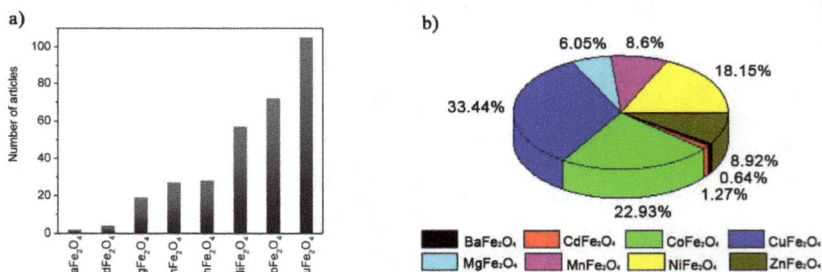

Figure 6. a) Graphical representation of the number of articles that investigated catalytic properties of listed ferrites ($CuFe_2O_4$, $CoFe_2O_4$, $NiFe_2O_4$, $ZnFe_2O_4$, $MnFe_2O_4$, $MgFe_2O_4$, $CdFe_2O_4$, $BaFe_2O_4$); b) 3D Color Pie Chart. Statistical data are obtained by search performed on 01.02.2019. for the first 40 pages of Google Scholar.

Below is given graphical illustration of interests of scientific community for investigation of nanocatalytic properties of ferrite spinels. Statistical data are obtained considering first 40 pages of Google scholar, collected in 2019 (accessed 01. February 2019). Figure 6 represents number of articles whose topic represents investigation of electrocatalytic properties of different nanoparticle ferrite spinels which shows catalytic activity ($CuFe_2O_4$, $CoFe_2O_4$, $NiFe_2O_4$, $ZnFe_2O_4$, $MnFe_2O_4$, $MgFe_2O_4$, $CdFe_2O_4$, $BaFe_2O_4$ (Fig. 6)).

Among all ferrite spinels, catalytic properties of nano-sized copper ferrite attracted the most of scientific attention. According to Fig.6, nano-sized copper ferrite catalytic behavior is investigated more than catalytic activity of any other ferrite (105 articles). Considering the number of articles written in order to investigate electrocatalytic activity of magnetic transition metal ferrites ($CoFe_2O_4$ and $NiFe_2O_4$), it is obvious that cobalt ferrite presents second ferrite with the most investigated catalytic properties (72 articles), while nickel ferrite is third (57 articles).

Interestingly, one of 3d transition element spinel ferrite that is not recognized in the literature as suitable for application in nano electrocatalysis is $CrFe_2O_4$. Although $CrFe_2O_4$ in combination with some other compounds built nanocomposites that could be used for different potential applications, in the literature is no comments about its potential usage for electrocatalysis. This obstacle could be overcome by deeper search of synthesis methods that could enable formation of nanocomposites containing $CrFe_2O_4$ for this purpose. Various synthesis approaches which enable preparation of materials based on chromium ferrite give evidence how structure and properties of nanomaterials could be altered and improved. One of the examples how alteration of synthesis influences structure and properties of nanocomposites containing chromium ferrite could be improved, could be found in doping chromium ferrite or preparation of chromium ferrite-based nanocomposite. Application of solution combustion method resulted in the preparation of cubic spinel structure nanocomposite - nickel doped chromium ferrite ($Ni_xCr_{1-x}Fe_2O_4$ (x=0, 0.25, 0.5, 0.75, 1)) [132]. On the other hand, the sol-gel method enables synthesis of nanocomposite based on $CrFe_2O_4$ and $BiFeO_3$, which is characterized by perovskite multiferroic structure [133]. Pandu noticed that the effect of synthesis with different dopants induces alterations in structure of composite nanomaterial, what simultaneously resulting in changes of its ferroelectric and dielectric properties. Consequently, author depicted to the possibility to change conductivity of the investigated material by reinforcing $CrFe_2O_4$_$BiFeO_3$ with carbon nanotubes [133]. Having in mind that modification of synthesis routes, as well as preparation of new materials, moves the boundaries in the nanomaterial science, it could be expected that in the future will be discovered nanocomposite material containing chromium ferrite, that could be appropriate for electrocatalysis.

That is only one of the possibilities, although different synthesis approaches or variation of synthesis conditions could be performed in order to obtain nanocomposites characterized by various properties, and in that way enabled tailoring of the nanomaterials potential application.

In the text below is listed some chosen applications of the most investigated spinel ferrite-based nanocomposites: $CuFe_2O_4$, $CoFe_2O_4$, and $NiFe_2O_4$.

3.1.1 Nanoelectrocatalytic applications of $CuFe_2O_4$-based nanocomposites

In the year 2015 Shahnavaz et al. found that an enzyme-free sensors, employed in order to detect glucose in alkaline conditions, could be prepared by using oxidative pyrrole chemical polymerization performed on the surface of core-shell nanoparticles, consisted of copper ferrite ($CuFe_2O_4$) in the core of nanoparticles, and polypyrrole (PPy) shell [134]. The nanocomposite material showed very high electrocatalytic activity, and it is recognized as a potential enzyme-free sensor [134].

Benvidi et al. showed that a highly sensitive voltammetric sensor, containing a carbon paste electrode (CPE) decorated with nanocomposite based on reduced graphene oxide (RGO) and copper ferrite ($CuFe_2O_4$) nanoparticles, could be used in order to detect and determine hydrogen peroxide (H_2O_2) [135]. $CuFe_2O_4$ nanoparticles were synthesized by coprecipitation procedure, starting from $Cu(NO_3)_2 \cdot 3H_2O$ and $Fe(NO_3)_3 \cdot 9H_2O$ solutions, using NaOH as precipitated agents [136]. Deposited nanocomposite acts as a nanoelectrocatalyst in the reduction reaction of H_2O_2, and designed sensor possessed very good electrocatalytic activity. Electrochemical sensor was already successfully probed in the H_2O_2 assay in biological and pharmaceutical samples. Also, excellent results are obtained during the examination of mouthwash solution, hair dry cream, as well as in the milk and green tea [135].

Pan et al. investigated a series of the samples based on $CuFe_2O_4$ nanoparticles deposited on glassy carbon electrode, with the aim to investigate ability of the material for reduction of p-nitrophenol, potassium chromate ($K_2Cr_2O_4$), and sodium nitrate ($NaNO_2$). This study revealed the influence of the variation of synthesis procedure onto the effectiveness of electrocatalytic properties of copper ferrite: if the $CuFe_2O_4$ nanoparticles are prepared with cyclohexylamine, obtained material showed better catalytic activity for p-nitrophenol, while preparation of mentioned ferrite with sodium hydroxide and PVP exhibited the highest activity for both $K_2Cr_2O_4$ and $NaNO_2$ reduction [137].

Furthermore, copper nanoparticles [137] as well as copper [138,139] supported on metal-oxide, presents active electrocatalysts that could be used in WGS reaction. Estrella et al. investigated the properties of $CuFe_2O_4$ and Cu/Fe_3O_4 catalysts, performed to WGS reaction conditions [140]. Authors found that interactions between metal and oxygen increased the stability of Fe^{3+} and Cu^{2+} cations in the $CuFe_2O_4$ crystal lattice [140]. In the temperature range from 350 °C to 450 °C occurred a significant reduction of $CuFe_2O_4$, resulting in Cu^0 and Fe_3O_4. Investigated samples shows electrocatalytic activity for the production of H_2 [140]. A lack of similar systematic studies slows discovery and development of more efficient electrocatalysts.

3.1.2 Nanoelectrocatalytic applications of $CoFe_2O_4$-based nanocomposites

Yu et al. prepared electrochemical biosensor that could be used for detection of micro ribonucleic acid (RNA), by employing nano-sized electrocatalysis without substrate [141]. Investigated biosensor was based on a padlock exponential rolling circle amplification (P-ERCA) assay, and cobalt ferrite ($CoFe_2O_4$) nanoparticles-assisted non-substrate electrocatalysts. This type of nano electroatalysis has been enabled by introducing redox molecule terbium, deposited onto the graphene [141]. Cobalt ferrite nanoparticles are chosen for this application because they possess an intrinsic peroxidase-like activity with H_2O_2 substrate [142-144], as well as because they act as nanoelectrocatalyst for signal amplification and in the absence of H_2O_2 [145]. P-ERCA assay was added in order to increase the sensitivity of the biosensor. The biosensor was probed in order to detect miR-21, and experimental results showed that the sensor is characterized by very high specificity and sensitivity [141].

Ding et al. showed that mesoporous $CoFe_2O_4$ nanoparticles possess satisfactory properties to participate in H_2O_2 reduction reaction as electrocatalyst [145]. Cobalt ferrite nanoparticles are obtained by a hard-templating synthesis assisted by aluminium oxide (Al_2O_3) [145]. Prepared nanoparticles are characterized by a superior mesoporous nanostructure with a high specific surface area (140.6 m^2g^{-1}) and volume (0.2410 cm^3g^{-1}), what enhance its electrocatalytic activity. Cobalt ferrite electrode was appropriate for HRR, showing good stability, and promising usage for H_2O_2-based alkaline fuel cells [145].

Liu et al. found that $CoFe_2O_4$ nanoparticles could be used as efficient electrocatalysts in OER and ORR processes [146]. Nanocomposite material was based on $CoFe_2O_4$ nanoparticles and biocarbon, and characterized by hierarchical structure and high surface area. The nanocomposite is obtained by facile biosynthesis method [146]. Experimental results revealed that synthesized material showed excellent catalytic activity, much higher in comparison with single phase, cobalt ferrite nanoparticles, or single phase -

biocarbon. Chronoamperometric measurements revealed that nanocomposite electrocatalyst possess high durability, that overcome durability of commercially used catalyst, Pt/C [146], what is attributed to two reasons; one is intensive coupling between ferrite nanoparticles and biocarbon, and the other is ascribed to hierarchical structure of prepared nanocomposite [146].

Yin et al. prepared $CoFe_2O_4$ nanoparticles characterized by a significant presence of oxygen vacancies and high specific active surface, by using hydrothermal synthesis [147]. Experimental investigation revealed that prepared nanoparticles could have a dual role: they showed remarkable long-term stability and improved electrocatalytic performance for HER and ORR, as well as they could be used as an air-cathode in Zn-air batteries [147].

3.1.3 Nanoelectrocatalytic applications of $NiFe_2O_4$-based nanocomposites

Abu-Dief et al. performed hydrothermal synthesis in order to obtain small (2-7 nm) $NiFe_2O_4$ nanoparticles [148]. Prepared nanoparticles are employed as catalysts in Claisen-Schmidt condensation reaction, which involves usage of acetylferrocene and aromatic/heterocyclic aldehydes, and enables obtaining of acetylferrocene chalcones in a high yields, for short time. Since the catalyst shows prominent magnetic properties, the authors showed that the catalyst could be recovered by the usage of an external magnetic field, and be reused a few times without losing its initial catalytic activity [148].

Ensafi et al. investigated nanocomposite material based on the nickel ferrite nanoparticles and carbon nanotubes, which could be used as a modified electrode [149]. Performed experimental research revealed that obtained material possesses promising properties, and could be used as electrocatalyst in the reaction of the sotalol oxidation. Investigation shows that prepared catalyst could be employed in order to detect sotalol in pharmaceutical and medical samples [149].

On the other hand, graphene doped with nitrogen and decorated with $NiFe_2O_4$ nanoparticles ($NiFe_2O_4$/N-Gr), is recognized as a novel nanohybrid, that showed promising catalytic properties for ORR in fuel cells [150]. Kiani et al. prepared nanocomposite by simple hydrothermal method, in order to obtain 20 nm in size nickel ferrite particles. The material showed very good catalytic properties, which is ascribed to high coupling of the graphene and nickel ferrite nanoparticles [150].

Moreover, Linghua et al. showed that nanocomposite based on graphitic carbon nitride and nickel ferrite nanoparticles possess excellent catalytic activity toward reaction of thermal decomposition of ammonium perchlorate [151].

3.2   Size and morphology of magnetic nanoparticles used in electrocatalysis

The transmission electron microscopy (TEM) presents the usual tool used for investigation of particle size, shape and morphology. Having in mind that these properties could influence the behavior of nanomaterials, in this part will be shortly listed some chosen examples of TEM micrographs of standard magnetic nanomaterial, applicable in electrocatalysis.

TEM analysis shows that magnetic nanoparticles of various sizes or morphologies could be used as electrocatalysts (Fig. 7). TEM micrographs of chosen examples of nanoelectrocatalysts are represented in Fig. 7; micrograph in the Fig. 7a) is recorded under the resolution of 200 nm, while the rest of micrographs is obtained with higher resolution, 50 nm.

Figure 7a) represents micrograph of the nanohybrid based on NiFe alloy nanoparticles loaded on graphene oxide (NiFe/GO) which could be used as electrocatalyst in electrochemical glucose sensing [152]. NiFe/GO material is prepared by a modified Hummers' method [153]. NiFe nanoparticles are irregular although nearly square in shape, with the sizes of 100 nm. Experimental results showed that deposition of NiFe/GO nanoparticles on the glassy carbon electrode enabled high selectivity and sensitivity for glucose, what is of importance for application in glucose sensing [154].

In the micrograph presented in Fig. 7b) is shown nano hybrid based on $Fe_3O_4\_Fe$ nanoparticles, that could be used as an electrocatalyst for the oxygen electrode of lithium-oxygen batteries [44]. Nanocomposite material is synthesized by low-cost method, via an electrical wire pulse process. Nanoparticles are nearly spherical in shape, 60 nm in size. Experimental results confirmed superior behavior of the electrode loaded with $Fe_3O_4\_Fe$ nanocatalysts: high discharge capacity (presented as a consequence of the good electrical conductivity of iron metal nanoparticles), low overpotential, long cycle stability, and fixed capacity regime [44].

Fig. 7c) showed micrograph of the nanocomposite based on cobalt ferrite nanoparticles and nitrogen doped activated carbon ($CoFe_2O_4$@N-AC), that could be used as an effective electrocatalyst for air cathode in microbial fuel cells [155]. Spinel $CoFe_2O_4$ nanoparticles are highly aggregated, 10-20 nm in size, synthesized by a hydrothermal method. Synthesized nanomaterial is characterized by large surface area, less total resistance, fast electron transport and very good electrocatalytic activity for ORR process [155].

The micrograph in Fig. 7d) presents the nanohybrid based on $NiFe_2O_4$ nanoparticles, cross-linked with multiwalled carbon nanotubes ($NiFe_2O_4$/MWCNT), which could be used as electrocatalyst for OER process [156]. Nickel ferrite nanoparticles are prepared by hydrothermal method. TEM shows that $NiFe_2O_4$ nanoparticles are 10-20 nm in size. Experimental investigation showed that prepared nanohybrid possesses much higher OER catalytic activity in comparison with single $NiFe_2O_4$, MCWNT, or commercial Pt/C electrocatalytic activity [165].

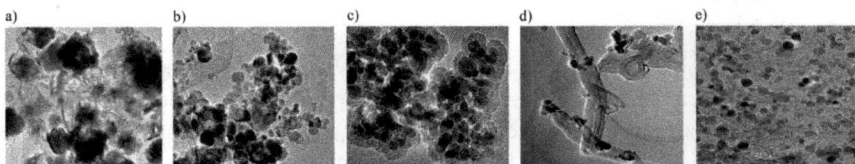

Figure 7. TEM micrographs of chosen nanoelectrocatalysts. Drafted by this author based on the literature data.

In Fig. 7e) is shown a micrograph of the nanohybrid based on $NiFe_2O_4$ nanoparticles and nitrogen doped graphene ($NiFe_2O_4$/N-Gr), which could be applied as an catalyst in ORR processes in fuel cells [157]. Nanohybrid is obtained by hydrothermal method, and nickel ferrite nanoparticles are 20 nm in size, flake in shape. Obtained material showed better electrocatalytic activity compared with both of each single phase that contribute to nanohybrid [157]. The origin of the nanocomposite improving the electrocatalysis properties is found in the nanocomposite structure, which consists of disordered crystal planes of nickel ferrite and the nitrogen-doped graphene at the interfaces. Disorder of crystal structure causes the appearance of further defects, which could be advantageous for oxygen molecules [157].

Fig. 7e) and c) presents one of the examples how various nanocomposite materials ($NiFe_2O_4$/N-Gr and $CoFe_2O_4$@N-AC) consisting different magnetic ferrite nanoparticles, could be used for the same purpose - as an electrocatalysts in ORR processes. The same synthesis method (hydrothermal method and nitrogen doping), but performed with altered synthesis conditions, resulted in preparation of different nanocomposites which could be used as ORR electrocatalyst (Fig. 7 c)-e)).

As a literature review revealed, based on Fig.7, it can be concluded that experimental characterization of the prepared nanoparticles, significantly differed in size and morphology, depicted to the fact that there is no limitation in size and morphology of the nanoparticles which could be used as potential electrocatalyst.

4. Influence of the synthesis parameters on the properties of nanocomposite materials of importance for catalysis

Although nanocomposite materials showed promising behavior for application in electrocatalysis, their overall potential for this purpose is still not sufficiently investigated, due to the lack of understanding relation between alteration of the synthesis parameters and structure of the synthesized nanocomposite material. Due to the fact that magnetic ordering influences crystal structure, as well as considering relation between magnetic, electric and catalytic properties, it is obvious that adjusting synthesis conditions enable tailoring properties of the materials via preparation of variously structured nanocomposites.

Having in mind that copper ferrite presents the most investigated material that could be used as one phase of nanocomposite materials applied in electrocatalysis, in the text below will be given example how variation of the copper and iron precursor concentration influences structural and magnetic properties of $CuFe_2O_4$-based nanocomposites obtained by the same coprecipitation procedure. Noteworthy, here should be mentioned that synthesis used for samples preparation were performed in atmospheric conditions, resulting in the simultaneous formation of a certain amount of CuO phase as a second phase, since stoichiometric $CuFe_2O_4$ cannot be synthesized by conventional techniques [158,159].

The most common usage of the nanocomposite containing $CuFe_2O_4$-CuO is as catalyst in water gas shift reaction [140]. A literature review revealed that active catalysts for WGS reaction presents extended surfaces of copper nanoparticles supported on metal oxides [160]. Noteworthy, Estrella found that an active role in WGS catalysis plays the oxide component of WGS catalysts [140], and the catalytic activity is strongly impacted by the properties of the support [161], which is determined by the synthesis conditions.

On the other hand, results in literature recognize $CuFe_2O_4$-Fe_2O_3-CuO nanocomposites as a material of importance for preparation of supercapacitor electrode in energy storage devices, due to the fact that this composite possess enhanced electrochemical activity [162]. It is important to notice that structural and magnetic properties of this material were not in detailed examined in the published study [162].

One of the parameters of high importance for determination of the materials structure is precursor concentration: higher precursor concentrations lead to higher rates of the reaction [163]. Accordingly, precursor concentration is varied in order to get deeper insight in its impact on the structure of nanocomposite materials. Since another pertinent parameter that determine materials structure presents formation mechanism, below will be commented formation mechanism of the examined samples.

On the other hand, with the aim to get deeper insight in the structural properties and magnetic behavior of these materials, further will be given preliminary characterization of the nanocomposite samples based on $CuFe_2O_4$ and CuO nanoparticles, prepared by coprecipitation method in the presence of NH_4OH as precipitated agent, which catalytic behavior has yet to be examined. Due to the importance of ferric hydroxide and copper oxide phases for understanding formation mechanism, samples consisted of 2L-Fhyd and CuO will also be discussed. It should be noted that the inability of determination of the precise 2L-Fhyd structure shadow insight in the formation of different intermediate species which presents a part of copper ferrite formation mechanism [164,165]. Preliminary results of the examination of the copper ferrite fundamental properties showed that even today rhw influence of synthesis parameter on the copper ferrite structure is not entirely understood. Investigation confirmed that precursor concentration presents synthesis parameter that strongly affects copper ferrite size and structure, which is of importance for further tailoring of $CuFe_2O_4$ properties and improvement of its electrocatalytic application.

4.1 Main structural properties of the phases within investigated samples

In the next part is given a brief overview on the main structural properties of the presented phases containing copper cations in the examined samples.

CuO possess the monoclinic structure, space group C2/c [166]; structure could be imagine as consisted of two zig-zag copper-oxide chains, that are propagate in [101] and [10$\bar{1}$] directions [167], Figure 8.

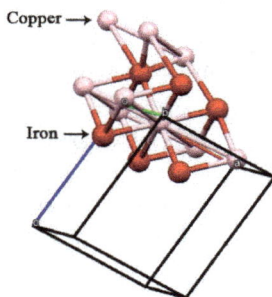

Figure 8. Representation of CuO structural fragment in the unit cell with axes.

$CuFe_2O_4$ crystallizes in two structures: $c-CuFe_2O_4$ and $t-CuFe_2O_4$ [168]. Thermal treatment initiates structural transformation of $CuFe_2O_4$ phase through cooperative Jahn-Teller effect [159]. Literature revealed that $t-CuFe_2O_4 \rightarrow c-CuFe_2O_4$ as well as $c-CuFe_2O_4 \rightarrow t-CuFe_2O_4$ structural conversion is possible with temperature increase [169-172].

$c-CuFe_2O_4$ presents inverse-spinel ferrite, characterized by face-centered cubic structure (space group: Fd3m) (Fig. 9)).

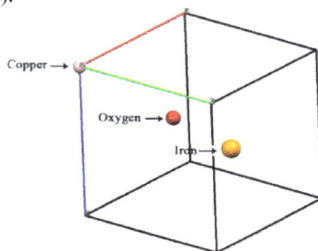

Figure 9. Representation of $c-CuFe_2O_4$ unit cell with axes.

Crystallographic data revealed that spinel ferrite $t-CuFe_2O_4$ belongs to body center cubic structure, with space group $I4_1/amd$ [173]. A unit cell with its axes is represented in Fig. 10.

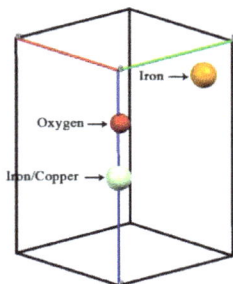

Figure 10. Representation of $t-CuFe_2O_4$ unit cell with axes.

4.2 Influence of the variation of Cu^{2+} precursor concentration on the $CuFe_2O_4$-based nanocomposite properties

Here is discussed the influence of the changing of the Cu^{2+} precursor amount during the coprecipitation synthesis, on the magneto-structural properties of the synthesized $CuFe_2O_4$-based nanocomposite.

Samples were prepared by using Cu^{2+} and Fe^{3+} cations in milimolar ratio 1:3 (S1), 3.1:3 (S2) [174,175]. Both samples were synthesized by the identical synthesis procedure, with the only difference in the amount of Cu^{2+} precursor involved in the coprecipitation. Synthesis started by dissolving $Fe(NO_3)_3 \cdot 9H_2O$ and $Cu(NO_3)_2 \cdot 3H_2O$ salts in water. As coprecipitation agent was used NH_4OH, in order to bring pH to 8. Samples were aged for 1h and rinsed with distilled water for a few times. Afterward, precipitates were dried and annealed for 3h at 1100 °C [174, 175].

XRD measurements showed that native powders of both samples consisted of 2L-Fhyd and CuO nanoparticles (Figure 11). Refinement in program Maud [176] revealed that the size of 2L-Fhyd coherent scattering domains in both samples was ~ 3 nm, while S1 and S2 crystallite sizes were similar, with the difference that sample S1 is consisted of slightly smaller CuO crystallites: d_{cr}^{S1} (CuO) ~ 23 nm and d_{cr}^{S2} (CuO) ~ 26 nm [174, 175].

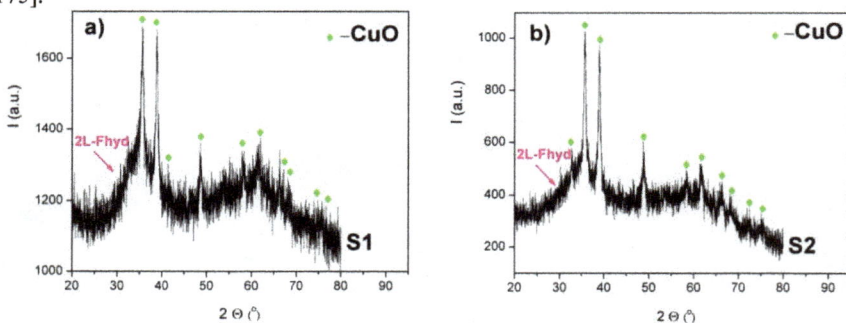

Figure 11. Diffraction pattern of the native samples: a) S1; b) S2 [175].

Fig. 12.a) showed that the sample S1 annealed at 1100 °C contained α-Fe_2O_3, $CuFe_2O_4$ and CuO phases, while sample S2 showed absence of the α-Fe_2O_3 phase (Fig. 12.b)).

Considering XRD pattern of the sample S1, it could be concluded that the particles of the biggest crystallite sizes within this sample presents hematite particles (~ 50 nm), and only slightly smaller in size c-$CuFe_2O_4$ nanoparticles: 46 nm [174, 175]. CuO nanoparticles are represented in such a small yield, which disables estimation of its crystallite size. Sample S2 showed presence of the c-CuFe$_2$O$_4$ nanoparticles characterized by the same crystallite size: 46 nm, and higher amount of CuO phase (crystallite size:~ 50 nm).

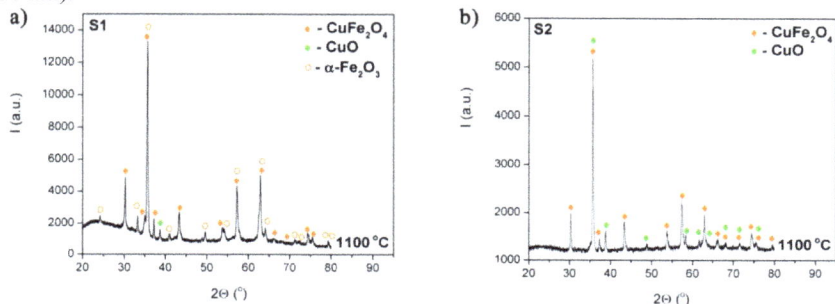

Figure 12. Diffraction pattern of the samples annealed at 1100 °C: a) S1; b) S2 [175].

Literature revealed that CuFe$_2$O$_4$ formation occurred via incorporation of iron cations in copper oxide lattice [169]. Since CuO could be characterized by electronic quasi degeneracy [177], addition of hole or electron into CuO structure resulted in the pseudo Jahn-Teller effect [177]. Iron cation enter CuO lattice by positioning in free hole or by replacement of the copper cation. CuO appeared as a second phase in a system due to the oxidation of released copper cations.

On the other hand, sample S1 annealed at 1100 °C contained a hematite phase (although presence of copper cations suppress presence of hematite phase [92]), because an iron precursor was presented in excess in the synthesis procedure. In this sample hematite is formed due to the fact that the pH of the solution was increased up to 8; if the pH solution would achieve much higher (10-14) or much lower (2-5) values, instead hematite goethite would be formed [178]. Noteworthy, the hematite phase is formed during annealing procedure via formation of defective 2L-Fhyd, which consists in the structural properties common with 2L-Fhyd and hematite structure. At certain temperature, defective 2L-Fhyd is converted into hematite phase (Figure 12a)). Simultaneously, very similar wt% of copper and iron precursors used for the preparation of the sample S2 consequently resulted in the absence of a hematite phase (Fig. 12b)), since the Cu^{2+} cations have a supressing effect on the hematite formation.

On the other hand, the fact that c-CuFe$_2$O$_4$ nanoparticles (synthesized by using the same concentration of iron precursor), presented in different nanocomposite materials, were characterized by the same crystallite size (46 nm), accented the impact of Fe^{3+} cations on the size of $CuFe_2O_4$ nanoparticles.

If we consider the iron-containing phases in investigated samples, it is obvious that the same amount of Fe^{3+} cations is distributed in two compounds within the sample S1 (c-$CuFe_2O_4$ and α-Fe_2O_3), while in the sample S2, the same amount of iron cations was built in the CuO lattice forming c-$CuFe_2O_4$, as the only iron-containing phase. This experimental result depicting to the fact that obtained copper ferrite phases in examined samples are significantly differed in stoichiometry: sample S1 contained c-$CuFe_2O_4$ phase which structure is characterized by the higher number of structural defects, compared to the c-$CuFe_2O_4$ phase within the sample S2 [174, 175].

Electron paramagnetic (EPR) measurements were conducted with the aim to confirm strong influence of variation of the synthesis conditions onto magnetic state of the investigated samples (Fig.13).

Non-annealed powders of the samples S1 and S2 are represented in Fig. 13 a) and b). Curves shape depicted to the presence of SPM behavior of the samples; having in mind phase composition of the samples and considering the fact that 2L-Fhyd nanoparticles are always SPM due to very small sizes of coherent scattering domains, it is obviously that CuO nanoparticles obtained by this method are also SPM.

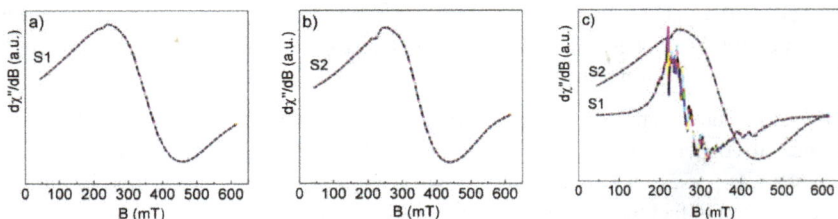

Figure 13. EPR spectra of the samples: a) S1; b) S2; c) samples S1 and S2 annealed at 1100 °C [175].

Noteworthy, magnetic state characteristic for the sample S2 has not been significantly changed after annealing at high temperature, although phase composition is altered by disappearing 2L-Fhyd and appearance of c-$CuFe_2O_4$ nanoparticles, Fig. 13. Contrary, sharp changes in the magnetic behavior of the sample S1 are observed (Fig. 13c)). Since the only difference in the chemical composition of the samples S1 and S2 after annealing process presents appearance of hematite nanoparticles, alteration of magnetic behavior (curve asymmetry, small intensity lines, various width of the right and left shoulders) could be explained in the term of the appearance of non-SPM nanohematite [174, 175].

4.3 Influence of the variation of Fe^{3+} precursor concentration on the $CuFe_2O_4$-based nanocomposite properties

Below is discussed the influence of the variation of the Fe^{3+} amount in the synthesis on the final structure of the $CuFe_2O_4$-based nanocomposite.

The same coprecipitation synthesis was performed in order to prepare samples, using Cu^{2+} and Fe^{3+} cations in milimolar ratio 3:3 (S1), 3:2.4 (S2) [179,180]. Hydrolysis of a Cu^{2+} and Fe^{3+} precursors was performed up to ph=8, by using different quantities of NH_4OH: 4.6 ml (sample S1) and 7.3 ml (sample S2). After precipitates were washed and dried, samples were annealed 3h at 1100 °C.

X-ray diffraction (XRD) measurements are presented in the Fig. 14. Both samples contains 2L-Fhyd, but only native powder of the sample S2 contains and CuO phase, appeared as a consequence of the fact that higher amount of copper precursor forced copper oxide formation [179, 180].

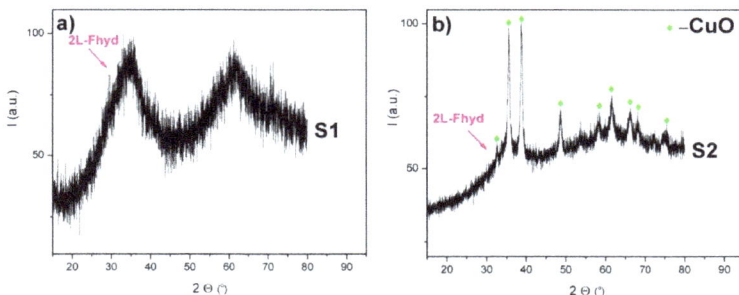

Figure 14. Diffraction pattern of the native samples: a) S1; b) S2 [180].

Annealing process at 1100 °C resulted in the formation of various $CuFe_2O_4$ structures (Figure 15).

Refinement revealed that annealing of the samples at 1100 °C led to the formation of t-$CuFe_2O_4$ nanoparticles within the sample S1 (d_{cr}=89 nm), and c-$CuFe_2O_4$ in the sample S2 (d_{cr}=170 nm). Although in the literature is found that there is certain temperature ranges within which both copper ferrite phases appeared together, in this study is shown that individual c-$CuFe_2O_4$ and t-$CuFe_2O_4$ phases (without traces of the other copper ferrite phases), could be obtained at temperature which overestimates temperature range where are these phases observed simultaneously. These experimental results confirmed the possibility of preparation of c- and t-$CuFe_2O_4$ phases (designated in literature as "low-" and "high-" temperature copper ferrite phase) at the same temperature, which is explained in the term of differences in formation mechanism. Excess of Cu^{2+} cations in the comparison with Fe^{3+} cations concentration resulted in the accelerated crystallization of c-$CuFe_2O_4$ [179,180] (sample S2), appeared as the consequence of the increased rate of the overall S1 sample formation, in comparison with the rate of the overall sample S2. In spite the fact that decreased concentration of the iron cations leads to the decreased rate of the sample S2 reaction mechanism, lower iron concentration favors t-$CuFe_2O_4$→c-$CuFe_2O_4$ transition, simultaneously. Consequently, accelerated Jahn-Teller effect enable starting of the t-$CuFe_2O_4$→c-$CuFe_2O_4$ structural transformation at lower temperatures, because Jahn-Teller effect presents a statistical phenomenon, and presence of the higher amount of Cu^{2+} cations increased probability that the net alignment will be reached [179,180]. Absence of the α-Fe_2O_3 phase within samples S1 and S2 is explained by the influence of a significant amount of Cu^{2+} cation in the synthesis route, which omitted formation of hematite.

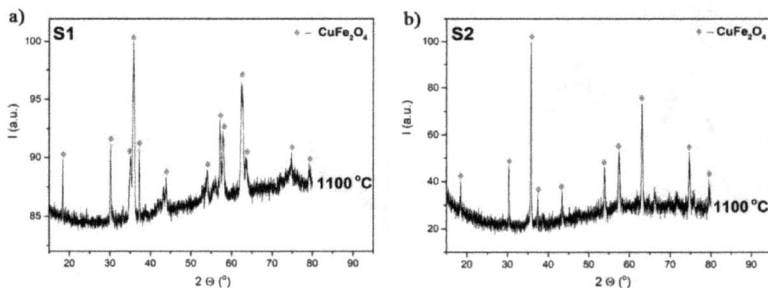

Figure 15. Diffraction pattern of the samples annealed at 1100 °C: a) S1; b) S2; c) S3 [180].

Conclusions

The main message of this chapter is to emphasize the significance of the magnetic nanocomposite materials for application in electrocatalysis. Today, the most attractive area of scientific examination presents the search for alternative techniques of hydrogen production (water splitting), and biomedicine; for this reason, in this chapter have been commented potential application of magnetic nanomaterials as electrocatalysts in these scientific fields. Some applications of magnetic-based nanocomposite materials as electrocatalysts are listed. Structure and morphology of magnetic nanomaterials used as potential electrocatalysts are also dicussed. Finally, since the literature revealed that various synthesis approaches could significantly improve electrocatalytic application of synthesized material, influence of the synthesis parameters on the properties of materials of importance for electrocatalysis is discussed on the example of $CuFe_2O_4$-based nanocomposites.

Acknowledgement

The research was carried out thanks to the support of the Ministry of Education, Science and Technology Development, Republic of Serbia (Project No. III 45015).

References

[1] F.P. Bowden, E. Rideal, The electrolytic behaviour of thin films. Part I. Hydrogeneral, Proc. Roy. Soc. A 120 (1928) 59-79. https://doi.org/10.1098/rspa.1928.0135

[2] S.H. Jordanov, P. Paunovic, O. Popovski, A. Dimitrov, D. Slavkov, Electrocataysts in the last 30 years - from precious metals to cheaper but sophisticated complex systems, Bull. Chem. Technol. Macedonia 23 (2004) 101–112.

[3] J.M. Leger, F. Hahn, Contribution of In-situ Infrared Reflectance Spectroscopy in the Study of Nanostructured Fuel Cell Electrodes, in: S.-G. Sun, P. A. Christensen, A. Wieckowski (Eds.), In-situ Spectroscopic Studies of Adsorption at the Electrode and Electrocatalysis, E-Publishing Inc., New York, 2007, pp. 1-36. https://doi.org/10.1016/B978-044451870-5/50004-X

[4] S. Blundell, Magnetism in Condensed Matter, Oxford University Press. Inc., New York, 2003. https://doi.org/10.1119/1.1522704

[5] C. Kittel, Introduction to Solid State Physics, John Willey&Sons, New York, 1996.

[6] L.D. Leslie-Pelecky, R.D. Rieke, Magnetic properties of nanostructured materials, Chem. Mater. 8 (1996) 1770–1783. https://doi.org/10.1021/cm960077f

[7] H. Zhong, C.A. Campos-Roldan, Y. Zhao, S. Zhang, Y. Feng, N. Alonso-Vante, Recent advances of cobalt-based electrocatalysts for oxygen electrode reactions and hydrogen evolution reaction, Catalysts 8 (2018) 1-43. https://doi.org/10.3390/catal8110559

[8] A.D. Handoko, S. Deng, Y. Deng, A.F.W. Cheng, K.W. Chan, H.R. Tan, Y. Pan, E.S. Tok, C.H. Sow, B.S. Yeo, Enhanced activity of H_2O_2-treated copper(ii) oxide

nanostructures for the electrochemical evolution of oxygen, Catal. Sci. Technol. 6 (2016) 269–274. https://doi.org/10.1039/C5CY00861A

[9] E. Antolini, Structural parameters of supported fuel cell catalysts: The effect of particle size, inter-particle distance and metal loading on catalytic activity and fuel cell performance, Appl. Catal. B 181 (2016) 298–313. https://doi.org/10.1016/j.apcatb.2015.08.007

[10] L.M. Dai, Y.H. Xue, L.T. Qu, H.J. Choi, J.B. Baek, Metal-free catalysts for oxygen reduction reaction, Chem. Rev. 115 (2015) 4823–4892. https://doi.org/10.1021/cr5003563

[11] B.I. Kharisov, O.V. Kharissova, H.V. Rasika Dias, U.O. Mendez, I.G. de la Fuente, Y. Pena, A.V. Dimas, Iron-based nanomaterials in the catalysis, in: N. Luis (Ed.), Advanced catalytic materials - Photocatalysis and other current trends, IntechOpen, London, 2016, pp. 1-29. https://doi.org/10.5772/61862

[12] J.H. Wang, W.Q. Liu, Z.C. Xing, A.M. Asiri, X.P. Sun, Recent progress in cobalt-based heterogeneous catalysts for electrochemical water splitting, Adv. Mater. 28 (2016) 215–230. https://doi.org/10.1002/adma.201502696

[13] W. Vielstich, A. Lamm, H.A. Gasteiger, Handbook of fuel cells: Fundamentals, technology, applications, John Willey&Sons, New York, 2003.

[14] T.R. Cook, D.K. Dogutan, S.Y. Reece, Y. Surendranath, T.S. Teets, D.G. Nocera, Solar energy supply and storagefor the legacy and nonlegacy worlds, Chem. Rev. 110 (2010) 6474–6502. https://doi.org/10.1021/cr100246c

[15] H.B. Gray, Powering the planet with solar fuel, Nat. Chem. 1 (2009) 1-7. https://doi.org/10.1038/nchem.137

[16] A. Kudo, Y. Miseki, Heterogeneous photocatalyst materials for water splitting, Chem. Soc. Rev. 38 (2009) 253–278. https://doi.org/10.1039/B800489G

[17] N.S. Lewis, D.G. Nocera, Powering the planet: Chemical challenges in solar energy utilization, Proc. Natl. Acad. Sci.103 (2006) 15729–15735. https://doi.org/10.1073/pnas.0603395103

[18] Y.Y. Liang, Y.G. Li, H.L. Wang, H.J. Dai, Strongly coupled inorganic/nanocarbon hybrid materials for advanced electrocatalysis, J. Am. Chem. Soc. 135 (2013) 2013–2036. https://doi.org/10.1021/ja3089923

[19] M.G. Walter, E.L. Warren, J. R. McKone, S.W. Boettcher, Q.X. Mi, E.A. Santori, N.S. Lewis, Solar water splitting cells, Chem. Rev. 110 (2010) 6446–6473. https://doi.org/10.1021/cr1002326

[20] H.L. Wang, H.J. Dai, Strongly coupled inorganic–nanocarbon hybrid materials for energy storage, Chem. Soc. Rev. 42 (2013) 3088–3113. https://doi.org/10.1039/c2cs35307e

[21] M. Gong, D.Y. Wang, C.C. Chen, B.J. Hwang, H. Dai, A mini review on nickel-based electrocatalysts for alkaline hydrogen evolution reaction, Nano Res. 9 (2016) 28-46. https://doi.org/10.1039/c2cs35307e

[22] G.W. Crabtree, M.S. Dresselhaus, M.V. Buchanan, The hydrogen economy, Phys. Today 57 (2004) 39–44. https://doi.org/10.1063/1.1878333

[23] M.S. Dresselhaus, I.L. Thomas, Alternative energy technologies, Nature 414 (2001) 332–337. https://doi.org/10.1038/35104599

[24] J.J. Spivey, Catalysis in the development of clean energy technologies, Catal. Today 100 (2005)171-176. https://doi.org/10.1016/j.cattod.2004.12.011

[25] P. Haussinger, R. Lohmuller, A.M. Watson, Hydrogen, in: F. Ullmann (Ed.), Ullmann's Encyclopedia of Industrial Chemistry, Wiley-VCH: Werlag GmbH& Co. KGaA, Berlin, 2002, pp. 281-304.

[26] M. Carmo, D.L. Fritz, J. Mergel, D. Stolten, A comprehensive review on PEM water electrolysis, Int. J. Hydrogen Energy 38 (2013) 4901–4934. https://doi.org/10.1016/j.ijhydene.2013.01.151

[27] M. Gong, H.J. Dai, A mini review of nife-based materials as highly active oxygen evolution reaction electrocatalysts, Nano Res. 8 (2015) 23–39. https://doi.org/10.1007/s12274-014-0591-z

[28] J.D. Holladay, J. Hu, D.L. King, Y. Wang, An overview of hydrogen production technologies, Catal. Today 139 (2009) 244–260. https://doi.org/10.1016/j.cattod.2008.08.039

[29] K. Zeng, D.K. Zhang, Recent progress in alkaline water electrolysis for hydrogen production and applications, Prog. Energy Combust. Sci. 36 (2010) 307–326. https://doi.org/10.1016/j.pecs.2009.11.002

[30] P. Millet, A. Godula-Jopek, Fundamentals of Water Electrolysis, in: A. Godula-Jopek (Ed.), Hydrogen Production: Electrolysis, Wiley-VCH: Verlag GmbH & Co. KGaA, Berlin, 2015, pp. 33-62. https://doi.org/10.1002/9783527676507.ch2

[31] H. Schafer, S. Sasaf, L. Walder, K. Kuepper, S. Dinklage, J. Wollschlager, L. Schneider, M. Steinhart, J. Hardeged, D. Daum, Stainless steel made to rust: A robust water-splitting catalyst with benchmark characteristics, Energy Environ. Sci. 8 (2015) 2685-2697. https://doi.org/10.1039/C5EE01601K

[32] D. Chen, Z.M. Baiyee, Z. Shao, F. Ciucci, Nonstoichiometric oxides as low-cost and highly-efficient oxygen reduction/evolution catalysts for low-temperature electrochemical devices, Chem. Rev. 115 (2015) 9869–9921. https://doi.org/10.1021/acs.chemrev.5b00073

[33] S.J. Trasatti, Electrocatalysis by oxides — Attempt at a unifying approach, J. Electroanal. Chem. 111 (1980) 125−131. https://doi.org/10.1016/0368-1874(80)80246-2

[34] Y. Lee, J. Suntivich, K.J. May, E.E. Perry, Y. Shao-Horn, Synthesis and activities of rutile IrO_2 and RuO_2 nanoparticles for oxygen evolution in acid and alkaline solutions, J. Phys. Chem. Lett. 3 (2012) 399-404. https://doi.org/10.1021/jz2016507

[35] F.M. Sapountzi, J.M. Gracia, C.J.Weststrate, H. Fredriksson, J.W. Niemantsverdriet, Electrocatalysis for the generation of hydrogen, oxygen and synthesis gas, Prog. Energy Combust. Sci. 58 (2017) 1-35. https://doi.org/10.1016/j.pecs.2016.09.001

[36] J. Xu, Y. Yang, Y.W. Li, Recent development in converting coal to clean fuels in China, Fuel 152 (2015) 122–30. https://doi.org/10.1016/j.fuel.2014.11.059

[37] Synthesis gas chemistry (Syngaschem), fundamental research projects, 2019 (accessed 05 March 2019). https://doi.org/10.31025/2611-4135/2019.13798

[38] T. Lim, J.W. Niemantsverdriet, J. Gracia, Layered antiferromagnetic ordering in the most active perovskite catalysts for oxygen evolution reaction, ChemCatChem 8 (2016) 2968-2974. https://doi.org/10.1002/cctc.201600611

[39] M.K. Masud, Md. N. Islam, Md. H. Haque, S. Tanaka, V. Gopalan, G. Alici, N.T. Nguyen, A.K. Lam, Md. S. A. Hossain, Y.Yamauchi, M.J.A. Shiddiky, Gold-loaded nanoporous superparamagnetic nanocubes for catalytic signal amplification in detecting miRNA, Chem. Commun. 53 (2017) 8231-8234. https://doi.org/10.1039/C7CC04789D

[40] M.K. Masud, S. Yadav, Md. N. Islam, N.T. Nguyen, C. Salomon, R. Kline, H. R. Alamri, Z.A. Alothman, Y. Yamauchi, Md.S.A. Hossain, M.J.A. Shiddiky, Gold-loaded nanoporous ferric oxide nanocubes with peroxidase-mimicking activity for electrocatalytic and colorimetric detection of autoantibody, Anal. Chem. 89 (2017) 11005-11013. https://doi.org/10.1021/acs.analchem.7b02880

[41] J. Tang, D. Tang, Non-enzymatic electrochemical immunoassay using noble metal nanoparticles: A review, Microchim. Acta 182 (2015) 2077-2089. https://doi.org/10.1007/s00604-015-1567-8

[42] C. Xu, J. Xie, D. Ho, C. Wang, N. Kohler, E.G. Walsh, J.R. Morgan, Y.E. Chin, S. Sun, $Au–Fe_3O_4$ Dumbbell Nanoparticles as Dual-Functional Probes, Angew. Chem. Int. Ed. 47 (2008) 173 –176. https://doi.org/10.1002/anie.200704392

Materials Research Forum LLC
https://doi.org/10.21741/9781644900611-2

[43] L. Liu, D. Deng, W. Sun, X. Yang, S. Yang, S. He, electrochemical biosensors with electrocatalysts based on metallic nanomaterials as signal labels, Int. J. Electrochem. Sci. 13 (2018) 10496-10513. https://doi.org/10.20964/2018.11.47

[44] S. Lee, G.H. Lee, H.J. Lee, M.A. Dar, D.W. Kim, Fe-based hybrid electrocatalysts for nonaqueous lithium-oxygen batteries, Sci. Rep. 7 (2017) 1-9. https://doi.org/10.1038/s41598-016-0028-x

[45] Z. Zhang, J. Hao, W. Yang, B. Lu, J. Tang, Modifying candle soot with FeP nanoparticles into high-performance and cost-effective catalysts for the electrocatalytic hydrogen evolution reaction, Nanoscale 7 (2015) 4400-4408. https://doi.org/10.1039/C4NR07436J

[46] A. Rinaldi, B. Mecheri, V. Garavaglia, S. Liococcia, P.D. Nardo, E. Traversa, Engineering materials and biology to boost performance of microbial fuel cells: A critical review, Energy Environ. Sci. 1 (2008) 417-429. https://doi.org/10.1039/b806498a

[47] G. Lu, Y. Zhu, L. Lu, K. Xu, H. Wang, Y. Jin, Z.J. Ren, Z. Liu, W. Zhang, Iron-rich nanoparticle encapsulated, nitrogen doped porous carbon materials as efficient cathode electrocatalyst for microbial fuel cells, J. Power Sources 315 (2016) 302-307. https://doi.org/10.1016/j.jpowsour.2016.03.028

[48] B. Seyyedi, Bio-inspired iron metal–carbon black based nano-electrocatalyst for the oxygen reduction reaction, Pigm. Resin Technol. 46 (2017) 267-275. https://doi.org/10.1108/PRT-07-2016-0081

[49] M. Saruyama, S. Kim, T. Nishino, M. Sakamoto, M. Haruta, H. Kurata, S. Akiyama, T. Yamada, K. Domen, T. Teranishi, Phase-segregated $NiP_x@FeP_yO_z$ core@shell nanoparticles: ready-to-use nanocatalysts for electro- and photo-catalytic water oxidation through in situ activation by structural transformation and spontaneous ligand removal, Chem. Sci. 9 (2018) 4830-4839. https://doi.org/10.1039/C8SC00420J

[50] L. Yu, J. F. Yang, B. Y. Guan, Y. Lu, X.W. Lou, Hierarchical hollow nanoprisms based on ultrathin ni-fe layered double hydroxide nanosheets with enhanced electrocatalytic activity towards oxygen evolution, Angew. Chem. Int. Ed. 57 (2018) 172-176. https://doi.org/10.1002/anie.201710877

[51] W. Ma, R. Ma, C. Wang, J. Liang, X. Liu, K. Zhou, T. Sasaki, A superlattice of alternately stacked Ni-Fe hydroxide nanosheets and graphene for efficient splitting of water, ACS Nano 9 (2015) 1977-84. https://doi.org/10.1021/nn5069836

[52] T. Shodiya, O. Schmidt, W. Peng, N. Hotz, Novel nano-scale $Au/\alpha\text{-}Fe_2O_3$ catalyst for the preferential oxidation of CO in biofuel reformate gas, J. of Catalysis 300 (2013) 63-69. https://doi.org/10.1016/j.jcat.2012.12.027

[53] L. Ji, Z. Guo, T. Yan, H. Ma, B. Du, Y. Li, Q. Wei, Ultrasensitive sandwich-type electrochemical immunosensor based on a novel signal amplification strategy using highly loaded palladium nanoparticles/carbon decorated magnetic microspheres as signal labels, Biosens. Bioelectron. 68 (2015) 757-762. https://doi.org/10.1016/j.bios.2015.02.010

[54] T. Zheng, Q. Zhang, S. Feng, J.J. Zhu, Q. Wang, H. Wang, Robust nonenzymatic hybrid nanoelectrocatalysts for signal amplification toward ultrasensitive electrochemical cytosensing, J. Am. Chem. Soc. 136 (2014) 2288−2291. https://doi.org/10.1021/ja500169y

[55] G.Y. Zhang, S.Y. Deng, W.R. Cai, S. Cosnier, X.J. Zhang, D. Shan, Magnetic zirconium hexacyanoferrate (ii) nanoparticle as tracingtag for electrochemical DNA assay, Anal. Chem. 87 (2015) 9093−9100. https://doi.org/10.1021/acs.analchem.5b02395

[56] D. Wu, H. Fan, Y. Li, Y. Zhang, H. Liang, Q. Wei, Ultrasensitive electrochemical immunoassay for squamous cell carcinoma antigen using dumbbell-like Pt-Fe$_3$O$_4$ nanoparticles as signal amplification, Biosens. Bioelectron. 46 (2013) 91–96. https://doi.org/10.1016/j.bios.2013.02.014

[57] H. Behret, H. Binder, G. Sandstede, Electrocatalytic oxygen reduction with thiospinels and other sulphidesof transition metals, Electrochim. Acta 20 (1975) 111–117. https://doi.org/10.1016/0013-4686(75)90047-X

[58] D. Baresel, W. Sarholz, P. Scharner, J. Schmitz, Transition Metal Chalcogenides as Oxygen Catalysts for Fuel Cells, Ber. Bunsen-Ges. 78 (1974) 608–618.

[59] R.A. Sidik, A.B. Anderson, Co$_9$S$_8$ as a catalyst for electroreduction of O$_2$: Quantum Chemistry Predictions, J. Phys. Chem. B 110 (2006) 936–941. https://doi.org/10.1021/jp054487f

[60] H.L. Wang, Y.Y. Liang, Y.G. Li, H.J. Dai, Co$_{1-x}$S-Graphene Hybrid: A high-performance metal chalcogenide electrocatalyst for oxygen reduction, Angew. Chem. Int. Ed. Engl. 50 (2011) 10969–10972. https://doi.org/10.1002/anie.201104004

[61] X.D. Jia, S.J. Gao, T.Y. Liu, D.Q. Li, P.G.Y.J. Feng, Controllable synthesis and bi-functional electrocatalytic performance towards oxygen electrode reactions of Co$_3$O$_4$/N-RGO Composites, Electrochim. Acta 2626 (2017) 104–112. https://doi.org/10.1016/j.electacta.2016.12.191

[62] Y.J. Feng, N. Alonso-Vante, Structure phase transition and oxygen reduction activity in acidic medium ofcarbon-supported cobalt selenide nanoparticles, ECS Trans. 25 (2009) 167–173. https://doi.org/10.1149/1.3210568

[63] Y.J. Feng, N. Alonso-Vante, Carbon-supported $CoSe_2$ nanoparticles for oxygen reduction Reaction in Acid Medium, Fuel Cells 10 (2010) 77–83. https://doi.org/10.1002/fuce.200900038

[64] G. Wu, H. T. Chung, M. Nelson, K. Artyushkova, K. L. More, C. M. Johnston, P. Zelenay, Graphene-enriched Co_9S_8-N-C non-precious metal catalyst for oxygen reduction in alkaline media, ECS Trans. 41 (2011) 1709–1717. https://doi.org/10.1149/1.3635702

[65] R.D. Apostolova, E.M. Shembel, I. Talyosef, J. Grinblat, B. Markovsky, D. Aurbach, Study of electrolytic cobalt sulfide Co_9S_8 as an electrode material in lithium accumulator prototypes, Russ. J. Electrochem. 45 (2009) 311–319. https://doi.org/10.1134/S1023193509030112

[66] C. Zhao, D.Q. Li, Y.J. Feng, Size-controlled hydrothermal synthesis and high electrocatalytic performance of CoS_2 nanocatalysts as non-precious metal cathode materials for fuel cells, J. Mater. Chem. A1 (2013) 5741–5746. https://doi.org/10.1039/c3ta10296c

[67] J.X. Wang, H. Inada, L.J. Wu, Y.M. Zhu, Y.M. Choi, P. Liu, W.P. Zhou, R.R. Adzic, Oxygen reduction onwell-defined core-shell nanocatalysts: particle size, facet, and Pt shell thickness effects, J. Am. Chem. Soc. 131 (2009) 17298–17302. https://doi.org/10.1021/ja9067645

[68] Y.C. Wang, T. Zhou, K. Jiang, P.M. Da, Z. Peng, J. Tang, B.A. Kong, W.B. Cai, Z.Q. Yang, G.F. Zheng, Reduced mesoporous Co_3O_4 nanowires as efficient water oxidation electrocatalysts and supercapacitor electrodes, Adv. Energy Mater. 4 (2014) 1400696–1400702. https://doi.org/10.1002/aenm.201400696

[69] L. Fenglei, Q. Wang, S.M. Choi, Y. Yin, Noble-metal-free electrocatalysts for oxygen evolution, Small 15 (2018) 1-9. https://doi.org/10.1002/smll.201804201

[70] Y.S. Xia, H.X. Dai, H.Y. Jiang, L. Zhang, Three-dimensional ordered mesoporous cobalt oxides: highly active catalysts for the oxidation of toluene and methanol, Catal. Commun. 11 (2010) 1171–1175. https://doi.org/10.1016/j.catcom.2010.07.005

[71] H.M. Li, X. Qian, C.L. Zhu, X.X. Jiang, L. Shao, L.X. Hou, Template synthesis of $CoSe_2/Co_3Se_4$ nanotubes: Tuning of their crystal structures for photovoltaics and hydrogen evolution in alkaline medium, J. Mater. Chem. A 5 (2017) 4513–4526. https://doi.org/10.1039/C6TA10718D

[72] P. Chen, W. Xia, F. Yang, A. Kostka, Interaction of cobalt nanoparticles with oxygen- and nitrogen-functionalized carbon nanotubes and impact on nitrobenzene hydrogenation catalysis, ACS Catal. 4 (2014) 15-19. https://doi.org/10.1021/cs500173t

[73] Z. Haihong, R. Tian, X. Gong, Y. Feng, Advanced bifunctional electrocatalyst generated through cobalt phthalocyaninetetrasulfonate intercalated Ni_2Fe-layered double hydroxides for a laminar flow unitized regenerative micro-cell, J. Power Sources 361 (2017) 21-30. https://doi.org/10.1016/j.jpowsour.2017.06.057

[74] M. Darayi, T. Csesznok, I. Sarusi, A. Kukovecs, I. Kiricsi, Beneficial effect of multi-wall carbon nanotubes on the graphitization of polyacrylonitrile (PAN) coating, Proc. Appl. Cer. 4 (2010) 59-62. https://doi.org/10.2298/PAC1002059D

[75] X.F. Dai, J.L. Qiao, X.J. Zhou, J.J. Shi, P. Xu, L. Zhang, J.J. Zhang, Effects of heat-treatment and pyridine addition on the catalytic activity of carbon-supported cobalt-phthalocyanine for oxygen reduction reaction in alkaline electrolyte, Int. J. Electrochem. Sci. 8 (2013) 3160–3175.

[76] S. Zhao, B. Rasimick, W. Mustain, H. Xu, Highly Durable and Active Co_3O_4 Nanocrystals Supported on Carbon Nanotubes as Bifunctional Electrocatalysts in Alkaline Media, Appl. Catal. B 203 (2017) 138–145. https://doi.org/10.1016/j.apcatb.2016.09.048

[77] S.Y. Liu, L. J. Li, H.S. Ahn, A. Manthiram, Delineating the roles of Co_3O_4 and N-doped carbon nanoweb (CNW) in Bifunctional Co_3O_4/CNW catalysts for oxygen reduction and oxygen evolution reactions, J. Mater. Chem. A 3(2015) 11615–11623. https://doi.org/10.1039/C5TA00661A

[78] Z.S. Wu, W.C. Ren, L. Wen, L.B. Gao, J.P. Zhao, Z.P. Chen, G.M. Zhou, F. Li, H.M. Cheng, Graphene anchored with Co_3O_4 nanoparticles as anode of lithium ion batteries with enhanced reversible capacity and cyclic performance, ACS Nano 4 (2010) 3187–3194. https://doi.org/10.1021/nn100740x

[79] N.P. Subramanian, S. P. Kumaraguru, H. Colon-Mercado, H. Kim, B.N. Popov, T. Black, D.A. Chen, Studies on Co-based catalysts supported on modified carbon substrates for PEMFC cathodes, J. Power Sources 157 (2006) 56–63. https://doi.org/10.1016/j.jpowsour.2005.07.031

[80] M. Khan, M.N. Tahir, S. F. Adil, H.U. Khan, M.R.H. Siddiqui, A.A. Al-Warthan, W. Tremel, Graphene based metal and metal oxide nanocomposites: Synthesis, properties and their applications, J. Mater. Chem. A 3 (2015) 18753–18808. https://doi.org/10.1039/C5TA02240A

[81] M. Kuang, Q. Wang, P. Han, G.F. Zheng, Cu, Co-embedded N-enriched mesoporous carbon for efficient oxygen reduction and hydrogen evolution reactions, Adv. Energy Mater. 7 (2017) 1700193–1700200. https://doi.org/10.1002/aenm.201700193

[82] Y. Zhu, Y. Wang, S. Liu, R. Guo, Z. Li, Facile and controllable synthesis at an ionic layer level of high performance NiFe-based nanofilm electrocatalysts for the oxygen

evolution reaction in alkaline electrolyte, Electrochem. Commun. 86 (2018) 38-42.
https://doi.org/10.1016/j.elecom.2017.11.008

[83] E.J. Popczun, J.R. McKone, C.G. Read, A.J. Biacchi, A.M. Wiltrout, N.S. Lewis,
R.E. Schak, Nanostructured nickel phosphide as an electrocatalyst for the hydrogen
evolution reaction, J. Am. Chem. Soc. 135 (2013) 9267−9270.
https://doi.org/10.1021/ja403440e

[84] W. Li, D. Xiong, X. Gao, W.G. Song, F. Xia, L. Liu, Self-supported Co-Ni-P ternary
nanowire electrodes for highly efficient and stable electrocatalytic hydrogen evolution
in acidic solution, Catal. Today 2 (2017) 122–129.
https://doi.org/10.1016/j.cattod.2016.09.007

[85] A.E. Mauer, D.W. Kirk, S.J. Thorpe, The role of iron in the prevention of nickel
electrode deactivation in alkaline electrolysis, Electrochim. Acta 52 (2007) 3505–
3509. https://doi.org/10.1016/j.electacta.2006.10.037

[86] M.K. Bates, Q. Jia, N. Ramaswamy, R.J. Allen, S. Mukerjee, Composite Ni/NiO-
Cr_2O_3 catalyst for alkaline hydrogen evolution reaction, J. Phys. Chem. C 119 (2015)
5467–5477. https://doi.org/10.1021/jp512311c

[87] E.S. Davydova, S. Mukerjee, F. Jaouen, D.R. Dekel, Electrocatalysts for hydrogen
oxidation reaction in alkaline electrolytes, ACS Catal. 8 (2018) 6665–6690.
https://doi.org/10.1021/acscatal.8b00689

[88] D.R. Dekel, Review of cell performance in anion exchange membrane fuel cells, J.
Power Sources 375 (2018)158–169. https://doi.org/10.1016/j.jpowsour.2017.07.117

[89] E.S. Davydova, J. Zaffran, K. Dhaka, M.C. Torokel, D.R. Dekel, Hydrogen
oxidation on Ni-based electrocatalysts: The effect of metal doping, Catalysts 8 (2018)
1-19. https://doi.org/10.20944/preprints201809.0532.v1

[90] S.L. Altmann, C.A. Coulson, W. Hume-Rothery, On the relation between bond
hybrids and the metallic structures, Proc. Roy. Soc. A 240 (1957) 145-159.
https://doi.org/10.1098/rspa.1957.0073

[91] S.J. Thomson, G.A. Harvey, The electrical conductivity of supported metal catalysts,
J. Catalysts, 22 (1971) 359-363. https://doi.org/10.1016/0021-9517(71)90207-7

[92] R.M. Cornell, U. Schwertmann, The iron oxides: Structure, properties, reactions,
occurrence and uses, John Wiley & Sons, New York, 2000.

[93] V.N. Nikolic, Preparation and Characterization of Fe_2O_3-SiO_2 Nanocomposite for
Biomedical Application, in: Y. Maeda (Ed.), Hematite, IntechOpen, London, 2019, pp.
1-22. https://doi.org/10.5772/intechopen.81926

[94] E. Roduner, Size matters: why nanomaterials are different, Chem. Soc. Rev. 35 (2006) 583–592. https://doi.org/10.1039/b502142c

[95] J.Y. Buzare, J.C. Fayet, The Fe^{3+} - O^{2-} pair in the diamagnetic AMF_3 perovskites: A sensitive probe for EPR investigations of structural phase changes, Solid State Commun. 21 (1977) 1097-1100. https://doi.org/10.1016/0038-1098(77)90315-5

[96] Z. Ali, I. Khan, I. Ahmad, M. Salman Khan, S.J. Asadabadi, Theoretical studies of the paramagnetic perovskites $MTaO_3$ (M = Ca, Sr and Ba), Mater. Chem. Phys. 162 (2015) 308-315. https://doi.org/10.1016/j.matchemphys.2015.05.072

[97] Y. Su, Y. Sui, J.G. Cheng, J.S. Zhou, X. Wang, Y. Wang, J.B. Goodenough, Critical behavior of the ferromagnetic perovskites $RTiO_3$ (R = Dy, Ho, Er, Tm, Yb) by magnetocaloric measurements, Phys. Rev. B 87, 195102 (2013). https://doi.org/10.1103/PhysRevB.87.195102

[98] T. Tajiri, S. Maruoka, H. Deguchi, S. Takagi, M. Mito, Y. Ishida, S. Kohiki, Superparamagnetic behavior of $La_{1-x}Sr_xMnO_3$ nanoparticles in the MCM-41 molecular sieve, Phys. B Condens Matter. 329 (2003) 860–861. https://doi.org/10.1016/S0921-4526(02)02576-0

[99] C.R. Sankar, P.A. Joy, Superspin glass behavior of a nonstoichiometric lanthanum manganite $LaMnO_{3.13}$, Phys. Rev. B 72, 13 (2005): 132407. https://doi.org/10.1103/PhysRevB.72.132407

[100] D.T.T. Nguyet, N.P. Duong, T. Satoh, L.N. Anh, T.D. Hien, Temperature-dependent magnetic properties of yttrium iron garnet nanoparticles prepared by citrate sol–gel, J. Alloys Compd. 541 (2012) 18–22. https://doi.org/10.1016/j.jallcom.2012.06.122

[101] L. Hadidi, E. Davari, D.G. Ivery, J.G.C. Veinot, Microwave-assisted synthesis and prototype oxygen reduction electrocatalyst application of N-doped carbon-coated Fe_3O_4 nanorods, Nanotechnology 28 (2017) 1-10. https://doi.org/10.1088/1361-6528/aa5716

[102] T. Tatarchuk, M. Bououdina, J. Judith Vijaya, L.J. Kennedy, Spinel Ferrite Nanoparticles: Synthesis, Crystal Structure, Properties, and Perspective Applications. In: International Conference on Nanotechnology and Nanomaterials; Lviv, Ukraine, 2016. https://doi.org/10.1007/978-3-319-56422-7_22

[103] Chemistry, LibreTexts. https://chem.libretexts.org/Bookshelves/Inorganic_Chemistry/Book%3A_Inorganic_C hemistry_(Wikibook)/Chapter_08%3A_Ionic_and_Covalent_Solids_-_Structures/8.6%3A_Spinel%2C_perovskite%2C_and_rutile_structures, 2019 (accessed 05 March 2019).

[104] C.F. Macrae, P.R. Edgington, P. McCabe, E. Pidcock, G.P. Shields, R. Taylor, M. Towler, J. van de Streek, Mercury: Visualization and analysis of crystal structures, J. Appl. Cryst. 39 (2006) 453–457. https://doi.org/10.1107/S002188980600731X

[105] ICSD Inorganic Crystals Structure Database, Release 2014/2, FIZ Karlsruhe, Eggenstein-Leopoldshafen, Germany.

[106] A.K. Kundu, Magnetic Perovskites: Synthesis, Structure & Physical Properties, Springer Nature, 2016.

[107] R.H. Mitchell, M.D. Welch, A.R. Chakhmouradian, Nomenclature of the perovskite supergroup: A hierarchical system of classification based on crystal structure and composition, Miner. Magazine 81 (2017) 411–461. https://doi.org/10.1180/minmag.2016.080.156

[108] Z. Yavari, M. Noroozifar, M. Khorasani-Motlagh, The improvement of methanol oxidation using nano-electrocatalysts, J. Exp. Nanosci. 11 (2016) 798-815. https://doi.org/10.1080/17458080.2016.1185805

[109] M. Risch, Perovskite electrocatalysts for the oxygen reduction reaction in alkaline media, Catalysts 7 (2017) 1-31. https://doi.org/10.3390/catal7050154

[110] K. Miyazaki, K. Kawaita, T. Abe, T. Fukutsuma, Single-step synthesis of nano-sized perovskite-type oxide/carbon nanotube composites and their electrocatalytic oxygen-reduction activities, J. Mater. Chem. 21 (2011) 1913-1919. https://doi.org/10.1039/C0JM02600J

[111] W. Eerenstein, N.D. Mathur, J. F. Scott, Multiferroic and magnetoelectric materials, Nature 442 (2006) 759-765. https://doi.org/10.1038/nature05023

[112] Y. Kiros, A.R. Paulraj, $La_{0.1}Ca_{0.9}MnO_3/Co_3O_4$ for oxygen reduction and evolution reactions (ORER) in alkaline electrolyte, J. Solid State Electrochem. 22 (2018) 1697–1710. https://doi.org/10.1007/s10008-017-3862-2

[113] Garnet Group ($X_3Y_2Z_3O_{12}$). http://ruby.colorado.edu/~smyth/G5200/10Garnets.PDF, 2019 (accessed 05 March 2019).

[114] University of Colorado, Mineral Structure Data, Garnet. http://ruby.colorado.edu/~smyth/min/garnet.html, 2019 (accessed 05 March 2019).

[115] Magnetism in Gemstones, Kirk Feral, 2011. https://www.gemstonemagnetism.com/understanding_garnets_through_magnetism.html, 2019 (accessed 05 March 2019).

[116] Z.D. Gordon, T. Yang, G.B.G. Morgado, C.K. Chan, Preparation of Nano- and Microstructured Garnet $Li_7La_3Zr_2O_{12}$ Solid Electrolytes for Li-Ion Batteries via

Cellulose Templating, ACS Sustainable Chem. Eng. 4 (2016) 6391−6398. https://doi.org/10.1021/acssuschemeng.6b01032

[117] J.K. Burdett, G.L. Price, S.L. Price, Role of the crystal-field theory in determining the structures of spinels, J. Am. Chem. Soc. 104 (1982) 92–95. https://doi.org/10.1021/ja00365a019

[118] L. Shen, P.E. Laibinis, T.A. Hatton, Bilayer surfactant stabilized magnetic fluids: Synthesis and interactions at interfaces, Langmuir 15 (1999) 447–453. https://doi.org/10.1021/la9807661

[119] R. Singh, G. Thirupathi, Manganese-zinc spinel ferrite nanoparticles and ferrofluids, in: M. Seehra (Ed.), Magnetic spinels-synthesis, properties and applications, IntechOpen, London, 2017, pp. 140–159. https://doi.org/10.5772/66522

[120] N. Sanpo, C. Wen, C.C. Berndt, J. Wang, Antibacterial properties of spinel ferrite nanoparticles, in: A. Mendez-Vilas (Ed.), Microbial pathogens and strategies for combating them: Science, technology and education, formatex research centre, Badajoz, 2013, pp. 239–250.

[121] L. Zhang, Y. Wu, Sol–gel synthesized magnetic MnFe$_2$O$_4$ spinel ferrite nanoparticles as novel catalyst for oxidative degradation of methyl orange, J. Nanomater. 6 (2013) 1–6. https://doi.org/10.1155/2013/640940

[122] F. Waag, B. Gokce, C. Kalapu, G. Bendt, S. Salamon, J. Landers, U. Hagemann, M. Heidelmann, S. Schulz, H. Wende, N. Hartmann, Adjusting the catalytic properties of cobalt ferrite nanoparticles by pulsed laser fragmentation in water with defined energy dose, Sci. Rep. 7 (2017) 13161. https://doi.org/10.1038/s41598-017-13333-z

[123] S. Joshi, V.B. Kamble, M. Kumar, A.M. Umarji, G. Srivastava, Nickel substitution induced effects on gas sensing properties of cobalt ferrite nanoparticles, J. Alloys Compd. 654 (2016) 460–466. https://doi.org/10.1016/j.jallcom.2015.09.119

[124] Q. Zafar, M.I. Azmer, A.G. Al-Sehemi, M.S. Al-Assiri, A. Kalam, K. Sulaiman, Evaluation of humidity sensing properties of TMBHPET thin film embedded with spinel cobalt ferrite nanoparticles, J. Nanopart. Res. 18(2016) 186-195. https://doi.org/10.1007/s11051-016-3488-9

[125] Y. Peng, Z. Wang, W. Liu, H. Zhang, W. Zuo, H. Tang, F. Chen, B. Wang, Size- and shape-dependent peroxidase-like catalytic activity of MnFe$_2$O$_4$ nanoparticles and their applications in highly efficient colorimetric detection of target cancer cells, Dalton Trans. 44 (2015) 12871–12877. https://doi.org/10.1039/C5DT01585E

[126] S. Reddy, B.K. Swamy, U. Chandra, K.R. Mahathesha, T.V. Sathisha, H. Jayadevappa, Synthesis of MgFe$_2$O$_4$ nanoparticles and MgFe$_2$O$_4$ nanoparticles/CPE for

electrochemical investigation of dopamine, Anal. Methods 3(2011) 2792–2796. https://doi.org/10.1039/c1ay05483j

[127] E. Cespedes, J.M. Byrne, N. Farrow, S. Moise, V.S. Coker, M. Bencsik, J.R. Lloyd, N.D. Telling, Bacterially synthesized ferrite nanoparticles for magnetic hyperthermia applications, Nanoscale 6 (2014) 12958–12970. https://doi.org/10.1039/C4NR03004D

[128] H. Choi, S. Lee, T. Kouh, S.J. Kim, C.S. Kim, E. Hahn, Synthesis and characterization of Co-Zn ferrite nanoparticles for application to magnetic hyperthermia, J. Korean Phys. Soc.70 (2017) 89–92. https://doi.org/10.3938/jkps.70.89

[129] C. Shu, H. Qiao, Tuning magnetic properties of magnetic recording media cobalt ferrite nano-particles by co-precipitation method. In: Symposium on Photonics and Optoelectronics; Wuhan, China. 2009. https://doi.org/10.1109/SOPO.2009.5230167

[130] A. Samavati, A.F. Ismail, Antibacterial properties of copper-substituted cobalt ferrite nanoparticles synthesized by co-precipitation method, Particuology 30 (2017) 158–163. https://doi.org/10.1016/j.partic.2016.06.003

[131] A.H. Ashour, A.I. El-Batal, M.I.A. Abdel Maksoud, G.S. El-Sayyad, S. Labib, E. Abdeltwab, M.M. El-Okr, Antimicrobial activity of metal-substituted cobalt ferrite nanoparticles synthesized by sol–gel technique, Particuology 40 (2018) 141-151. https://doi.org/10.1016/j.partic.2017.12.001

[132] A.S. Khader, M.S. Shariff, F. Nayeem, J. Basavaraja, H. Mandanakumara, M.S. Thyagaraj, Structural and dielectric properties of Ni^{2+} doped Chromium Ferrite by Solution Combustion method, JCPS 9 (2016) 993-997.

[133] R. Pandu, $CrFe_2O_4$-$BiFeO_3$ perovskite multiferroic nanocomposites – A review, Mat. Sci. Res. India 11 (2014) 128-145. https://doi.org/10.13005/msri/110206

[134] Z. Shahnavaz, F. Lorestani, W.P. Meng, Y. Alias, Core-shell–$CuFe_2O_4$/PPy nanocomposite enzyme-free sensor for detection of glucose, J Solid State Electrochem. 19 (2015) 1223–1233. https://doi.org/10.1007/s10008-015-2738-6

[135] A. Benvidi, M.T. Nafar, S. Jahanbani, M.D. Tezerjani, M. Rezaeinsaab, S. Dalirnasab, Developing an electrochemical sensor based on a carbon paste electrode modified with nano-composite of reduced graphene oxide and $CuFe_2O_4$ nanoparticles for determination of hydrogen peroxide, Mater. Sci. Eng. C 75 (2017) 1435-1447. https://doi.org/10.1016/j.msec.2017.03.062

[136] S. Tao, F. Gao, X. Liu, O.T. Sorensen, Preparation and gas-sensing properties of $CuFe_2O_4$ at reduced temperature, Mater. Sci. Eng. B 77 (2000) 172-176. https://doi.org/10.1016/S0921-5107(00)00473-6

[137] J.A. Rodriguez, P. Liu, J. Hrbek, J. Evans, M. Perez, Water gas shift reaction on Cu and Au nanoparticles supported on CeO_2(111) and ZnO(0001): intrinsic activity and importance of support interactions, AngewChemInt Ed Engl. 46 (2007) 1329-32. https://doi.org/10.1002/anie.200603931

[138] C.T. Campbell, B.E. Koel, H_2S/Cu(111): A model study of sulfur poisoning of water-gas shift catalysts, Surf. Sci.186 (1987) 393-398. https://doi.org/10.1016/S0039-6028(87)80384-9

[139] A.A. Gokhale, J.A. Dumesic, M. Mavrikakis, On the mechanism of low-temperature water gas shift reaction on copper, J. Am. Chem. Soc. 130 (2008) 1402-1414. https://doi.org/10.1021/ja0768237

[140] M. Estrella, L. Barrio, G. Zhou, X. Wang, Q. Wang, W. Wen, J.C. Hanson, A.I. Frenkel, J.A. Rodriguez, Characterization of $CuFe_2O_4$ and Cu/Fe_3O_4 Water-Gas Shift Catalysts, J. Phys. Chem. C 113 (2009) 14411–14417. https://doi.org/10.1021/jp903818q

[141] M.A. Ansari, A. Baykal, S. Asiri, S. Rehman, Synthesis and characterization of antibacterial activity of spinelchromium-substituted copper ferrite nanoparticles for biomedical application, JIOPM 28 (2018) 2316–2327. https://doi.org/10.1007/s10904-018-0889-5

[142] S.A. Morrison, C.L. Cahill, E.E. Carpenter, S. Calvin, V.G. Harris, Preparation and characterization of MnZn–ferrite nanoparticlesusing reverse micelles, J. Appl. Phys. 93 (2003) 7489–7491. https://doi.org/10.1063/1.1555751

[143] S.G. Ali, M.A. Ansari, H.M. Khan, M. Jalal, A.A. Mahdi, S.S. Cameotra, Antibacterial and antibiofilm potential of green synthesized silver nanoparticles against imipenem resistant clinical isolates of *P. aeruginosa*, BioNanoSci. 8 (2018) 544–553. https://doi.org/10.1007/s12668-018-0505-8

[144] J.N. Payne, H.K. Waghwani, M.G. Connor, W. Hamilton, S. Tockstein, H. Moolani, F. Chavda, V. Badwaik, M.B. Lawrenz, R. Dakshinamurthy, Novel synthesis of kanamycin conjugated gold nanoparticles with potent antibacterial activity, Front. Microbiol. 7 (2016) 607-617. https://doi.org/10.3389/fmicb.2016.00607

[145] R. Ding, L. Lv, L. Qi, M. Jia, H. Wang, A facile hard-templating synthesis of mesoporous spinel $CoFe_2O_4$ nanostructures as promising electrocatalysts for the H_2O_2 reduction reaction, RSC Adv. 4 (2014) 1754-1760. https://doi.org/10.1039/C3RA45560B

[146] S. Liu, W. Bian, Z. Yang, J. Tian, C. Jin, M. Shen, Z. Zhou, R. Yang, A facile synthesis of $CoFe_2O_4$/biocarbon nanocomposites as efficient bi-functional

electrocatalysts for the oxygen reduction and oxygen evolution reaction, J. Mater. Chem. A 2 (2014) 18012-18019. https://doi.org/10.1039/C4TA04115A

[147] J. Yin, L. Shen, Y. Li, M. Lu, H. Sun, P. Xia, $CoFe_2O_4$ nanoparticles as efficient bifunctional catalysts applied in Zn-air battery, J. Mater. Res. 33 (2018) 590-600. https://doi.org/10.1557/jmr.2017.404

[148] A.M. Abu-Dief, I. F. Nassar, W. H. Elsayed, Magnetic $NiFe_2O_4$ nanoparticles: efficient, heterogeneous and reusable catalyst for synthesis of acetylferrocene chalcones and their anti-tumor activity, Appl. Organometal. Chem. 30 (2016) 917–923. https://doi.org/10.1002/aoc.3521

[149] A.A. Ensafi, A.R. Allafchian, B. Rezaei, R. Mohammadzadeh, Characterization of carbon nanotubes decorated with $NiFe_2O_4$ magneticnanoparticles as a novel electrochemical sensor: Application for highly selective determination of sotalol using voltammetry, Mater. Sci. Eng. C 33 (2013) 202–208. https://doi.org/10.1016/j.msec.2012.08.031

[150] M. Kiani, J. Zhang, J. Fan, H. Yang, G. Wang, J. Chen, R. Wang, Spinel nickel ferrite nanoparticles supported onnitrogen doped graphene as efficient electrocatalyst for oxygen reduction in fuel cells, Mater. Express 7 (2017) 261-273.

[151] Google Patents, $G-C_3N_4/NiFe_2O_4$ composite material, as well as preparation method and application thereof. https://patentimages.storage.googleapis.com/72/ae/71/02dafc71ab6026/CN104646044A.pdf, 2019 (accessed 01 February 2019). https://doi.org/10.1166/mex.2017.1376

[152] J. Li, W. Tang, J. Huang, J. Jin, J. Ma, Polyethyleneimine decorated graphene oxide-supported $Ni_{1-x}Fe_x$ bimetallic nanoparticles as efficient and robust electrocatalysts for hydrazine fuel cells, Catal. Sci. Technol. 3 (2013) 3155–3162. https://doi.org/10.1039/c3cy00487b

[153] W.S. Hummers, R.E. Offeman, Preparation of graphitic oxide, J. Am. Chem. Soc. 80 (1958) 1339-1348. https://doi.org/10.1021/ja01539a017

[154] Z.-P. Deng, Y. Sun, Y.-C. Wang, J.-D. Gao, A NiFe Alloy Reduced on Graphene Oxide for Electrochemical Nonenzymatic Glucose Sensing, Sensors 18 (2018) 3972. https://doi.org/10.3390/s18113972

[155] Q. Huang, P. Zhou, H. Yang, L. Zhu, H. Wu, In situ generation of inverse spinel $CoFe_2O_4$ nanoparticles onto nitrogen doped activated carbon for an effective cathode electrocatalyst of microbial fuel cells, Chem. Eng. J. 325 (2017) 466–473. https://doi.org/10.1016/j.cej.2017.05.079

[156] P. Li, R. Ma, Y. Zhou, Y. Chen, Q. Liu, G. Peng, Z. Liang, J. Wang, Spinel nickel ferrite nanoparticles strongly crosslinked with multiwalled carbon nanotubes as a biefficient electrocatalyst for oxygen reduction andoxygen evolution, RSC Adv. 5 (2015) 73834. https://doi.org/10.1039/C5RA14713A

[157] M. Kiani, J. Zhang, J. Fan, H. Yang, G. Wang, J. Chen, R. Wang, Spinel nickel ferrite nanoparticles supported on nitrogen doped graphene as efficient electrocatalyst for oxygen reduction in fuel cells, Mater. Express 7 (2017) 261-273. https://doi.org/10.1166/mex.2017.1376

[158] X. Tang, A. Manthiram, J.B. Goodenough, Copper Ferrite Revisited, J Solid State Chem. 79 (1989) 250-262. https://doi.org/10.1016/0022-4596(89)90272-7

[159] N. Nanba, S. Kobayashi, Semiconductive Properties and Cation Distribution of Copper Ferrites $Cu_{1-\delta}Fe_{2+\delta}O_4$, Japan J. Appl. Phys. 17 (1978) 1819-1824. https://doi.org/10.1143/JJAP.17.1819

[160] J.A. Rodriguez, P. Liu, J. Hrbek, J. Evans, M. Perez, Water gas shift reaction on Cu and Au nanoparticles supported on $CeO_2(111)$ and $ZnO(0001)$: Intrinsic activity and importance of support interactions, Angew. Chem. Int. Ed. 46 (2007) 1329-1338. https://doi.org/10.1002/anie.200603931

[161] J. B. Park, J. Graciani, J. Evans, D. Stacchiola, S. Ma, P. Liu, A. Nakamura, J. Fernandez-Sanz, J. Hrbek, J. A. Rodriguez, High catalytic activity of $Au/CeO_x/TiO_2(110)$ controlled by the nature of the mixed-metal oxide at the nanometer level, Proc. Natl.Acad. Sci. U.S.A. 106 (2009) 4975-4985. https://doi.org/10.1073/pnas.0812604106

[162] R. Khan, M. Habib, M. A. Gondal, A. Khalil, Z.U. Rehman, Z. Muhammad, Y.A. Halem, C.Wang, C.Q. Wu, L. Song, Facile synthesis of $CuFe_2O_4$–Fe_2O_3 composite for high-performance supercapacitor electrode applications, Mater. Res. Express 4 (2017) 1-7. https://doi.org/10.1088/2053-1591/aa8dc4

[163] H. Yang, J. Yan, Z. Lu, X. Cheng, Y. Tang, Photocatalytic activity evaluation of tetragonal $CuFe_2O_4$ nanoparticles for the H_2 evolution under visible light irradiation, J. Alloys Compd. 476 (2009) 715–719. https://doi.org/10.1016/j.jallcom.2008.09.104

[164] V.A. Drits, B.A. Sakharov, A.L. Salyn, A. Manceau, Structural model for ferrihydrite, Clay Miner. 28 (1993) 185-207. https://doi.org/10.1180/claymin.1993.028.2.02

[165] Y. Cudennec, A. Lecerf, The transformation of ferrihydrite into goethite or hematite, revisited, J. Solid State Chem. 179 (2006) 716–722. https://doi.org/10.1016/j.jssc.2005.11.030

[166] S. Asbrink, L.J. Norrby, A refinement of the crystal structure of copper (ii) oxide with a discussion of some exceptional E.s.d.'s, ActaCrystallogr. B 26 (1969) 319-328.

[167] A.B. Kuzmenko, D. van der Marel, P.J.M. van Benthum, E.A. Tishchenko, C. Presura, A.A. Bush, Infrared spectroscopic study of CuO: Signatures of strong spin-phonon interaction and structural distortion, Phys. Rev. B 63 (2000) 094303. https://doi.org/10.1103/PhysRevB.63.094303

[168] A.M. Balagurov, I.A. Bobrikov, M.S. Maschenko, D. Sangaa, V.G. Simkin, Structural phase transition in $CuFe_2O_4$ spinel, Crystallogr. Rep. 58 (2013) 710-717.

[169] N.H. Li, S.L. Lo, C.Y. Hu, C.H. Hsieh, C.L. Chen, Stabilization and phase transformation of $CuFe_2O_4$ sintered from simulated copper-laden sludge, J. Hazard. Mater. 190 (2011) 597–603. https://doi.org/10.1016/j.jhazmat.2011.03.089

[170] M.M. Rashad, R.M. Mohamed, M.A. Ibrahim, L.F.M. Ismail, E.A.Abdel-Aal, Magnetic and catalytic properties of cubic copper ferrite nanopowders synthesized from secondary resources, Adv. Powder Technol. 23 (2012) 315–323. https://doi.org/10.1016/j.apt.2011.04.005

[171] Z.H. Xiao, S.H. Jin, J.H. Wang, C.H. Liang, Magnetism and phase transformation of Cu-Fe composite oxides prepared by the sol-gel route, Hyperfine Interact. 217 (2013) 151–156. https://doi.org/10.1007/s10751-012-0662-z

[172] S. Pongpadung, T. Kamwanna, V. Amornkitbamrung, Effect of fabrication method on the structural and the magnetic properties of copper ferrite, J. Korean Phys. Soc. 68 (2016) 697–704. https://doi.org/10.3938/jkps.68.697

[173] A. Kyono, S.A. Gramsch, Y. Nakamoto, M. Sakata, M. Kato, T. Tamura, T. Yamanaka, High-pressure behavior of cuprospinel $CuFe_2O_4$: influence of the Jahn-Teller effect on the spinel structure, Am. Mineral. 100 (2015) 1752-1761. https://doi.org/10.2138/am-2015-5224

[174] V.N. Nikolic, M.M. Vasic, D. Kisic, Influence of the Fe^{3+} cation on the formation mechanism and crystallite size of $CuFe_2O_4$ nanoparticles, 26th Conference of the Serbian Crystallographic Society; Srebrno Jezero, Serbia. 2019

[175] V.N. Nikolic, M.M. Vasic, D. Kisic, Observation of c-$CuFe_2O_4$ nanoparticles of the same crystallite size indifferent nanocomposite materials: The influence of Fe^{3+} cations, J. Solid State Chem. 275 (2019) 187-196. https://doi.org/10.1016/j.jssc.2019.04.007

[176] L. Lutterotti, Total pattern fitting for the combined size–strain–stress–texture determination in thin film diffraction, Nucl. Instrum. Methods B 268 (2010) 334–340. https://doi.org/10.1016/j.nimb.2009.09.053

[177] A.S. Moskvin, N.N. Loshkareva, Y.P. Sukhorukov, M.A. Sidorov, A.A. Samokhvalov, Characteristic features of the electronic structure of copper oxide (CuO): Initiation of the polar configuration phase and middle-IR optical absorption, Zh. Eksp. Teor. Fiz.105 (1994) 967-993.

[178] J. Zhao, F.E. Huggins, Z. Feng, G.P. Huffman, Ferrihydrite: surface structure and its effects on phase transformation, Clays and Clay Minerals 42 (1994) 737-745. https://doi.org/10.1346/CCMN.1994.0420610

[179] V.N. Nikolic, M.M. Milic, Evolution of $CuFe_2O_4$ and α-Fe_2O_3 structural properties initiated by thermal treatment. In: 25th Conference of the Serbian Crystallographic Society; Bajina Basta, Serbia. 2018

[180] V.N. Nikolic, M. Vasic, M.M. Milic, Observation of low- and high-temperature $CuFe_2O_4$ phase at 1100 °C: The influence of Fe^{3+} ions on $CuFe_2O_4$ structural transformation, Cer. Inter. 44 (2018) 21145-21152. https://doi.org/10.1016/j.ceramint.2018.08.157

Magnetochemistry - Materials and Applications
Materials Research Foundations **66** (2020) 87-129

Materials Research Forum LLC
https://doi.org/10.21741/9781644900611-3

Chapter 3

Magnetic Nanomaterials for Separations

Sunil Kumar and Rashmi Madhuri[*]

Department of Applied Chemistry, Indian Institute of Technology (Indian School of Mines), Dhanbad, Jharkhand 826 004, India

Abstract

Magnetic nanomaterials have low cost, high removal capacity, reusability, high surface-volume ratio, and high reactivity towards eluent compounds, therefore, are highly suitable for separation purposes. Their only drawback associated with magnetic nanoparticles (MNPs) is their poor stability in harsh conditions. Therefore, in general, for their wider applications modified MNPs are used. Magnetic nanomaterials were modified or coated with various kinds of materials to protect their outer surface from strong pH, high temperature, etc. Here, in this chapter, we focused on the modification of MNPs and their role in separation science. For example, modifications with inorganic materials (like silica, metal, and carbon) improve the reactivity of MNPs and reduce their agglomeration. Similarly, modification with organic materials (like surfactants and polymer) may improve stability of MNPs and avoid their agglomeration. Here, we have included the role of modified MNPs for successful separation of protein, nucleic acid, heavy metal ions, dyes, etc.

Keywords

Separation, Magnetic Nanoparticles, Metal Ions, Carbonaceous Nanomaterials, Surfactants, Polymers

Contents

1.   Introduction

Nanomaterials are very popular materials in recent year due to their unique physical and chemical properties along with their small size [1]. Nanomaterials are materials having a size ranging from 1 nm to 100 nm. They can be divided by different classification methods, like based on dimension, they can be classified in three categories: one dimensional, two dimensional and three dimensional [2]. They can be also grouped into four different categories, depending on functionality at the surface of the materials, as carbon-based nanomaterials, inorganic nanomaterials, organic nanomaterials, nanocomposites [3]. The carbon-based nanomaterials comprise of single or multiwalled nanotubes, fullerenes, graphene, and nano-diamonds, which have remarkable properties like high mechanical, electrical and optical activity [4, 5]. However, organic compound modified nanomaterials i.e. organic nanomaterials can be applied to mechanical separation, fabrication of nanodevice sensor and retroviral transduction in animals [6]. The best application of nanocomposites is chitin nanomaterials, which have high mechanical strength, good thermal stability, and antibacterial properties and can be used as gas barrier [7]. Next is inorganic nanomaterials, consist of a composition of metal and their oxides, nitrite, sulphides, carbides, halides and alloys. They can be further classified on the basis of their intrinsic properties like magnetic, electronic, and optical compounds, ionic and superconducting materials, structural nanomaterials, ceramics, catalytic and porous nanomaterials [8, 9]. Among the different kind of nanomaterials, recently, magnetic nanomaterials have become very popular in laboratory as well as industries, owing to its low toxicity, high surface-volume ratio and easy separation of the pure compound by simple external magnet only.

Magnetic nanoparticles (MNPs) or magnetic nanomaterials (MNs) are a category of nanosized material which is operated by utilizing an external magnet. These materials or substance are usually consisting of two compounds. The magnetic material that can be nickel (Ni), iron (Fe) or cobalt (Co) and a chemical constituent which has functional groups. Use of magnetic nanomaterials has gained popularity since 2000 because of their non-hazardous properties, strong magnetization value, and an active surface, which can be applied for targeting, imaging, separation and therapeutic activities [10]. William Fullarton has applied first-time magnetic separation of iron ore in 1792, but the separation of related biological compounds by magnetic nanomaterials were reported successfully in 1970 [11]. In recent time, magnetic nanomaterials were applied in different fields like research and industrial community (chemical, medical and environmental) [12]. They have found to be suitable for magnetic hyperthermia [13], catalysis [14], environmental remediation [15], drug and gene delivery [16, 17], separation [18], sensing [19], biomedical field [20], and magnetic resonance imaging [21]. There are different types of magnetic nanomaterials made up of mainly three known metal such as iron, nickel, and cobalt which utilized in the formation of their oxides or metallic form. For example, iron oxide NPs have iron oxides wherein two constituents like maghemite (γ-Fe_2O_3) or magnetite (Fe_3O_4) are present [22].

The magnetic nanoparticles (MNPs) have attracted attention in recent years due to their enormous surface region and small size. Due to the large proportion of atoms exposed on their surface, they have high surface energy. According to the literature, magnetic anisotropic energy gets lowered with decrease in size of the particle to nanometer range [23]. Therefore, the thermal motion of magnetic dipole present in MNPs gets increased opposite the magnetic anisotropic energy. Till then, the thermal motion of magnetic dipole present in the MNPs becomes greater than the ordering of magnetic dipoles. In simple words, orientation of magnetic moments of the nanoparticles get suppressed over the energy generated by thermal motion of magnetic dipoles [23]. As a result, MNPs start behaving like superparamagnetic materials, which can be reversibly transit between pseudoparamagnetic characteristic and pseudoferromagnetic characteristics, with the influence of external magnets. The MNPs act as superparamagnetic materials which have large and stable magnetic moment and also behave like paramagnetic atoms, which give quick response towards external magnets and shows almost insignificant remanence and coercivity. The magnetic dipole is easily moved towards magnetic fields with insignificance remanence and coercivity is called a paramagnetic substance, but without magnetic field, the superparamagnetic substance retained their magnetic properties, which can be applied for separation purposes [23]. Nowadays, different methods are popularly used for the synthesis of MNPs owing to their characteristic properties like large surface region, great bonding with contaminants and easy separation by external magnets. Figure 1 shows the role of various kind of magnetic nanomaterials in the field of separation science.

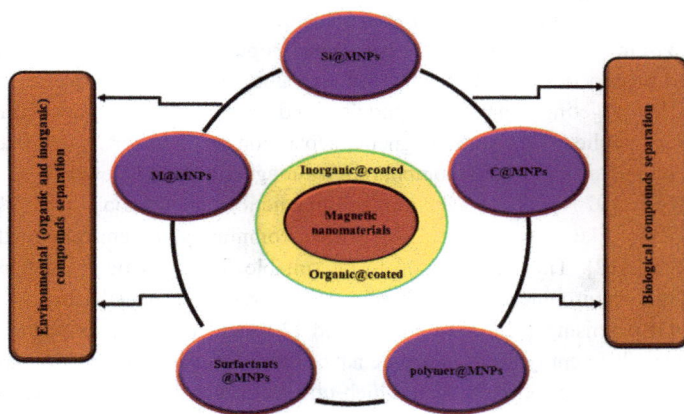

Figure 1. Schematic illustration showing application of magnetic nanomaterials in different areas.

2. Synthesis of MNPs

Synthesis process commonly used for magnetic nanomaterials can be divided in two broad classes: (1) physical methods and (2) chemical methods.

2.1 Physical methods used for the synthesis of MNPs

The physical methods used for the formation of MNPs are the top-down approach, in which the bulk material or macromolecule are reduced to nanosized materials. The physical methods used for synthesis of MNPs can further be divided in different types, based on the techniques used.

2.1.1 Mechanical milling method

Mechanical milling (MM) is the process, where solid-state reaction occurs. The technique allows the conversion of bulk materials into homogeneous reinforcement nanomaterials by the interaction between balls and particles in terms of plastic distortion, constant breaking and cold welding of the compounds. The preparation of fine magnetic nanomaterials was depended on various parameters such as reaction time, rate of reaction, ball-to-powder weight ratio and monitor machine. From this technique, nanoscale alloys, composites and intermetallic compounds can be synthesized successfully in the last few years [24, 25]. Balaraju et al. have synthesized $NiO-MnO_2$ nanomaterials by using the mechanical milling method [26]. The proposed materials have been characterised by field emission scanning electron microscopy (FE-SEM), Vibrating sample magnetometer (VSM), X-ray diffractometer (XRD). Scherer's relation is used for measurement of the size of nanoparticles, which is occurred between 25-35 nm. The prepared magnetic nanocomposite has excellent magnetic moment calculated at 100 K [26].

2.1.2 Vapour deposition method

Vapour deposition method is well known method of synthesizing MNPs. It can be further classified in two types i.e. physical vapour deposition and chemical vapour deposition. In physical vapour deposition or thin film deposition method the solid and liquid materials are converted into vapour state, which is condensed and adsorbed at a particular surface. Pham et al. have synthesized cobalt iron oxide thin Films by atomic film coating method in the presence of reactive compounds such as $Co(TMHD)_2$ and $Fe(TMHD)_3$, [here TMHD stands for bis(2,2,6,6-tetramethyl-3,5-heptanedionato)], and atomic oxygen as the metal and oxidation sources, respectively [27]. It is deposited on to the surface of $SrTiO_3$ (001) as required temperature at 200 °C. The magnetic behaviour was judge by two properties, one is the thickness of the film, and another is annealing temperature. The saturation magnetization was obtained in the range of 260 to 550 emu cm^{-3} and magnetic coercivity ranges from 0.2 to 2.2 kO [27].

In chemical vapour deposition (CVD) mixed reaction plays the leading role to deposit solid and volatile compounds from a volatile precursor through chemical reactions on the substrate [28]. There are various kind of CVD methods based on different process used such as laser chemical vapour deposition (CVD), photochemical vapour deposition (PVD), atmospheric pressure chemical vapour deposition chemical beam epitaxy (AP-CVD), chemical vapour infiltration (CVI), and plasma-enhanced chemical vapour deposition (PE-CVD), metal-organic chemical vapour deposition (MO-CVD), low-pressure chemical vapour deposition (LP-CVD) and plasma-assisted chemical vapour deposition (PA-CVD) [29]. Alijani et al. has synthesized magnetic carbon nanotubes via one step process modified with diatomite earth compounds by chemical vapor deposition method. The synthesized materials have shown superparamagnetic properties, which are used for separation of M^{2+} ions from aqueous medium [30].

2.1.3 Electrical explosion of wires (EEW) method

The electrical explosion of wire is an old technique for nanomaterials production, but is most popular because of its simplicity and cost-effectiveness. The synthesis of MNPs by EEW, depends on two parameters, firstly, metal have large energy density for complete vaporisation of metal and another is the simultaneous interaction between metal vapour and plasma, which will control the crystal growth and results in synthesized MNP. The formation of magnetic materials from this technique depends on some basic properties like physical transformation of wire, high temperature, high pressure, strong wave, type of irradiation material and current rate etc. [31, 32]. For example, Beketov et al. has developed iron and nickel magnetic nanoparticle by electrical explosion of wire method [33]. Lee et al. has evaluated the one-pot synthesis of $Fe_3O_4/Fe/MWCNT$ from this technique, in which Fe wire was vaporised and prepared MNPs get functionalised by MWCNT in distilled water at a normal temperature [34].

2.2   Chemical methods used for the synthesis of MNPs

Chemical methods used for synthesis of MNPs are more popular than physical methods owing to their easy process, low cost and flexibility in modifications. The well-known chemical methods used for synthesis of MNPs are discussed below.

2.2.1 Co-precipitation method

The synthesis of MNPs by co-precipitation method is reacted in aqueous solution with a magnetic precursor, which acquire the primary medium. For the synthesis, reducing agent (like [NaBH$_4$], ascorbic acid [35, 36]), inorganic salts as precursor (like ferric chloride, cobalt nitrate, nickel acetate etc.) [37] and organic molecule (as a stabilizing agent, like citrate, oleic acid) [37, 38] are mostly required. The size and shape of MNPs synthesized via co-precipitation method depended on various parameters such as the concentration of precursor, stirring speed, reactants ratio, pH and temperature of a medium [40-41]. The process is very simple, where metal salt as precursor will be added in aqueous solution, followed by addition of reducing agent and stabilizing agents. Whole mixture will be stirred for certain time and particular temperature to prepared desired MNPs. Chandane et al. has prepared iron oxide NPs via the co-precipitation method in existence of NaBH$_4$ (reducing agent). The prepared MNPs has shown excellent superparamagnetic properties having 0.83506G coercivity (Hc) and 11.51emu g^{-1} saturation magnetization (Ms) [35].

2.2.2 Thermal decomposition method

In this method, metallic organic precursor will decompose at high temperature (~350 °C) and high boiling point organic solvent (like polyethylene glycol) with suitable surfactants compounds (sodium oleate, oleic acid, sodium dodecylamine sulfonate etc.) were used. The method gives magnetic nanomaterials with monitored size and shape [42, 43]. Several types of MNPs have been synthesized from this method such as Fe$_3$O$_4$ [44], CoFe$_2$O$_4$ [45], Au@Fe$_3$O$_4$ [46] etc. Using this method, synthesis of Mn$_x$Ga$_{1-x}$Fe$_2$O$_4$ (x=0–1) NPs was reported in the presence of tetraethylene glycol, which shows good superparamagnetic properties [47].

2.2.3 Hydrothermal method

Among the various reported conventional technique for preparation of magnetic nanomaterials, the hydrothermal method has attracted much attention in recent years. The hydrothermal process can be accomplished in a closed stainless-steel autoclave with Teflon lining in the presence of the elevated temperature and pressure. Hydrothermal method is cost-effective gives pure product, energy storing, have nucleation control, comparatively pollution free, have high dispersion and high rate of reaction as well as maintain the morphology of resulting MNPs [48]. For example, using this method, NiFe$_2$O$_4$ NPs has been synthesized by utilizing glycerol as surfactants and NaOH act as a precipitating constituents [49].

2.2.4 Microemulsion method

Microemulsion is a thermodynamically steady isotropic dispersion between the two non-miscible oil and water phases in the presence of a surfactant. Here, surfactant molecule makes a monolayer interface between water and oil layer and plays a wide role in synthesis. Therefore, the characteristics of prepared MNPs using this method depends upon on the type and the morphology of the surfactant. Lu et al. have been synthesized a bifunctional magnetic-fluorescent nanomaterial i.e. $Fe_3O_4@SiO_2$-PDI-PAA/Ca^{2+} by depositing the Perylene diimide Perylene diimide (PDI) modified star polymer (PAA) (Polyacrylic acid) on to the $Fe_3O_4@SiO_2$ core-shell nanomaterials by reverse micro emulsion method [50]. The nanocomposite was characterised by different techniques like transmission electron microscopy, fluorescence spectroscopy, ultraviolet-visible spectroscopy and vibrating sample magnetometer. The prepared nanocomposite exhibits a robust red emission and superparamagnetic properties at average temperature.

3.   Modification or functionalization of magnetic nanoparticles

The stability of MNPs is a necessary condition to apply them in various areas, where, the bare or unmodified MNPs failed to show their proper and full activity. The unmodified or bare magnetic nanoparticles i.e. without any modification have some drawbacks like their easy agglomeration/aggregation, high risk of oxidation in acidic or basic solutions and easy release of metal ion from MNPs. To resolve this problem, MNPs are functionalized or modified with suitable functional groups, which will not only solve these problems but enhance their properties as well.

One of the easiest procedures to answer this problem is coating or modification of nanoparticle's surface with a compound, functional group or layer, which can isolate them from the damaging environment. Surface coating of MNPs with polymers, organic compounds, inorganic moieties, even with nanoparticles/materials are very common in recent time and such modifications not only protect the MNPs from damage but also enhance their properties. During modification, we have to keep in mind that selection of functionalizing agent or coating layer should always be chosen according to the properties they can present, their overall impact, cost as well as application. Therefore, use of biocompatible, stable, easily available and low-cost materials are always preferred for modification of MNPs. Darwish et al. have prepared oleic acid-stabilized magnetite nanoparticles in the alkaline medium by co-precipitation method. The prepared MNPs shows good specific absorption rate (214.95 W/g) and therefore applied for hyperthermia [51]. MNPs are usually modified or functionalized to target very specific applications and can be used in multiple fields. Here, we have confined our search limited to the role of MNPs in separation science only.

4.   What is separation?

Conversion of a mixture of solutions or chemicals in their individual components or more distinct product mixtures is called as separation [52]. The scientist working in this field called it as separation science, because, it is the most important and foremost process, prior to obtaining pure form of any substance/compound/chemical.

Separation of compounds was reported via various techniques such as centrifugation [53], crystallization [54], chromatography [55], distillation/fractional distillation [56], sublimation [57] and extraction. But, the role of MNPs or magnetically driven separation is one of the wide potential application areas in terms of cost, selectivity and specificity. The magnetic nanomaterials are well known materials with properties like high removal capacity, high reactivity towards eluents compounds, high separation ability, regeneration feature and reusability [58]. In general, separation performed by using MNPs follows a similar pattern and can be easily demonstrated by the simple schematic representation (Figure 2). As shown in the scheme, firstly, in the sample of analyte, MNPs were added. Then, the analyte gets adsorbed on the surface of the MNPs. Finally, an external magnet was placed near to the beaker or vial containing analyte adsorbed MNPs. The MNPs get attracted towards the magnet and finally suitable solvent is added to elute the analyte. The pure analyte gets eluted in the solvent and separated MNPs can be collected, washed and prepared for re-use for another separation.

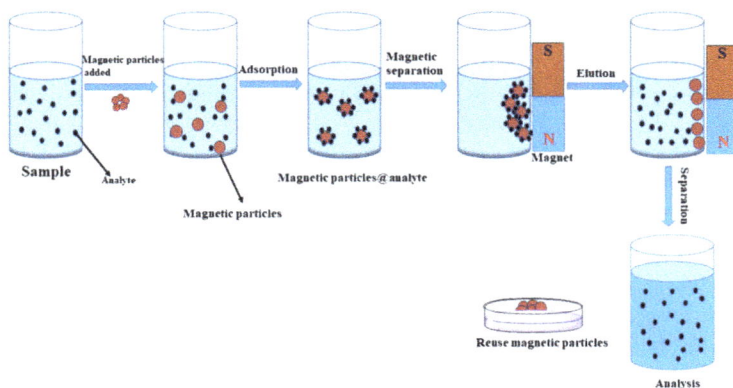

Figure 2. General protocol used for separation of analyte using magnetic nanomaterials with the help of external magnet {modified from reference 5, with permission}.

Magnetochemistry - Materials and Applications   Materials Research Forum LLC
Materials Research Foundations 66 (2020) 87-129   https://doi.org/10.21741/9781644900611-3

A number of literatures has been reported, MNPs are successfully employed for separation of variety of compounds/chemicals/contaminants etc. For example, Rahbar et al. have developed zirconia and magnetite (Fe_3O_4) nanoparticles modified with chitosan and applied it for MSPE (micro-solid-phase extraction) of organophosphorus pesticides from fruit juice and aqueous samples preceeding to gas chromatography-mass detection [59]. They have used the concentration range 0.1–500 ng/mL and found limit of detection (LOD) is 0.034 ng/mL. Similarly, in another literature, protein separation was done by the hydrophobic silica modified MNPs bonded with alkyl group [60]. Some different metal ions like Ag^+ [61], Pb^{2+}, Cd^{2+} [62], other compounds like carvedilol, amlodipine besylate, and losartan from plasma sample [63], industrial dyes like methylene blue by carboxyl functionalized magnetic nanoparticle [64], chiral compounds by teicoplanin-modified hybrid magnetic mesoporous silica nanoparticles [65], lysozyme from serum by magnetic@poly(styrene-sulfonate-N-isopropylacrylamide) [66] have been separated using modified MNPs. Similarly, MNPs were also used for selective detection and removal of 17b-estradiol from milk sample [67], methylprednisolone acetate [68], histamine [69], Hb [70], 4-tert-octylphenol, penconazole and chlorpyrifos-ethyl [71]. As we can see, for separation of different kind of compounds different kind of modification is needed to the MNPs. In the next section, we have discussed the role of different kind of modified MNPs used for separation of various type of compounds.

5.   Role of magnetic nanomaterials in separation

As we have discussed in introduction section that a large variety of applications has already been envisioned for MNPs. The potential and versatility of MNPs in the field of separation science begin from their rapid and simple separation using simple external magnets, which can eliminate the monotonous and expensive separation processes usually applied in chemistry.

5.1   Separation by silica modified magnetic nanomaterials

Use of magnetic materials are increasing day by day because they have the proper size, shape, stable and dispersible in aqueous media [72]. The bare magnetic materials suffer the lack of stability and low-efficiency. Silica modification of MNPs is a very popular and easy step to sort out these problems. Silica coated MNPs accept much observation, owing to their low cost, high stability and non-swelling properties [73]. Silica is stable in acidic medium and behaves as inert platform in the presence of redox process, which generally used for protection of inner core of MNPs and surface functionalization of magnetic materials [74]. Pei et al. has synthesized $Fe_3O_4@SiO_2$–MPS nanocomposite via the sol–gel method, which have excellent hydrophobicity with contact angle (CA)

Materials Research Forum LLC

https://doi.org/10.21741/9781644900611-3

towards water $137 \pm 3°$, while CA for oil is found nearly $0°$. The prepared nanocomposite has shown good superparamagnetic property with saturation magnetisation of 44.24 emu/g [75].

In another method, high-performance liquid chromatography- ultraviolet visible spectroscopy (HPLC-UV) technique was used for separation of four amines including putrescine, histamine, cadaverine and tyramine in Iranian Kashk by solid phase extraction process [76]. But prior to that sample preparation was done using silica modified magnetic nanomaterials. The separation process was affected by some parameters like removal of solvent, weight of mesoporous sorbent, removal time and contact time (CT). For the four amines, recovery was found in the range of 97.16-104.5%. Therefore, the silica modified MNPs successfully used to separate four amines from reported local Iranian compound. Wang et al. have synthesized multifunctional magnetic core shell nanostructure by immobilising the cadmium-tellurium QDs and Rhodamine 6G-deprived receptor on to the surface of the magnetic mesoporous silica nanocomposite [74]. Synthesis protocol and separation mechanism are shown in Figure 3. The linearity over the concentration range and detection limit for separation of Hg^{2+} from the environmental, biological, and polluted areas is found as 0.7 to $90×10^{-8}$ mol/L and $2.5×109^{-9}$ mol/L.

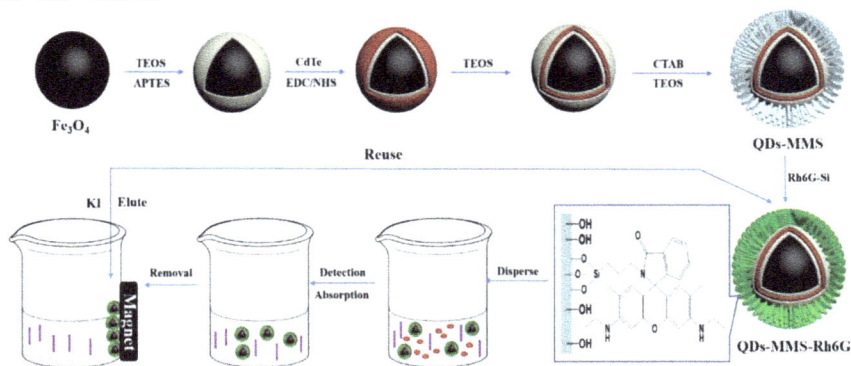

Figure 3. Schematic illustration showing synthesis of QDs-MMS-Rh6G for separation of mercury ion, reproduced with the permission taken from [74].

A similar method was applied by Wang et al. to synthesized silica@Fe_3O_4 via the Stöber method [73]. They have used tetraethyl orthosilicate as silica precursor, which have several advantages such as equal size and shape, protected from acid and basic medium, thermal robust and moderate cytotoxic. The resulting nanomaterials were successfully used for separation of congo red dye from aqueous medium. In another method, silica@MNPs modified with quaternary ammonium coated on to materials through sorption/desorption method was formed and used for removal of uranium (VI) from aqueous solutions [77]. In this experiment, optimal adsorption properties i.e. contact time and pH were used. Kinetic and thermodynamic variable was observed using non-linear regression. The prepared material has exhibit the adsorption capacity of ~87 mg/g in 120 minutes of equilibrium time. Chen et al. has synthesized silica@MNPs and applied it to sensitive, specific, and rapid immunochromatographic strips to separate nucleic acid from cronobacterlysate [78]. The detection limit was reported 105 and 106 cfu/mL in pure culture and powder infant formula. It confirms that magnetic materials have potential for microbial separation from food items or clinical diagnosis also. Javaheri et al. has synthesized Fe_3O_4@SiO_2@MPS@poly(4-vinylpyridine) (MPS = 3-(trimethoxysilyl) propyl methacrylate) nanocomposite via co-precipitation method, which is applied for detection of nitrate ion aqueous medium. The large adsorption capacity was found 80.6 mg/g for nitrate ions [79].

5.2   Separation by alumina modified magnetic nanomaterials

Alumina or aluminium oxide (Al_2O_3) is another popular option to prevent the agglomeration of MNPs, increase the stability and sorption capacity compared with the naked magnetic materials [80]. Modification with aluminium oxide, leads to several advantages including low price, high stability, high biocompatibility, large surface area, high conductance, and simple synthetic procedure [81]. Some of the literatures showing Al_2O_3 coated MNPs towards separation of analytes are discussed below. Chai et al. has synthesized the sulfate@Fe_3O_4/Al_2O_3 NPs with high adsorption capacity and low-cost separation of fluoride from drinking water by ion the exchange method. Here, the sulfate ion present in the NPs get exchanged with the fluoride ion and used for removal of fluoride ion [82].

In another work, uniform size of aluminium oxide was wrapped at the surface of magnetic nanomaterial, which is not only use to separate the phosphopeptides from the biological sample but also showed magnetic properties, which help to the isolation of pure compounds by the use of an external magnet [83]. Sun et al. have prepared Fe_3O_4/Al_2O_3 NPs for detection and separation of sulfonamides in different soil samples via magnetic solid-phase extraction [84]. The modified MNPs was added in the soil sample, ultrasonicated and separated by an external magnet. The recovery range were found between 71% and 93% and LOD were found between 0.37 and 6.74 ng g^{-1}. In another way, Chen et al. have investigated a pigeon ovalbumin-bound Fe_3O_4@Al_2O_3

MNPs for separation of uropathogenic Escherichia *coli* [85], shown in Figure 4. The bacteria separated by the bonding with modified MNPs were observed by matrix-assisted laser desorption/ionization mass spectrometry (MALDI/MS). The detection limit for E. *coli* was found to be $\sim 9.60 \times 10^4$ cfu/mL (0.5 mL).

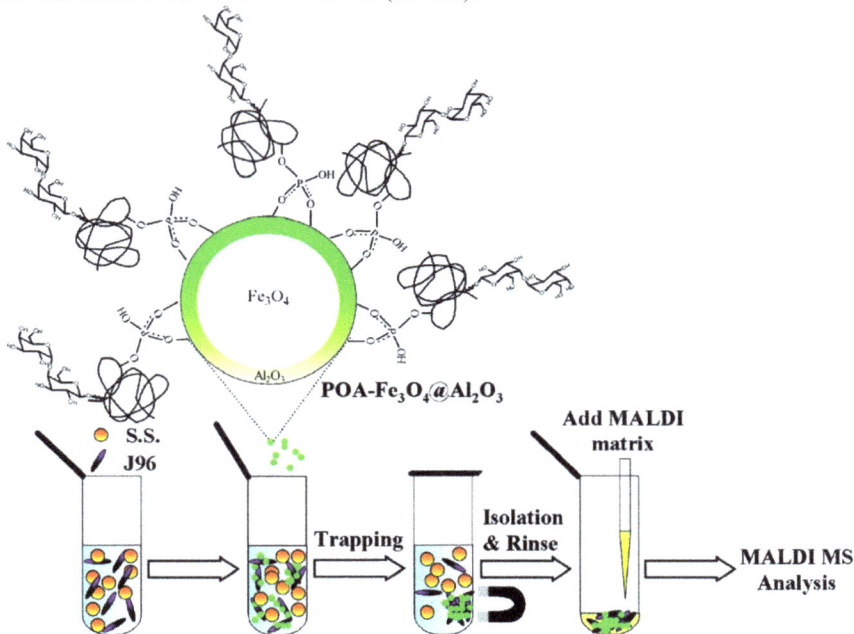

Figure 4. Schematic mechanism showing trapping target bacteria using POA-Fe$_3$O$_4$@Al$_2$O$_3$ NPs, reproduced with the permission taken from [85].

5.3   Separation by zirconium modified magnetic nanomaterials

Zirconium oxide (ZrO_2) has Lewis acid sites at their surface, which possess the ability to combine with Lewis bases. This combination has many advantages for the separation of pure compound [86]. So, combination of ZrO_2 and MNPs not only protect the MNPs from the environment but also gives better bonding between material and contaminants/compounds.

Wang et al. has used magnetic core-shell composite i.e. Fe_3O_4 coated with ZrO_2 for separation of phosphate in water effect of magnetic field. The adsorption of phosphate on the sorbent was explained by using the phosphate adsorption kinetics and Freundlich adsorption model [87]. In another work, glucose modified magnetic zirconium-based metal-organic structure materials were applied for separation of Co(II) from the aqueous medium under the optimal experimental condition [88]. The maximum adsorption capacity was obtained as 178.6 mg/g at 288 K, 222.2 mg/g at 298 K, and 270.3 mg/g at 308 K for Co (II). Zhang et al. have prepared a multifunctional magnetic zirconium hexacyanoferrate (ZrHCF) nanoparticles in which the magnetic beads and zirconium hexacyanoferrate (II) (ZrHCF) acted as core-shell nanomaterials [89]. The Zr (IV) in ZrHCF is easily reacted with phosphonate groups present in DNA and used for their magnetic separation (Figure 5). The proposed sensor showed high sensitivity towards DNA in the linear range of 1.0 fM to 1.0 nM and showed a detection limit of 0.43 fM.

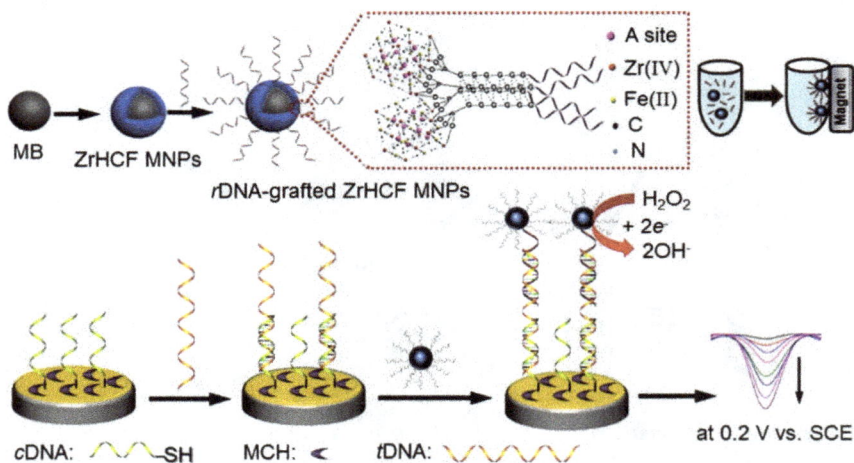

Figure 5. Schematic representation for separation of DNA by ZrHCF MNPs, reproduced with the permission taken from [89].

5.4 Separation by silver modified magnetic nanomaterials

Silver based materials are always at the top most priority of scientists, and researchers owing to their low cost, low toxicity, good stability and no adverse effect [90]. Das et al. have developed $Fe_3O_4@$ polypyrrole (PPy)-mercaptoacetic acid (MAA)/Ag nanomaterials via the two-step procedure. First, the magnetic material was modified by

PPy and MAA monomer. After polymerization, the combination of Fe_3O_4@PPy-MAA was mixed with $AgNO_3$ leads to deposition of AgNPs on their surface under the normal temperature. The final material Fe_3O_4@PPy-MAA/Ag was applied to the detection of 4-nitrophenol and organic dyes (methyl orange and methylene blue) from wastewater at normal temperature [91].

Veisi et al. have synthesized Fe_3O_4@tannic acid (TA)/Ag nanomaterials for potential detection and separation of Rhodamine B, 4-nitrophenol and Methylene blue [92]. Here, tannic acid was responsible for functionalization of phenolic, hydroxyl, and carbonyl functional group on surface of Fe_3O_4 NPs. By attaching these functional groups (FG) on the surface of MNPs, their surface activities were improved in terms of stability. Wang et al. have developed silver@ MnZn ferrite nanocomposite as functional antimicrobial agents, which is essential for bacterial inactivation in biological fluids [93]. The prepared magnetic nanocomposite is not only able to differentiate the GramVE+bacteria's and GramVE-negative bacteria's (but also employed to separate blood platelets. The separation procedure used in this work shown is shown in Figure 6.

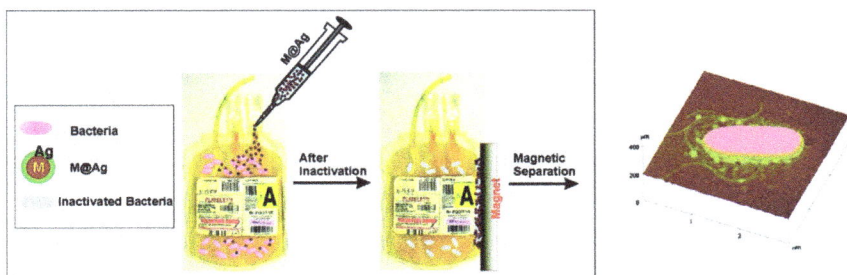

Figure 6. Schematic representation of silver-coated magnetic core@shell nanoparticles (m@ag) as a functional antimicrobial agent for inactivation of bacteria in blood platelets and the subsequent magnetic separation to remove m@ag from the platelets. Right side is AFM image of single bacterial cell, reproduced with the permission taken from [93].

5.5   Separation by gold modified magnetic nanomaterials

Modification of MNPs with gold materials is also very common, which provides robustness, optical properties and high surface area to the resulting material [94]. Some of the Au modified MNPs are discussed below. Wang et al. have prepared $MnFe_2O_4$@Au nanoparticles and combined it with antibody of Staphylococcus aureus (S. aureus) to capture and separate the bacterial strain via surface-enhanced Raman scattering method [95]. From this method, detection limit for Staphylococcus aureus was found as 10 cell/mL. The whole process used for separation and synthesis of nanoparticles are shown in Figure 7. In a similar work, non-destructive separation method is used for identification and detection of diribonucleoside from the DNA-protein composition by using gold-iron oxide nanomaterials, which is prepared by the coprecipitation method in the existence of citrate substance [96]. Au-Fe_3O_4 nanomaterials have several advantages like robust bonding of single-stranded DNA with gold surface, good recovery, and easy separation of DNA (diribonucleoside) from DNA-protein composition using external magnet.

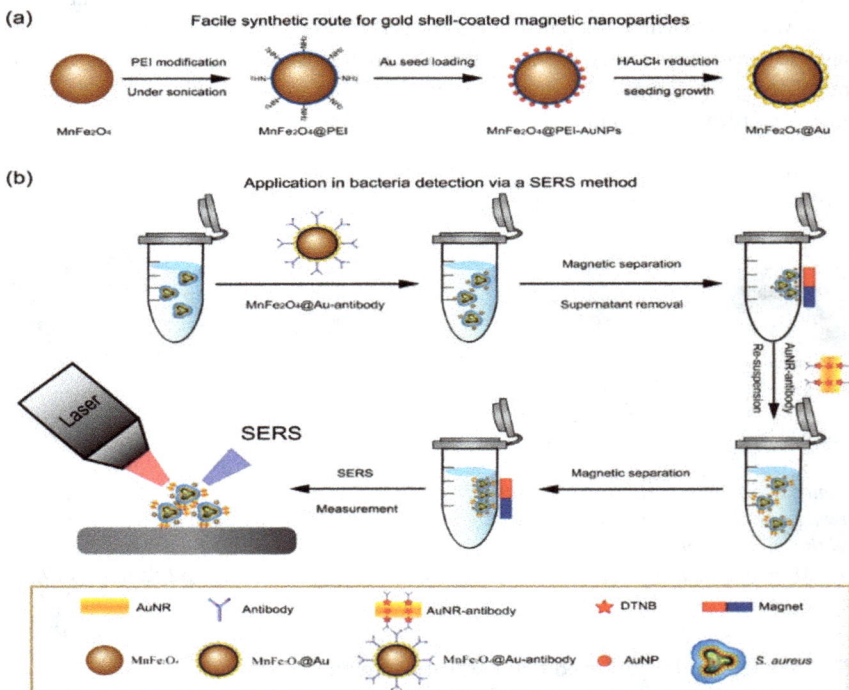

Figure 7. Synthesis of AuMNPs and their application in removal of bacteria by step-wise process, reproduced with the permission taken from [95].

The presence of antibiotic in water leads to a serious problem for human health. To resolve this problem, gold nanoparticles fabricated magnetic beads (AuNPs/MBs) were synthesized by Luo et al. [97]. The fluorescence aptamer sensor modified with AuNPs/MBs provides good limit of detection of 0.07 ng/mLin the linear range of 0.1-100 ng/mL. Khun et al. has developed iron ferrite magnetic particle with chitosan and gold modification and coated onto the glass electrode to separate the glucose from human serum. In this method, glucose oxidase was immobilised on the surface of MNPs without using cross linker and used for potentiometric estimation of glucose in the concentration range of 1.0×10^{-6}- 3.0×10^{-2} M. The proposed sensor exhibited a sensitivity of 27.3 ± 0.8 mV/decade with reactive time of 7.0 s [98].

5.6   Separation by manganese oxide modified magnetic nanomaterials

The wide use of manganese oxide (MnO_2) in combination with MNPs can be attributed to their low cost, eco-friendly nature and large surface area, and strong oxidising/adsorptive properties [99]. As a first example showing the role of $MnO_2@MNPs$, we have taken the work carried out by Wen et al. [100]. They have used $MnO_2@Fe_3O_4$ nanomaterials for adsorption of arsenic from aqueous to solid state by three-techniques i.e. X-ray absorption spectroscopy (XAS), high-performance liquid chromatography-Inductively Coupled plasma-mass spectrometry (HPLC-ICP/MS), and X-ray photoelectron spectroscopy (XPS). Arsenic is toxic elements and found in air, soil and polluted water. When the concentration of arsenic increases in drinking water, it can affect human health because of their carcinogenic properties. The prepared $MnO_2@Fe_3O_4$ nanocomposites was used for separation of arsenic from different samples.

Kang et al. has devolved $Fe_3O_4@MnO_2$ for the decomposition of organic dye (methylene blue) and separated them by two processes: (1) advanced oxidation processes and (2) adsorptive bubble separation [101]. In the process mixture of H_2O_2 and $Fe_3O_4@MnO_2$ has produced bubbles that used to stabilize the layers of surfactants (cetyltrimethyl ammonium chloride (CTAC) or sodium dodecyl sulfate (SDS). The proposed technique has successfully separated methylene blue dye from river water industrial wastewater. In another work, $MnFe_2O_4$ nanocomposite was synthesized to separate compounds (selenite and selenate) in aqueous medium, discovered by Gonzalez et al. by using dynamic reaction cell inductively-coupled plasma-mass spectrometry (DRC-ICP-MS) and batch techniques. The binding capacity was obtained as 6573.76 and 769.23 mg Se kg^{-1} of nanomaterials [102].

5.7   Separation by titanium oxide and zinc oxide modified magnetic nanomaterials

The titanium oxide (TiO_2) acts as semiconductor and have photocatalytic activity also, therefore commonly applied in sunscreens, catalysts, paints, cosmetics, wastewater treatment processes, plastics etc. [103]. The combination of MNPs with TiO_2 are very commonly used for separation of metal ions. For example, Mousavi et al. have synthesized the (Fe_3O_4@TiO_2-CN) and utilized it for removal of Ni^{+2} and Pb^{+2} ions from polluted water. The maximum adsorption capacity for lead and nickel were obtained as 92.6 and 75.76 mg L^{-1} and the adsorption of M^{+2} ions followed the Langmuir monolayer adsorption isotherm [104].

Bi et al. have developed Fe_3O_4@C@TiO_2-NTs nanocomposite through oil-phase cyclic magnetic adsorption technique [105]. Herein, the separation of Pb (II) was performed using the prepared materials by the reduction of Fe ion into Fe^0 from the wastewater collected from different industries (i.e. printing industry and dye industry). Habila et al. has synthesized the Fe_3O_4@SiO_2@TiO_2 nanocomposite for detection of heavy metals via ICP-MS method. The saturation magnetisation calculated from the standard temperature hysteresis loop for iron oxide and Fe_3O_4@SiO_2@TiO_2 were found 72 and 40 emu/g, respectively, which revealed the separation capacity of prepared nanomaterials [106]. From this, LODs were calculated as, 0.041, 0.082, 0.066 and 0.049 μg/L for cadmium (II), lead (II), copper (II), and zinc (II) respectively, while the LOQs were found to be, 0.12, 0.25, 0.20 and 0.15 μg/L for cadmium (II), lead (II), Copper (II), and Zinc (II).

Similarly, MNPs modified with ZnO was also used for separation of variety of compounds. For example, Abazari et al. have evaluated the performance of Fe_3O_4-ZnO nanocomposites for degradation and separation of methylene blue and Procion Red MX-5B. They magnetic nanomaterials retained their catalytic properties, after several use [107]. In another work, 3D Fe_2O_3@ZnO core-shell nanomaterials were fabricated by Zhang et al. and applied for selective separation of ciprofloxacin. They have compared the rate of reaction for core-shell materials and their individual components. It was found that Fe_2O_3@ZnO has revealed 4 to 20 times higher first order kinetic rate constant than of pure ZnO and iron oxide, respectively [108].

5.8   Separation by carbon modified magnetic nanomaterials

The carbon-based materials have large surface area, higher conductivity and high porosity. In addition, they have low cost, are easy in preparation and required less time for synthesis as well as processing [109]. MNPs modified with carbon-based nanomaterials as single-wall carbon nanotubes and multi-wall carbon nanotubes, graphene, mesoporous carbon etc. are widely used in various fields of research [110]. For example, Yan et al. has prepared the titanium oxide modified magnetic carbon nanotubes nanocomposites (MagCNTs@TiO$_2$) via solvothermal reaction and applied them for detection of phosphopeptides [111]. The peptides were separated by magnetic process and detected by mass spectrometry (MS).

Chromium have carcinogenic properties and have an adverse effect to human health and aquatic animals [112]. For their removal, Fe$_3$O$_4$/C nanocomposites were synthesized by taking FeCl$_3$, glucose and sodium acetate as precursors using one-step hydrothermal process in which glucose plays the main role for reduction of Fe ion and stabilization of the final compound. The prepared nanocomposite was used for separation of toxic Cr (VI) ion as well as Rhodamine B organic pollutant from wastewater [113]. You et al. have synthesized the Fe$_3$O$_4$@graphene nanocomposite via one-step hydrothermal process by using graphene oxide, FeCl$_3$, and D-glucose as reducing agent [114]. The material was successfully applied for separation of arsanilic acid and showed the adsorption capacity of 313.7±11.2 mg/g in 15 minutes at pH 3-6 from wastewater sample. Mercury (Hg) is extremely harmful, even at their low concentration to human health according to the Agency for Toxic Substances and Disease Registry (ATSDR). So, their removal is necessary. For this, the polyethylenimine (PEI) functionalized magnetic amorphous carbon thin film nanocomposite (Fe$_3$O$_4$-PEI-ACTF N) was prepared [114]. The prepared nanocomposites exhibit the adsorption capacity of 714 mg/g for Hg (II).

Du et al. have prepared a novel carbon@magnetic honeycomb sorbent for example Hkust−Fe$_3$O$_4$, which have several important features as higher superparamagnetic properties, excellent macro/mesoporous structure, high stability and high adsorption capacity [115]. Due to these properties, the carbon-based magnetic honeycomb material used for separation of Rhodamine B. The Fe$_3$O$_4$@CNC nanocomposite was used for separation of lysozyme from the white part of the egg [116]. The materials exhibited remarkable adsorption capacity for lysozyme i.e. 860 ± 14.6 mg g^{-1} with recovery of 98%. Some of the other examples showing their role of magnetic nanomaterials in separation science are listed in Table 1.

Table 1. Modified magnetic nanomaterials for separation of biologically important compounds.

S.N.	Analyte	Magnetic materials	Detection technique	Linear range (ng/L)	Limit of Detection (LOD) ng/L	Recovery (%)	Precision (%)	Ref.
1.	Biogenic amines	m-MCM-41	HPLC-UV	5×10^4– 1.5×10^8	8000 - 16000	94.40 - 104.0	1.02-4.14	76
2.	16SrRNA	Si@MNP	Immunochromaticgraphic strip	10^8 cfu mL^{-1}	10^6 cfu mL^{-1}	100	-	78
3.	Fluoride	Sulfate-doped Fe_3O_4/Al_2O_3	Ion chromatography	-	70.4×10^6	70-90	-	82
4.	Sulfonamides	Fe_3O_4/Al_2O_3 NPs	MSPE	10^3–10^5	1300	71-90	2.1-7.9	84
5.	S. aureus	AuMNPs	SERS	$10-10^5$ cfu mL^{-1}	10 cfu mL^{-1}	-	-	95
6.	Cu (II)	Fe_3O_4@SiO_2@TiO_2	ICP-MS	5×10^6 – 1.2×10^9	66-82	87 - 110	0.01 - 0.9	106
7.	Heme proteins	Fe_3O_4@Al_2O_3	MGCE	3.14×10^{11} - 6.8×10^{13}	9.8×10^{10}	95 - 97	2.91 -2.46	149
8.	Pesticides	m-ZrO_2@Fe_3O_4	GC–MS/MS	-	0.02 ng/g	69.8 - 117.1	3.4 -16.5	150
9.	CHIKV	Au@MNPs	CV	48.7- 48.7×10^6	48.7	-	-	151

MGCE = magnetic glassy carbon electrode, MSPE = magnetic solid-phase extraction, GC–MS/MS = Gas chromatography-tandem mass spectrometry, SERS = surface-enhanced Raman scattering, ICP-MS = inductively coupled plasma Mass Spectroscopy, HPLC-UV = High Performance Liquid Chromatography- Ultraviolet–visible spectroscopy, CV = cyclic voltammetry, m-MCM-41 = magnetic mesoporous silica nanoparticles, cfu = Colony-forming unit.

5.9 Separation by surfactants modified magnetic nanomaterials

Surfactants play a main role in terms of the stability of magnetic nanomaterials and controlled the oxidation of magnetic nanomaterials [117]. They are basically used to reduce the surface tension of aqueous solution consist of both hydrophilic and hydrophobic group [118]. Some common surfactants are octadecyl trimethylammonium chloride (OTC), cetyl trimethyl ammonium bromide (CTAB), and C_{14}-dialkyldimethylammonium amine-based surfactants C_8- to C_{18}-alkylamines, C_8-tetraalkylammonium, pyridinium compounds, cetyl trimethyl ammonium chloride (CTAC) and benzalkonium [119, 120]. Among the different kind of surfactants reported in the literature, CTAB has the highest popularity.

For example, Zhang et al. have synthesized Fe_3O_4 MNG@CTAB (MNG= magnetic nanoparticle graphene) nanocomposite via dispersive solid phase extraction-gas chromatography–mass spectrometry for separation of polycyclic aromatic hydrocarbon (PAHs) from seawater, collected from 56 different places and found the limits of detection ranged from 0.009 to 0.018 mg/L and the recovery range from 79.01 to 99.67% [121]. Similarly, in another work, CTAB modified magnetic nanomaterials (Fe_3O_4) was used to elaborate trace amount of polycyclic aromatic hydrocarbon (PAHs) via the solid phase extraction method from five occurring samples including river waters, rainwater, wastewater, and tap water, analysed by ultra-performance liquid chromatography (UPLC) with fluorescence detection (FLD) [122]. The synthesis process used for preparation of nanomaterial is shown in Figure 8. The LOD is ranged from 0.4 to 10.3 ng/L for PAHs.

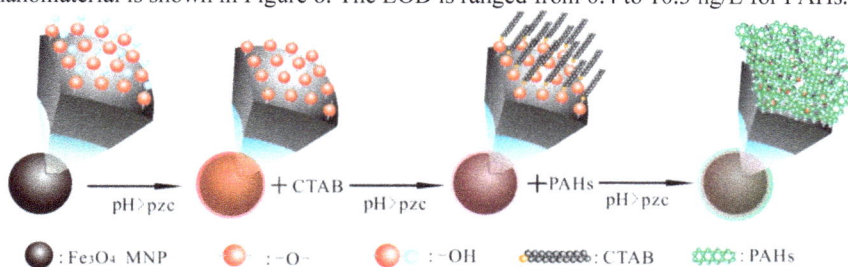

Figure 8. Schematic illustration showing mechanism of adsorption of PAHs using Fe_3O_4–CTAB MNPs, reproduced with the permission taken from [122].

Similarly, anionic surfactants are also used in modification of MNPs. The well-known anionic surfactant is Sodium dodecyl benzene sulfonate (SDBS), which can be retained in the environment meaning it has low biodegradability. Similarly, another common surfactant is sulfonate-based surfactant i.e. hexadecyldiphenyl oxide disulfonate (DPDS) and also known as "twin-head" surfactant, which is mainly used for groundwater purification [123]. Huang et al. have used three types of surfactants i.e. Si_4ASO_3Li, Si_4ASO_3Na, Si_4ASO_3K, and another is thio-ester oxidation process for modification of MNPs [124]. Here, tetrasiloxane chain represent the hydrophobic group, while sulfonate act as the hydrophilic group. The anionic silicon surfactants reduce the surface tension of water as 19.8 mN m^{-1} at the critical high concentration. Cai et al. have prepared cellulose@SDS-MMT (sodium dodecyl sulfonate-montmorillonite) using cellulose acetate as precursor and SDS as surfactant [125]. For the fabrication, firstly, the SDS-modified MMT is spin into composite nanofiber mats and cellulose acetate was converted into cellulose by alkaline hydrolysis process on the composite nanofiber materials. The fabricated nanocomposite was used to separate the heavy metal ions.

5.10 Separation by polymers modified magnetic nanomaterials

Modification of MNPs with polymeric substances have a great popularity in recent time, due to the possibility of the number of the functional group attached at the surface of materials that provide stability and avoid the agglomeration of MNPs [126, 127]. Various reported methods were noticed in previous years, where polymer coated magnetic materials were designed by different polymerization technique such as reversible addition-fragmentation chain transfer (RAFT), controlled radical polymerization (CRP), atom transfer radical polymerization (ATRP), and self-assembly methods etc.

In the RAFT method, polymerisation was performed with the incorporation of a one extra compound, which allowed the coating of polymers like bottle-brush polymers, multi-block copolymers, star polymers, surface coatings and hyperbranched polymer on the surface of MNPs [128]. Atom transfer radical polymerization is the class of CRP technique in which a fast-dynamic interaction occurs between radical, dormant alkyl halides, and metal catalyst. ATRP process can further be divided in different types i.e. activators regenerated by electron transfer (ARGET)-ATRP, electrochemically mediated ATRP (eATRP), initiators for continuous activator regeneration (ICAR)-ATRP, supplemental activator and reducing Agent (SARA)-ATRP, and photochemical-organic activated-ATRP [128].

Some of the literatures citing the application of polymer modified MNPs in separation science are discussed below. The catastrophic oil spills and oil spread present over water were harmful to the environment. Palchoudhary et al. have prepared water-soluble polyvinylpyrrolidone@iron oxide nanoparticle as a cost-effective, non-toxic and stable material for successful separation of oil from wastewater [129]. Ni et al. have prepared the glutathione-functionalized Fe_3O_4@polydopamine [130]. The modified MNPs shows good affinity towards the separation of protein molecules via interaction between glutathione s-transferase (GST) and glutathione (GSH).

Ghasemi et al. have developed Fe_3O_4@GAA (GAA = guanidine acetic acid) for separation of heavy metal including Zinc (II), Copper (II), Silver (I), and Mercury (II) from aqueous medium [131]. The optimum diameter of magnetic nanoparticle is observed in the range of 20–30 nm. The optimum pH = 7 and time of 10 minutes are responsible for maximum separation of analyte. The detection limit is found as 0.002, 0.01, 0.005, and 0.01 g/L and for Silver (I), Mercury (II), Zinc (II), and Copper (II), and respectively with recovery range from 95 to 100%. Similarly, Pb(II) has been separated by a novel magnetic nanocomposite i.e. Fe_3O_4@DAPF (2,3-diaminophenol and formaldehyde) from the aqueous medium by Venkateswarlu et al. [132]. The synthesis procedure used for nanomaterial is shown in Figure 9 (a), while bonding between metal ions and nanomaterial is shown in Figure 9 (b). The largest separation value found from this magnetic material is 83.3 mg/g in only 25s, due to their maximum saturation magnetization reported as i.e. 56.1 emu/g. In another work, $PFe_3O_4@NH_2$-MIL-125 magnetic nanocomposite (P = polymer) was prepared for separation of heavy metal ion Pb(II) from aqueous medium by Venkateswarlu et al. [133]. The separation of lead ions depends on different parameter such as pH, separation time, concentration of sample, and temperature. The large adsorption capacity was found as 561.7 mg g^{-1}, at optimum pH 5 and optimum temperature 298 K.

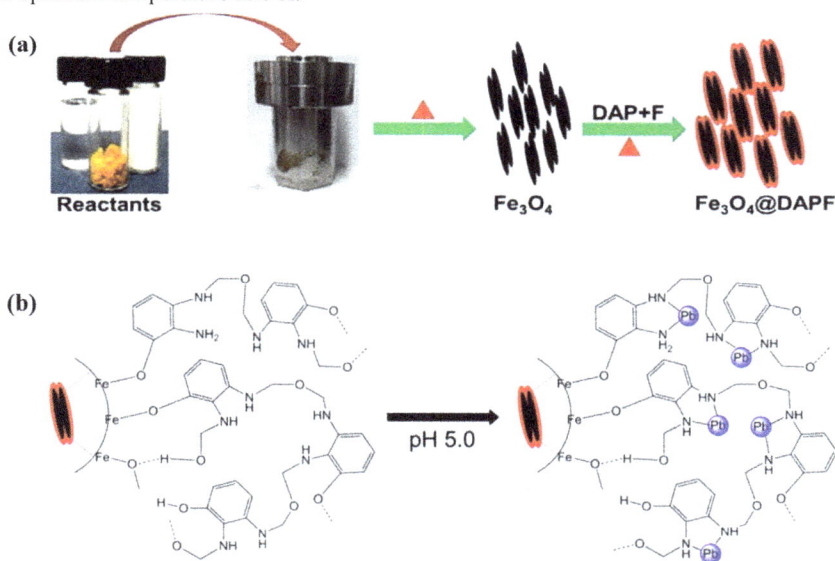

Figure 9. (a) Synthesis of $Fe_3O_4@DAPF$ CSFMNRs and (b) mechanism of trapping of Pb(II) to the nanomaterial at pH 5, reproduced with the permission taken from [132].

Similarly, polycyclic aromatic hydrocarbon (PAH) are also very popular toxic compounds. If, they get penetrated into the soil, it remains there for a long time, which is carcinogenic to human health through the food cycle [134]. To resolve this type of problem, $Fe_3O_4@$polyaniline-based nanocomposite was synthesized followed by two steps i.e. polymerization and oxidation process [135]. Using the nanomaterial, separation of polycyclic compounds such as fluoranthene, pyrene and benzopyrene were performed successfully.

5.11   Separation by magnetic molecularly imprinted polymer (MMIP)

Molecularly imprinted polymer (MIP) is a synthetic polymer, which is polymerised in the presence of template, monomer, and crosslinker. After polymerisation, template molecule gets extracted from the polymer matrices, which leaves cavities that are corresponding to the size and shape of the template molecule. This recognition site present on the surface of polymer matrices is known as a molecularly imprinted cavity and polymer containing such sites are available to bind with the template only, called as molecularly imprinted polymer (MIP). MIP are special kind of polymers, which can be applied for selective, sensitive and specific determination of template molecule from different types of matrices. But they have problems in terms of limited stability and poor adsorption. To resolve this issue and improve the performance of MIPs, in recent time, their combination with MNPs has become very popular [18]. Among different combination of MIPs with nanomaterials, their combination with magnetic nanomaterials are most popular and known as magnetic molecularly imprinted polymer (MMIP). The MMIPs has received great attention in recent years, due to excellent properties such as high adsorption capacity, high stability, highly selective for analyte, repeatability, low price and simple synthetic procedure [136, 137].

For example, Lu et al. have investigated MMIPs using $Fe_3O_4@SiO_2$ (basic materials) NPs by surface molecular imprinting technique, all processes are completed through template (diuron) monomer (α-methacrylic acid) crosslinker (trimethylolpropane trimethacrylate) initiator (azobisisobutyronitrile), and porogen (acetonitrile) [138]. The separation efficiency of diuron depends on different analytical parameters like pH of solvent, weight of reactive compounds and efficiency of analyte. The linear range is found from 0.02 to 10.0 mg L^{-1} with limit of detection is 0.012 mg/L. Hashemi et al. has developed MWCNTs based MMIPs and used it for estimation of melamine by charge transfer complexation process [139]. The linear range and limit of detection are found as 10.0–600.0 ng/mL and 3.0 ng/mL, respectively. This method has been successfully applied to the detection of melamine in milk and dry milk samples. Similar approach has been developed by Hashemi et al. where MIP has been fabricated on the surface of vinyl@MNPs, using bovine haemoglobin as template, N, N-methylenebisacrylamide as functional monomer [140]. The prepared MMIP was selectively used for determination of bovine haemoglobin obtained from a calf blood sample. Tang et al. has synthesized $Fe_3O_4@GO@MIP$ by self-assembly method and applied it for detection of interleukin-8 [141]. The proposed MMIPs has showed good limit of detection of 0.04 pM in the linear range of 0.1 to 10 pM. In another study, an ultrasound assisted MMIP is formed by Messaoud et al. [142]. The proposed method is used for detection of Bisphenol A in the tap and mineral water samples along with the linear range of 0.07 µM to 10 µM with LOD = 8.8 nM.

Table 2. Combination of organic materials coated MNPs for separation.

S. N.	Analyte	Magnetic materials	Detection technique	Linear range (ng/L)	LOD (ng/L)	Recovery (%)	Precision RSD (%)	Ref.
1.	Protein	Fe_3O_4@CNC	Dopamine ligand exchange	-	8.6×10^{11}	98	14.6	116
2.	Hg (II), Cu (II), Zn (II), Ag (I)	Fe_3O_4@GAA	ICP-OES	$20\text{-}55 \times 10^3$	10-2	95 - 100	0.1-0.5	131
3.	Diuron	Fe_3O_4@SiO$_2$-Polymer	HPLC	20×10^3-10×10^6	12×10^3	83.56-116.1	1.21-6.81	138
4.	Melamine	MIPs/MMW CNTs	Spectrophoto-metry	10 – 600	3.0	94.6 - 102.3	2.8-3.9	139
5.	IL-8	Fe_3O_4@GO @MIP	CV	0.25 - 25.04	0.1	92.89	2.67-3.17	141
6.	Bisphenol A	AuNPs/CBN Ps/SPCE	DPV	15.9×10^3 - 22.8×10^5	20.08×10^2	96.4–104.3	3.18- 8.3	142
7.	Hg^{2+}, Cd^{2+}, Ni^{2+}, Cu^{2+}	MIIPs	SPE	100–15×10^3	6.0 22.5	94.7-110.2	1.9-5.8	147
8.	Phthalate Esters	ALG@C$_{18}$-Fe_3O_4-TNs	SPE	$60\text{-}10^3$	11-46	84 -109	0.3 -11	152
9.	Pb^{2+}, Cu^{2+}	Fe_3O_4@SiO$_2$-NH-MFL	Ion exchange chromatogram -phy	-	1.5×10^{11} 7.07×10^{10}	-	-	153
10.	Hg (II)	MMIP	SPE-ICP	-	0.03	98.3	1.47	154

LOD = Limit of Detection, SPE = solid phase extraction, ALG@C$_{18}$-Fe_3O_4-TNs = alginate polymer cage C$_{18}$ magnetic titanate nanotube, Fe_3O_4@CNC = Iron oxide cellulose nanocrystal, Fe_3O_4@GAA = magneticfunctionalized by guanidine acetic acid, ICP-OES = inductively coupled plasma optical emission spectrometry, HPLC = High-performance liquid chromatography, MIPs/MMWCNTs = molecularly imprinted polymer coated on magnetic multiwalled carbon nanotubes, SPE-ICP = inductively coupled plasma-optical emission spectrometry, MIIPs = Multi-ion imprinted polymers, IL-8 = interleukin-8, AuNPs/CBNPs/SPCE = Gold nanoparticles/ carbon black nanoparticles/ Screen Printed Carbon Electrode, DPV = differential pulse voltammetry.

In addition to these, several types of heavy metals were also separated using different kind of MMIPs, such as In(III)@polymer (IIP) for In(III) [143], Cd(II) @ polymer for Cd(II) [144], Cr(VI)@polymer for Cr(VI) [145], (IIP)@N-(pyridin-2-ylmethyl) ethenamine@Fe_3O_4 nano-particles for Hg(II) ion [146]. Multi-ion imprinted magnetic polymers has also been designed for separation of Mercury (II), Cadmium (II), Nickel (II) and Copper (II) from wastewater samples with good recovery range 94.7-110.2% [147].

Chen et al. have synthesized MIP for the extraction of sulfonamides (SAs) from honey [148]. The separation of analyte along with polymer was done by the simple magnet, finally, sulfonamides (SAs) were separated from polymer by liquid chromatography-tandem mass spectrometry. The detection limit for sulfonamides (SAs) was observed in the range of 1.5-4.3 ng/g with recovery range of 67.1% to 93.6%. Some of the other organic material modified MNPs and their application in separation of different kind of compounds are shown in Table 2.

Conclusion

Recently, considerable attention has been given towards the synthesis of high-quality magnetic materials with good composition and tailored properties. For this, various new techniques have been developed which may leads to the synthesis of desirable size, shape, magnetic properties in the MNPs. Similarly, in the field of magnetic nanomaterial-based separation some new sample pre-treatment has also been introduced like magnetic solid phase separation (MSPE), in which materials are not packed in column and separation were performed by the use of external magnets only. We have discussed in this chapter that to improve the physicochemical properties of the magnetic nanomaterials, they are modified with suitable materials like silica, metal, carbon, surfactants, and polymers. But we still need some new materials which will not only improve the surface properties of MNPs but also improve their stability, non-swelling properties, sorption capacity, conductivity, porosity, and avoid the agglomeration/oxidation. Recently, the magnetic molecularly imprinted polymer has played an essential role for separation of pollutants and biological compounds. The functionalised magnetic nanomaterials or core-shell magnetic nanomaterials were successfully applied for separation of different organic and inorganic compounds (heavy metal ions) from various environmental matrices as well used for separation of biological components also like protein, cells, nucleic acid, antibiotic, virus etc. With the growing success of MNPs and day by day improvement in this area assure us that in the near future we will see some better, enhanced and more efficient magnetic nanomaterials for separation of variety of compounds with higher selectivity, sensitivity, reproducibility, repeatability, and reusability.

Author declaration

Mr. Kumar has given the major contribution in writing this book chapter along with drawing the Figures and Tables, taking the copyright permission etc.

References

[1] N. Sanvicens, M. Pilar Marco, Multifunctional nanoparticles-properties and prospects for their use in human medicine, Trends Biotechnol. 26 (2008) 425-433. https://doi.org/10.1016/j.tibtech.2008.04.005

[2] L.M. Rossi, N.J.S Costa, F.P. Silva, R. Wojcieszak, Magnetic nanomaterials in catalysis advanced catalysts for magnetic separation and beyond, Green Chem. 16 (2014) 2906-2933. https://doi.org/10.1039/c4gc00164h

[3] M. Chellappa, U. Vijayalakshmi, Fabrication of Fe_3O_4-silica core-shell magnetic nano-particles and its characterization for biomedical applications, Mater Today Proc. 9 (2019) 371-379. https://doi.org/10.1016/j.matpr.2019.02.166

[4] H. Shokrollahi, A. Khorramdin, G. Isapour, Magnetic resonance imaging by using nano-magnetic particles, J. Magn. Magn. Mater. 369 (2014) 176-183. https://doi.org/10.1016/j.jmmm.2014.06.023

[5] L. Chen, T. Wang, J. Tong, Application of derivatized magnetic materials to the separation and the preconcentration of pollutants in water samples, TrAC, Trends Anal. Chem. 7 (2011) 1095-1108. https://doi.org/10.1016/j.trac.2011.02.013

[6] N. Ali, H. Zaman, M. Bilal, M.S. Nazir, H. M. Iqbal, Environmental perspectives of interfacially active and magnetically recoverable composite materials-A review, Sci. Total Environ. 670 (2019) 523-538. https://doi.org/10.1016/j.scitotenv.2019.03.209

[7] T.H. Tran, H.L. Nguyen, D.S. Hwang, J.Y. Lee, H.G. Cha, J.M. Koo, S.Y. Hwang, J. Park, D.X. Oh, Five different chitin nanomaterials from identical source with different advantageous functions and performances, Carbohydr. Polym. 205 (2019) 392-400. https://doi.org/10.1016/j.carbpol.2018.10.089

[8] S. Mohamed M.S. Veeranarayanan, T. Maekawa, D.S. Kumar, External stimulus responsive inorganic nanomaterials for cancer theranostics, Adv. Drug Deliver Rev. 138 (2019) 18-40. https://doi.org/10.1016/j.addr.2018.10.007

[9] S.C. Tjong, Synthetic architecture of inorganic nanomaterials, Nanocrystalline Materials, Elsevier (2006).

[10] Z. Chen, C. Wu, Z. Zhang, W. Wu, X. Wang, Z. Yu, Synthesis, functionalization, and nanomedical applications of functional magnetic nanoparticles, Chin. Chem. Lett. 29 (2018) 1601-1608. https://doi.org/10.1016/j.cclet.2018.08.007

[11] M. R. Parker, The physics of magnetic separation, Contemp Phys. 18 (1977) 279-306. https://doi.org/10.1080/00107517708231486

[12] L. Mohammed, H. G. Gomaa, D. Ragab, J. Zhu, Magnetic nanoparticles for environmental and biomedical applications: A review, Particuology 30 (2017) 1-14. https://doi.org/10.1016/j.partic.2016.06.001

[13] M. Kubovcikova, M. Koneracka, O. Strbak, M. Molcan, V. Zavisova, I. Antal, I. Khmara, Poly-L-lysine designed magnetic nanoparticles for combined hyperthermia, magnetic resonance imaging and cancer cell detection, J. Magn. Magn. Mater. 475 (2019) 316-326. https://doi.org/10.1016/j.jmmm.2018.11.027

[14] K. Wang, X. Xu, L. Lu, A. Li, X. Han, Y. Wu, J. Miao, Y. Jiang, Magnetically recoverable $Ag/Bi_2Fe_4O_9$ nanoparticles as a visible-light-driven photocatalyst, Chem. Phys. Lett. 715 (2019) 129-133. https://doi.org/10.1016/j.cplett.2018.11.021

[15] Z. Wang, J. Guo, J. Ma, L. Shao, Highly regenerable alkali-resistant magnetic nanoparticles inspired by mussels for rapid selective dye removal offer high-efficiency environmental remediation, J. Mater. Chem. A 39 (2015) 19960-19968. https://doi.org/10.1039/C5TA04840K

[16] W. Song, D.A. Gregory, H.A. janabi, M. Muthana, Z. Cai, X. Zhao, Magnetic-silk/polyethyleneimine core-shell nanoparticles for targeted gene delivery into human breast cancer cells, Int. J. Pharm. 555 (2019) 322-336. https://doi.org/10.1016/j.ijpharm.2018.11.030

[17] A. Sharifi, S.Y. Motlagh, H. Badfar, Numerical investigation of magnetic drug targeting using magnetic nanoparticles to the Aneurysmal Vessel, J. Magn. Magn. Mater. 474 (2019) 236-245. https://doi.org/10.1016/j.jmmm.2018.10.147

[18] A. Ito, R. Teranishi, K. Kamei, M. Yamaguchi, A. Ono, S. Masumoto, Y. Sonoda, M. Horie, Y. Kawabe, M. Kamihira, Magnetically triggered transgene expression in mammalian cells by localized cellular heating of magnetic nanoparticles, J. Biosci. Bioeng. (2019). https://doi.org/10.1016/j.jbiosc.2019.03.008

[19] A.O. Adeoye, J.F. Kayode, B.I. Oladapo, S.O. Afolabi, Experimental analysis and optimization of synthesized magnetic nanoparticles coated with PMAMPC-MNPs for bioengineering application, St. Petersburg Polytechnical University Journal: Phys. Mathematics, 3 (2017) 333-338. https://doi.org/10.1016/j.spjpm.2017.10.003

[20] A. Sharma, A.I.Y. Tok, C. Lee, R. Ganapathy, P. Alagappan, B. Liedberg, Magnetic field assisted preconcentration of biomolecules for lateral flow assaying, Sens. Actuators, B 285 (2019) 431-437. https://doi.org/10.1016/j.snb.2019.01.073

[21] T. Chungcharoen, K Netjaibun, T. Pratabkong, P. Suwannasam, W. Limmun. Effects of inner angle of bowl, flow rate and speed on the efficiency of glycerol separation from the raw biodiesel using cylindrical bowl centrifuge, Energy Procedia 138 (2017) 405-410. https://doi.org/10.1016/j.egypro.2017.10.184

[22] A. Azzouz, S. K. Kailasa, S.S. Lee, A. J. Rascón, E. Ballesteros, M. Zhang, K.H. Kim, Review of nanomaterials as sorbents in solid-phase extraction for environmental samples, TrAC, Trends Anal. Chem. 113 (2018) 256-279. https://doi.org/10.1016/j.trac.2019.02.017

[23] L. Chen, C.H. Zhou, S. Fiore, D.S. Tong, H. Zhang, C.S. Li, S.F. Ji, W.H. Yu. Functional magnetic nanoparticle/clay mineral nanocomposites: Preparation, magnetism and versatile applications, Appl. Clay. Sci. 127 (2016) 143-163. https://doi.org/10.1016/j.clay.2016.04.009

[24] A. Canakci, T. Varol, S. Ozsahin, Analysis of the effect of a new process control agent technique on the mechanical milling process using a neural network model: measurement and modelling, Measurement 46 (2013) 1818-1827. https://doi.org/10.1016/j.measurement.2013.02.005

[25] J. Alonso, J.M. Barandiarán, L.F. Barquín, A.G-Arribas. Magnetic nanoparticles, synthesis, properties, and applications, In magnetic nanostructured materials, Elsevier, 2018, 1-40. https://doi.org/10.1016/B978-0-12-813904-2.00001-2

[26] B. Balaraju, S. Kaleemulla, C. Krishnamoorthi, Structural and magnetic properties of $NiO-MnO_2$ nanocomposites prepared by mechanical milling, J. Magn. Magn. Mater. 464 (2018) 36-43. https://doi.org/10.1016/j.jmmm.2018.05.039

[27] C.D. Pham, J. Chang, M.A. Zurbuchen, J. P. Chang, Magnetic properties of $CoFe_2O_4$ thin films synthesized by radical-enhanced atomic layer deposition, ACS Appl. Mater. Interfaces 9 (2017) 36980-36988. https://doi.org/10.1021/acsami.7b08097

[28] D.M. Mattox, Handbook of physical vapor deposition (PVD) processing. William Andrew, 2010. https://doi.org/10.1016/B978-0-8155-2037-5.00008-3

[29] P.M. Martin, Deposition technologies, an overview. Handbook of deposition technologies for films and coatings, Third Edition, Elsevier Inc., 2010, 1-31. https://doi.org/10.1016/B978-0-8155-2031-3.00001-6

[30] H. Alijani, M. H. Beyki, Z. Shariatinia, M. Bayat, F. Shemirani, A new approach for one step synthesis of magnetic carbon nanotubes/diatomite earth composite by chemical vapor deposition method: Application for removal of lead ions, Chem. Eng. J. 253 (2014) 456-463. https://doi.org/10.1016/j.cej.2014.05.021

[31] C. Peng, J. Wang, N. Zhou, G. Sun. Fabrication of nanopowders by electrical explosion of a copper wire in water, Curr. Appl. Phys. 16 (2016) 284-287. https://doi.org/10.1016/j.cap.2015.12.009

[32] F. Lv, H. Qi, P. Liu, J. Liu, Molecular dynamics simulation of the thermal pulse explosion of metal nanowire, AIP Advances 8 (2018) 075307. https://doi.org/10.1063/1.5037662

[33] I.V Beketov, A.P. Safronov, A.V. Bagazeev, A. Larrañaga, G.V. Kurlyandskaya, A.I. Medvedev, In situ modification of Fe and Ni magnetic nanopowders produced by the electrical explosion of wire, J. Alloys Compd. 586 (2014), S483-S488. https://doi.org/10.1016/j.jallcom.2013.01.152

[34] K. Song, W. Kim, CY. Suh, D. Shin, K.S. Ko, K. Ha, Magnetic iron oxide nanoparticles prepared by electrical wire explosion for arsenic removal, Powder Technol. 246 (2013) 572-574. https://doi.org/10.1016/j.powtec.2013.06.023

[35] P. Chandane, J. Ladke, C. Jori, S. Deshmukh, S. Zinjarde, M. Chakankar, H. Hocheng, U. Jadhav, Synthesis of magnetic Fe_3O_4 nanoparticles from scrap iron and use of their peroxidase like activity for phenol detection, J Environ. Chem Eng. 7 (2019), 103083. https://doi.org/10.1016/j.jece.2019.103083

[36] Y.V.M. Reddy, B. Sravani, S. Agarwal, V.K. Gupta, G. Madhavi, Electrochemical sensor for detection of uric acid in the presence of ascorbic acid and dopamine using the poly (DPA)/SiO_2@Fe_3O_4 modified carbon paste electrode, J. Electroanal. Chem. 820 (2018) 168-175. https://doi.org/10.1016/j.jelechem.2018.04.059

[37] Z. Chen, C. Wu, Z. Zhang, W. Wu, X. Wang, Z. Yu, Synthesis, functionalization, and nanomedical applications of functional magnetic nanoparticles, Chin. Chem. Lett. 29 (2018) 1601-1608. https://doi.org/10.1016/j.cclet.2018.08.007

[38] S. Nigam, K. C. Barick, D. Bahadu, Development of citrate-stabilized Fe_3O_4 nanoparticles: conjugation and release of doxorubicin for therapeutic applications, J. Magn. Magn. Mater. 323 (2011) 237-243. https://doi.org/10.1016/j.jmmm.2010.09.009

[39] M. SA. Darwish, Effect of carriers on heating efficiency of oleic acid-stabilized magnetite nanoparticles, J. Mol. Liq. 231 (2017) 80-85. https://doi.org/10.1016/j.molliq.2017.01.094

[40] R. Valenzuela, M. C. Fuentes, C. Parra, J. Baeza, N. Duran, S. K. Sharma, M. Knobel, J. Freer, Influence of stirring velocity on the synthesis of magnetite nanoparticles (Fe_3O_4) by the co-precipitation method, J. Alloys Compd. 488 (2009) 227-231. https://doi.org/10.1016/j.jallcom.2009.08.087

[41] L.B. Mello, L.C. Varanda, F.A. Sigoli, I.O. Mazali, Co-precipitation synthesis of (Zn-Mn)-co-doped magnetite nanoparticles and their application in magnetic hyperthermia, J. Alloys Compd. 779 (2019) 698-705. https://doi.org/10.1016/j.jallcom.2018.11.280

[42] J. Sánchez, D. Alicia C-Hernández, J C.E. Bocardo, J.M.A. Robles, P.Y.R. Rodríguez, R.A.J. Terán, P.B-Pérez, L.E.D-L. Prado, Synthesis of MnxGa1− xFe$_2$O$_4$ magnetic nanoparticles by thermal decomposition method for medical diagnosis applications, J. Magn. Magn. Mater. 427 (2017): 272-275. https://doi.org/10.1016/j.jmmm.2016.10.098

[43] A. Ahab, F. Rohman, F. Iskandar, F. Haryanto, I. Arif, A simple straightforward thermal decomposition synthesis of PEG-covered Gd_2O_3 (Gd_2O_3@ PEG) nanoparticles, Adv. Powder Technol. 27 (2016) 1800-1805. https://doi.org/10.1016/j.apt.2016.06.012

[44] X.H. Yi, F.X. Wang, X.D. Du, H. Fu, C.C. Wang, Highly efficient photocatalytic Cr (VI) reduction and organic pollutants degradation of two new bifunctional 2D Cd/Co-based MOFs, Polyhedron 152 (2018) 216-224. https://doi.org/10.1016/j.poly.2018.06.041

[45] M. Chen, L.L. Shao, J.J Li, W.J Pei, M.K. Chen, X.H Xie. One-step hydrothermal synthesis of hydrophilic Fe_3O_4/carbon composites and their application in removing toxic chemicals, RSC Adv. 6 (2016) 35228-35238. https://doi.org/10.1039/C6RA01408A

[46] J.E Szulejko, K.H. Kim, R.J.C. Brown, M.S. Bae, Review of progress in solvent-extraction techniques for the determination of polyaromatic hydrocarbons as airborne pollutants, TrAC, Trends Anal. Chem. 61 (2014) 40-48. https://doi.org/10.1016/j.trac.2014.07.001

[47] J. Sánchez, D. Alicia C-Hernández, J.C. E-Bocardo, J.M. A-Robles, P.Y.R. Rodríguez, R. A. J-Terán, P. B-Pérez, L.E. D-L. Prado, Synthesis of MnxGa1− xFe$_2$O$_4$ magnetic nanoparticles by thermal decomposition method for medical diagnosis applications, J. Magn. Magn. Mater. 427 (2017): 272-275. https://doi.org/10.1016/j.jmmm.2016.10.098

Materials Research Forum LLC
https://doi.org/10.21741/9781644900611-3

[48] H. Emadi, M. S-Niasari, A. Sobhani, Synthesis of some transition metal (M: 25Mn, 27Co, 28Ni, 29Cu, 30Zn, 47Ag, 48Cd) sulfide nanostructures by hydrothermal method, Adv. Colloid Interface Sci. 246 (2017) 52-74. https://doi.org/10.1016/j.cis.2017.06.007

[49] D.K. Dinkar, B. Das, R. Gopalan, B.S. Dehiya, Effects of surfactant on the structural and magnetic properties of hydrothermally synthesized $NiFe_2O_4$ nanoparticles, Mater. Chem. Phys. 218 (2018), 70-76. https://doi.org/10.1016/j.matchemphys.2018.07.020

[50] Y. Lu, Y. Zheng, S. You, F. Wang, Z. Gao, J. Shen, W. Yang, M. Yin, Bifunctional magnetic-fluorescent nanoparticles: synthesis, characterization, and cell imaging, ACS Appl. Mater. Interfaces 7 (2015): 5226-5232. https://doi.org/10.1021/am508266p

[51] M.S.A. Darwish, N.H.A Nguyen, A. Ševců, I. Stibor, S. K. Smoukov. Dual-modality self-heating and antibacterial polymer-coated nanoparticles for magnetic hyperthermia, Mater. Sci. Eng., C. 63 (2016) 88-95. https://doi.org/10.1016/j.msec.2016.02.052

[52] K.M Yenkie, W.Z. Wu, R.L. Clark, B.F. Pfleger, T.W. Root, C.T. Maravelias, A roadmap for the synthesis of separation networks for the recovery of bio-based chemicals: matching biological and process feasibility, Biotechnol. Adv. 34 (2016) 1362-1383. https://doi.org/10.1016/j.biotechadv.2016.10.003

[53] T. Chungcharoen, K. Netjaibun, T. Pratabkong, P. Suwannasam, W. Limmun, Effects of inner angle of bowl, flow rate and speed on the efficiency of glycerol separation from the raw biodiesel using cylindrical bowl centrifuge, Energy Procedia 138 (2017) 405-410. https://doi.org/10.1016/j.egypro.2017.10.184

[54] X. Jiang, W. Xiao, G. He, Falling film melt crystallization (III): Model development, separation effect compared to static melt crystallization and process optimization, Chem. Eng. Sci. 117 (2014): 198-209. https://doi.org/10.1016/j.ces.2014.06.027

[55] Y. Xu, X. Lv, G. Yang, J. Zhan, M. Li, T. Long, C.T. Ho, S. Li, Simultaneous separation of six pure polymethoxyflavones from sweet orange peel extract by high performance counter current chromatography, Food Chem. 292 (2019) 160-165. https://doi.org/10.1016/j.foodchem.2019.04.031

[56] A. Tripodi, D. Manzini, M. Compagnoni, G. Ramis, I. Rossetti, Alternative integrated distillation strategies for the purification of acetonitrile from ethanol ammoxidation, J. Ind. Eng. Chem. 59 (2018) 35-49. https://doi.org/10.1016/j.jiec.2017.10.003

[57] M. A Varfolomeev, V. B. Novikov, R. N. Nagrimanov, B. N. Solomonov, Modified solution calorimetry approach for determination of vaporization and sublimation

enthalpies of branched-chain aliphatic and alkyl aromatic compounds at T= 298.15 K,
J. Chem. Thermodyn. 91 (2015) 204-210 https://doi.org/10.1016/j.jct.2015.07.037

[58] Y.R. Zhang, S.Q. Wang, S.L. Shen, B.X. Zhao, A novel water treatment magnetic
nanomaterial for removal of anionic and cationic dyes under severe condition, Chem.
Eng. J. 233 (2013) 258-264. https://doi.org/10.1016/j.cej.2013.07.009

[59] S. Štěpánová, V. Kašička, Recent applications of capillary electromigration methods
to separation and analysis of proteins, Anal. Chim. Acta. 933 (2016) 23-42.
https://doi.org/10.1016/j.aca.2016.06.006

[60] J.H. Chang, J. Lee, Y. Jeong, J.H. Lee, I.J. Kim, S.E. Park, Hydrophobic partitioning
approach to efficient protein separation with magnetic nanoparticles, Anal. Biochem.
405 (2010) 135-137. https://doi.org/10.1016/j.ab.2010.05.027

[61] S.K Mwilu, E. Siska, R.B.N. Baig, R.S. Varma, E. Heithmar, K.R. Rogers,
Separation and measurement of silver nanoparticles and silver ions using magnetic
particles, Sci. Total Environ. 472 (2014) 316-323.
https://doi.org/10.1016/j.scitotenv.2013.10.077

[62] I. Ali, C. Peng, D. Lin, D. P. Saroj, I. Naz, Z.M. Khan, M. Sultan, M. Ali,
Encapsulated green magnetic nanoparticles for the removal of toxic Pb^{2+} and Cd^{2+}
from water development, characterization and application, J. Environ. Manage. 234
(2019) 273-289. https://doi.org/10.1016/j.jenvman.2018.12.112

[63] H, Heidari, B.L. Khosrowshahi, Magnetic solid phase extraction with carbon-coated
Fe_3O_4 nanoparticles coupled to HPLC-UV for the simultaneous determination of
losartan, carvedilol, and amlodipine besylate in plasma samples, J. Chromatogr. B.
1114-1115 (2019) 24-30. https://doi.org/10.1016/j.jchromb.2019.03.025

[64] Z. Jiaqi, D. Yimin, L. Danyang, W. Shengyun, Z. Liling, Z. Yi, Synthesis of
carboxyl-functionalized magnetic nanoparticle for the removal of methylene blue,
Colloids Surf. A 572 (2019) 58-66. https://doi.org/10.1016/j.colsurfa.2019.03.095

[65] J. Wu, P. Su, J. Huang, S. Wang, Y. Yang, Synthesis of teicoplanin-modified hybrid
magnetic mesoporous silica nanoparticles and their application in chiral separation of
racemic compounds, J. Colloid Interface Sci. 399 (2013) 107-114.
https://doi.org/10.1016/j.jcis.2013.02.045

[66] M. Zhang, J. Qiao, L. Qi. Dual-functional polymer-modified magnetic nanoparticles
for isolation of lysozyme, Anal. Chim. Acta 1035 (2018) 70-76.
https://doi.org/10.1016/j.aca.2018.07.019

[67] R. Gao, X. Cui, Y. Hao, L. Zhang, D. Liu, Y. Tang, A highly-efficient imprinted magnetic nanoparticle for selective separation and detection of 17β-estradiol in milk, Food Chem. 194 (2016) 1040-1047. https://doi.org/10.1016/j.foodchem.2015.08.112

[68] N. Baimani, P.A. Azar, S. W. Husain, H.A. Panahi, A. Mehramizi, Providing hyper-branched dendrimer conjugated with β-cyclodextrin based on magnetic nanoparticles for the separation of methylprednisolone acetate, J. Chromatogr. A 1571 (2018) 38-46. https://doi.org/10.1016/j.chroma.2018.08.005

[69] W.W. Ye, Y.T. Ding, Y. Sun, F. Tian, M. Yang, A Nanoporous Alumina Membrane Based Impedance Biosensor for Histamine Detection with Magnetic Nanoparticles Separation and Amplification, Procedia Eng. 27 (2017) 116-117. https://doi.org/10.1016/j.protcy.2017.04.051

[70] B. Sun, X. Ni, Y. Cao, G. Cao, Electrochemical sensor based on magnetic molecularly imprinted nanoparticles modified magnetic electrode for determination of Hb, Biosens. Bioelectron. 91 (2017) 354-358. https://doi.org/10.1016/j.bios.2016.12.056

[71] E. Akkaya, G.D. Bozyiğit, S. Bakirdere, Simultaneous determination of 4-tert-octylphenol, chlorpyrifos-ethyl and penconazole by GC-MS after sensitive and selective preconcentration with stearic acid coated magnetic nanoparticles, Microchem. J. 146 (2019) 1190-1194. https://doi.org/10.1016/j.microc.2019.01.077

[72] D. Xu, X. Ming, M. Gan, X. Wu, Y. Dong, D. Wang, H. Wei, F. Xu, Rapid detection of Cronobacter spp. in powdered infant formula by thermophilic helicase-dependent isothermal amplification combined with silica-coated magnetic particles separation, J. Immunol. Methods 462 (2018) 54-58. https://doi.org/10.1016/j.jim.2018.08.008

[73] P. Wang, X. Wang, S. Yu, Y. Zou, J. Wang, Z. Chen, N.S. Alharbi, Silica coated Fe_3O_4 magnetic nanospheres for high removal of organic pollutants from wastewater, Chem. Eng. J. 306 (2016) 280-288. https://doi.org/10.1016/j.cej.2016.07.068

[74] A.S Timin, A.V. Solomonov, A. Kumagai, A. Miyawaki, S. Yu Khashirova, A. Zhansitov, E.V. Rumyantsev, Magnetic polymer-silica composites as bioluminescent sensors for bilirubin detection, Mater. Chem. Phys. 183 (2016) 422-429. https://doi.org/10.1016/j.matchemphys.2016.08.048

[75] Y. Pei, Q. Han, L. Tang, L. Zhao, L. Wu, Fabrication and characterisation of hydrophobic magnetite composite nanoparticles for oil/water separation, Mater. Technol. 31 (2016) 38-43. https://doi.org/10.1179/1753555715Y.0000000024

[76] R. Molaei, H. Tajik, M. Moradi, Magnetic solid phase extraction based on mesoporous silica-coated iron oxide nanoparticles for simultaneous determination of biogenic amines in an Iranian traditional dairy product; Kashk, Food Control. 101 (2019) 1-8. https://doi.org/10.1016/j.foodcont.2019.02.011

[77] M.T. Aljarrah, M.S.A-Harahsheh, M. Mayyas, M. Alrebaki, In situ synthesis of quaternary ammonium on silica-coated magnetic nanoparticles and its application for the removal of uranium (VI) from aqueous media, J. Environ. Chem. Eng. 6, (2018) 5662-5669. https://doi.org/10.1016/j.jece.2018.08.070

[78] F. Chen, X. Ming, X.X. Chen, M. Gan, B.G. Wang, F. Xu, H. Wei, Immunochromatographic strip for rapid detection of Cronobacter in powdered infant formula in combination with silica-coated magnetic nanoparticles separation and 16S rRNA probe, Biosens. Bioelectron. 61 (2014) 306-313. https://doi.org/10.1016/j.bios.2014.05.033

[79] F. Javaheri, S. Hassanajili, Synthesis of $Fe_3O_4@SiO_2@MPS@P4VP$ nanoparticles for nitrate removal from aqueous solutions, J. Appl. Polym. Sci. 133 (2016) 71348-51154. https://doi.org/10.1002/app.44330

[80] S. B. Ulaeto J. K. Pancrecious, T. P. D. Rajan, B. C. Pai, Smart Coatings,In Noble Metal-Metal Oxide Hybrid Nanoparticles, Woodhead Publishing, 2019, 341-372. https://doi.org/10.1016/B978-0-12-814134-2.00017-6

[81] H.P. Peng, R.P. Liang, J.D. Qiu, Facile synthesis of $Fe_3O_4@Al_2O_3$ core-shell nanoparticles and their application to the highly specific capture of heme proteins for direct electrochemistry, Biosens. Bioelectron. 26 (2011) 3005-3011. https://doi.org/10.1016/j.bios.2010.12.003

[82] L. Chai, Y. Wang, N. Zhao, W. Yang, X. You, Sulfate-doped Fe_3O_4/Al_2O_3 nanoparticles as a novel adsorbent for fluoride removal from drinking water, Water Res. 47 (2013) 4040-4049. https://doi.org/10.1016/j.watres.2013.02.057

[83] Y. Li, Y. Liu, J. Tang, H. Lin, N. Yao, X. Shen, C. Deng, P. Yang, X. Zhang, $Fe_3O_4@Al_2O_3$ magnetic core-shell microspheres for rapid and highly specific capture of phosphopeptides with mass spectrometry analysis, J. Chromatogr. A 1172 (2007) 57-71. https://doi.org/10.1016/j.chroma.2007.09.062

[84] L. Sun, X. Sun, X. Du, Y. Yue, L. Chen, H. Xu, Q. Zeng, H. Wang, L. Ding, Determination of sulfonamides in soil samples based on alumina-coated magnetite nanoparticles as adsorbents, Anal. Chim. Acta. 665 (2010) 185-192. https://doi.org/10.1016/j.aca.2010.03.044

[85] J.C. Liu, P.J. Tsai, Y.C. Lee, Y-C. Chen, Affinity capture of uropathogenic Escherichia coli using pigeon ovalbumin-bound $Fe_3O_4@Al_2O_3$ magnetic nanoparticles, Anal. Chem. 80 (2008) 5425-5432. https://doi.org/10.1021/ac800487v

[86] X.T. Peng, L. Jiang, Y. Gong, X.Z. Hu, L. J. Peng, Y.Q. Feng, Preparation of mesoporous ZrO_2-coated magnetic microsphere and its application in the multi-residue analysis of pesticides and PCBs in fish by GC-MS/MS, Talanta 132 (2015) 118-125. https://doi.org/10.1016/j.talanta.2014.08.069

[87] W. Wang, H. Zhang, L. Zhang, H. Wan, S. Zheng, Z. Xu, Adsorptive removal of phosphate by magnetic $Fe_3O_4@C@ ZrO_2$, Colloids Surf., A 469 (2015) 100-106. https://doi.org/10.1016/j.colsurfa.2015.01.002

[88] G. Yuan, C. Zhao, H. Tu, M. Li, J. Liu, J. Liao, Y. Yang, J. Yang, N. Liu, Removal of Co (II) from aqueous solution with Zr-based magnetic metal-organic framework composite, Inorg. Chim. Acta. 483 (2018) 488-495. https://doi.org/10.1016/j.ica.2018.08.057

[89] G-Y. Zhang, S-Y. Deng, W-R. Cai, S. Cosnier, X-J. Zhang, D. Shan, Magnetic zirconium hexacyanoferrate (II) nanoparticle as tracing tag for electrochemical DNA assay, Anal. Chem. 87 (2015) 9093-9100. https://doi.org/10.1021/acs.analchem.5b02395

[90] J. López, J.M.A-Torres, L.A. Arce-Saldaña, A. Portillo-López, S. González-Martínez, J. S. Betancourt, M. E. Gómez, Ag nanoparticles embedded in a magnetic composite for magnetic separation applications, J. Alloys Compd. 786 (2019) 839-847. https://doi.org/10.1016/j.jallcom.2019.02.029

[91] R. Das, V.S. Sypu, H. K. Paumo, M. Bhaumik, V. Maharaj, A. Maity, Silver decorated magnetic nanocomposite ($Fe_3O_4@PPy-MAA/Ag$) as highly active catalyst towards reduction of 4-nitrophenol and toxic organic dyes, Appl. Catal. B 244 (2019) 546-558. https://doi.org/10.1016/j.apcatb.2018.11.073

[92] H. Veisi, S.B. Moradi, A. Saljooqi, P. Safarimehr, Silver nanoparticle-decorated on tannic acid-modified magnetite nanoparticles ($Fe_3O_4@TA/Ag$) for highly active catalytic reduction of 4-nitrophenol, Rhodamine B and Methylene blue, Mater. Sci. Eng. C 100 (2019) 445-452. https://doi.org/10.1016/j.msec.2019.03.036

[93] L. Wang, J. Luo, S. Shan, E. Crew, J. Yin, C-J. Zhong, B. Wallek, S. SS. Wong, Bacterial inactivation using silver-coated magnetic nanoparticles as functional antimicrobial agents, Anal. Chem. 83 (2011) 8688-8695. https://doi.org/10.1021/ac202164p

Materials Research Forum LLC
https://doi.org/10.21741/9781644900611-3

[94] D. Qi, H. Zhang, J. Tang, C. Deng, X. Zhang, Facile synthesis of mercaptophenylboronic acid-functionalized core– shell structure $Fe_3O_4@C@Au$ magnetic microspheres for selective enrichment of glycopeptides and glycoproteins, J. Phys. Chem. C 114, (2010) 9221-9226. https://doi.org/10.1021/jp9114404

[95] J. Wang, X. Wu, C. Wang, Z. Rong, H. Ding, H. Li, S. Li, Facile synthesis of Au-coated magnetic nanoparticles and their application in bacteria detection via a SERS method, ACS Appl. Mater. Interfaces 8 (2016) 19958-19967. https://doi.org/10.1021/acsami.6b07528

[96] A. C. Dutta, N. Agnihotri, R. Doong, A. De, Label-free and nondestructive separation technique for isolation of targeted DNA from DNA-protein mixture using magnetic $Au-Fe_3O_4$ nanoprobes, Anal. Chem. 89 (2017) 12244-12251. https://doi.org/10.1021/acs.analchem.7b03095

[97] Z. Luo, Y. Wang, X. Lu, J. Chen, F. Wei, Z. Huang, C. Zhou, Y. Duan, Fluorescent aptasensor for antibiotic detection using magnetic bead composites coated with gold nanoparticles and a nicking enzyme, Anal. Chim. Acta 984 (2017) 177-184. https://doi.org/10.1016/j.aca.2017.06.037

[98] K. Khun, Z. H. Ibupoto, J. Lu, M. S. AlSalhi, M. Atif, A. A. Ansari, M. Willander, Potentiometric glucose sensor based on the glucose oxidase immobilized iron ferrite magnetic particle/chitosan composite modified gold coated glass electrode, Sens. Actuators B 173 (2012) 698-703. https://doi.org/10.1016/j.snb.2012.07.074

[99] L. Jiang, Q. Ye, J. Chen, Z. Chen, Y. Gu. Preparation of magnetically recoverable bentonite-Fe_3O_4-MnO_2 composite particles for Cd (II) removal from aqueous solutions, J. Colloid Interface Sci. 513 (2018) 748-759. https://doi.org/10.1016/j.jcis.2017.11.063

[100] Z. Wen, Y. Zhang, Y. Wang, L. Li, R. Chen, Redox transformation of arsenic by magnetic thin-film MnO_2 nanosheet-coated flowerlike Fe_3O_4 nanocomposites, Chem. Eng. J. 312 (2017) 39-49. https://doi.org/10.1016/j.cej.2016.11.112

[101] Y.G. Kang, H. Yoon, C-S. Lee, E.J. Kim, Y.S. Chang, Advanced oxidation and adsorptive bubble separation of dyes using MnO_2-coated Fe_3O_4 nanocomposite, Water Res. 151 (2019) 413-422. https://doi.org/10.1016/j.watres.2018.12.038

[102] C. M. Gonzalez, J. Hernandez, Jason G. Parsons, J.L.G. Torresdey, A study of the removal of selenite and selenate from aqueous solutions using a magnetic iron/manganese oxide nanomaterial and ICP-MS, Microchem. J. 96, (2010) 324-329. https://doi.org/10.1016/j.microc.2010.05.005

[103] V. Iswarya, M. Bhuvaneshwari, N. Chandrasekaran, A. Mukherjee, Trophic transfer potential of two different crystalline phases of TiO_2 NPs from Chlorella sp. to Ceriodaphnia dubia, Aquat. Toxicol. 197 (2018) 89-97. https://doi.org/10.1016/j.aquatox.2018.02.003

[104] S.V. Mousavi, A. Bozorgian, N. Mokhtari, M.A. Gabris, H.R. Nodeh, W.A.W. Ibrahim. A novel cyanopropylsilane-functionalized titanium oxide magnetic nanoparticle for the adsorption of nickel and lead ions from industrial wastewater: Equilibrium, kinetic and thermodynamic studies, Microchem. J. 145 (2019) 914-920. https://doi.org/10.1016/j.microc.2018.11.048

[105] J. Bi, X. Huang, J. Wang, T. Wang, H. Wu, J. Yang, H. Lu, H. Hao, Oil-phase cyclic magnetic adsorption to synthesize $Fe_3O_4@C@TiO_2$-nanotube composites for simultaneous removal of Pb (II) and Rhodamine B, Chem. Eng. J. 366 (2019) 50-61. https://doi.org/10.1016/j.cej.2019.02.017

[106] M.A Habila, Z.A.A. Othman, A.M.E. Toni, J.P. Labis, M. Soylak, Synthesis and application of $Fe_3O_4@ SiO_2@TiO_2$ for photocatalytic decomposition of organic matrix simultaneously with magnetic solid phase extraction of heavy metals prior to ICP-MS analysis, Talanta, 154 (2016) 539-547. https://doi.org/10.1016/j.talanta.2016.03.081

[107] R. Abazari, A. R. Mahjoub, S. Sanati, Magnetically recoverable Fe_3O_4-ZnO/AOT nanocomposites: synthesis of a core-shell structure via a novel and mild route for photocatalytic degradation of toxic dyes, J. Mol. Liq. 223 (2016) 1133-1142. https://doi.org/10.1016/j.molliq.2016.09.038

[108] N. Li, J. Zhang, Y. Tian, J. Zhao, J. Zhang, W. Zuo, Precisely controlled fabrication of magnetic 3D γ-$Fe_2O_3@ZnO$ core-shell photocatalyst with enhanced activity: ciprofloxacin degradation and mechanism insight, Chem. Eng. J. 308 (2017) 377-385. https://doi.org/10.1016/j.cej.2016.09.093

[109] M. Chen, L.-L. Shao, J. J. Li, W-J. Pei, M-K. Chen, X-H. Xie, One-step hydrothermal synthesis of hydrophilic Fe_3O_4/carbon composites and their application in removing toxic chemicals, RSC Advances 6 (2016) 35228-35238. https://doi.org/10.1039/C6RA01408A

[110] Y. Zhang, R. Li, J. Fang, C. Wang, Z. Cai, Simultaneous determination of eighteen nitro-polyaromatic hydrocarbons in PM2.5 by atmospheric pressure gas chromatography-tandem mass spectrometry, Chemosphere 198 (2018) 303-310. https://doi.org/10.1016/j.chemosphere.2018.01.131

[111] Y. Yan, Z. Zheng, C. Deng, X. Zhang, P. Yang, Selective enrichment of phosphopeptides by titania nanoparticles coated magnetic carbon nanotubes, Talanta 118 (2014)14-20. https://doi.org/10.1016/j.talanta.2013.09.036

[112] X.H. Yi, F.X. Wang, X.D. Du, H. Fu, C.C. Wang, Highly efficient photocatalytic Cr (VI) reduction and organic pollutants degradation of two new bifunctional 2D Cd/Co-based MOFs, Polyhedron 152 (2018) 216-224. https://doi.org/10.1016/j.poly.2018.06.041

[113] M. Chen, L-L. Shao, J-J. Li, W.-J. Pei, M-K. Chen, X-H. Xie, One-step hydrothermal synthesis of hydrophilic Fe_3O_4/carbon composites and their application in removing toxic chemicals, RSC Advances 6 (2016) 35228-35238. https://doi.org/10.1039/C6RA01408A

[114] N. You, X.F. Wang, J.Y. Li, H.T. Fan, H. Shen, Q. Zhang, Synergistic removal of arsanilic acid using adsorption and magnetic separation technique based on $Fe_3O_4@$ graphene nanocomposite, J. Ind. Eng. Chem. 70 (2019) 346-354. https://doi.org/10.1016/j.jiec.2018.10.035

[115] C. Du, Y. Shui, Y. Bai, Y. Cheng, Q. Wang, X. Zheng, Y. Zhao, Bottom-up formation of carbon-based magnetic honeycomb material from metal-organic framework-guest polyhedra for the capture of Rhodamine B, ACS Omega 4 (2019) 5578-5585. https://doi.org/10.1021/acsomega.8b03664

[116] J. Guo, I. Filpponen, L.S. Johansson, P. Mohammadi, M. Latikka, M.B. Linder, R. H.A. Ras, O. J. Rojas, Complexes of magnetic nanoparticles with cellulose nanocrystals as regenerable, highly efficient, and selective platform for protein separation, Biomacromolecules 18 (2017) 898-905. https://doi.org/10.1021/acs.biomac.6b01778

[117] O.P Artykulnyi, V.I. Petrenko, L.A. Bulavin, O.I. Ivankov, M.V. Avdeev, Impact of poly (ethylene glycol) on the structure and interaction parameters of aqueous micellar solutions of anionic surfactants, J. Mol. Liq. 276 (2019) 806-811. https://doi.org/10.1016/j.molliq.2018.12.035

[118] S.Ö. Engin, H. Akbaş, M. Boz, Synthesis and physicochemical properties of double-chain cationic surfactants, J. Chem. Eng. Data. 61 (2015) 142-150. https://doi.org/10.1021/acs.jced.5b00367

[119] X. Gu, F. Zhang, Y. Li, J. Zhang, S. Chen, C. Qu, G. Chen, Investigation of cationic surfactants as clean flow improvers for crude oil and a mechanism study, J. Petrol. Sci. Eng. 164 (2018) 87-90. https://doi.org/10.1016/j.petrol.2018.01.045

[120] N. Timmer, P. Scherpenisse, J.L.M. Hermens, S.T.J. Droge, Evaluating solid phase (micro-) extraction tools to analyze freely ionizable and permanently charged cationic surfactants, Anal. Chim. Acta 1002 (2018) 26-38. https://doi.org/10.1016/j.aca.2017.11.051

[121] S. Zhang, W. Wu, Q. Zheng, Evaluation of modified Fe_3O_4 magnetic nanoparticle graphene for dispersive solid-phase extraction to determine trace PAHs in seawater, Anal. Methods 7 (2015) 9587-9595. https://doi.org/10.1039/C5AY02470F

[122] A. Middea, L.S. Spinelli, F.G. Souza Jr, R. Neumann, T.L. Fernandes, O.F.M. Gomes, Preparation and characterization of an organo-palygorskite-Fe_3O_4 nanomaterial for removal of anionic dyes from wastewater, Appl. Clay Sci. 139 (2017) 45-53. https://doi.org/10.1016/j.clay.2017.01.017

[123] T. Doura, F. Tamanoi, M. Nakamura, Miniaturization of thiol-organosilica nanoparticles induced by an anionic surfactant, J. Colloid Interface Sci. 526 (2018) 51-62. https://doi.org/10.1016/j.jcis.2018.04.090

[124] Y. Huang, L. Meng, M. Guo, P. Zhao, H. Zhang, S. Chen, J. Zhang, S. Feng, Synthesis, Properties, and Aggregation Behavior of Tetrasiloxane-Based Anionic Surfactants, Langmuir 34 (2018) 4382-4389. https://doi.org/10.1021/acs.langmuir.8b00825

[125] J. Cai, M. Lei, Q. Zhang, J.R. He, T. Chen, S. Liu, S-H. Fu, T.T. Li, G. Liu, P. Fei, Electrospun composite nanofiber mats of Cellulose@Organically modified montmorillonite for heavy metal ion removal: Design, characterization, evaluation of absorption performance, Compos Part A. Appl Sci Manuf. 92 (2017) 10-16. https://doi.org/10.1016/j.compositesa.2016.10.034

[126] H. Niu, S. Zhang, X. Zhang, Y. Cai, Alginate-polymer-caged, C18-functionalized magnetic titanate nanotubes for fast and efficient extraction of phthalate esters from water samples with complex matrix, ACS Appl. Mater. Interfaces 2 (2010) 1157-1163. https://doi.org/10.1021/am100010x

[127] C. Kaewsaneha, P. Tangboriboonrat, D. Polpanich, A. Elaissari, Multifunctional fluorescent-magnetic polymeric colloidal particles: Preparations and bioanalytical applications, ACS Appl. Mater. Interfaces 7 (2015) 23373-23386. https://doi.org/10.1021/acsami.5b07515

[128] B.D. Fairbanks, P.A. Gunatillake, L. Meagher, Biomedical applications of polymers derived by reversible addition-fragmentation chain-transfer (RAFT), Adv. Drug Delivery Rev. 91 (2015) 141-152. https://doi.org/10.1016/j.addr.2015.05.016

[129] S. Palchoudhury, J.R. Lead, A facile and cost-effective method for separation of oil-water mixtures using polymer-coated iron oxide nanoparticles, Environ. Sci. Technol. 48 (2014) 14558-14563. https://doi.org/10.1021/es5037755

[130] K. Ni, J. Yang, Y. Ren, D. Wei, Facile synthesis of glutathione-functionalized $Fe_3O_4@$ polydopamine for separation of GST-tagged protein, Mater. Lett. 128 (2014) 392-395. https://doi.org/10.1016/j.matlet.2014.04.124

[131] E. Ghasemi, A. Heydari, M. Sillanpää, Central composite design for optimization of removal of trace amounts of toxic heavy metal ions from aqueous solution using magnetic Fe_3O_4 functionalized by guanidine acetic acid as an efficient nano-adsorbent, Microchem. J. 147 (2019) 133-141. https://doi.org/10.1016/j.microc.2019.02.056

[132] S. Venkateswarlu, M. Yoon, Core-shell ferromagnetic nanorod based on amine polymer composite ($Fe_3O_4@$ DAPF) for fast removal of Pb (II) from aqueous solutions, ACS Appl. Mater. Interfaces. 7 (2015) 25362-25372. https://doi.org/10.1021/acsami.5b07723

[133] S. Venkateswarlu, A. Panda, E. Kim, M. Yoon, Biopolymer-Coated Magnetite Nanoparticles and Metal-Organic Framework Ternary Composites for Cooperative Pb (II) Adsorption, ACS Appl. NanoMater. 1 (2018) 4198-4210. https://doi.org/10.1021/acsanm.8b00957

[134] J.E Szulejko, K.H. Kim, R.J.C. Brown, M.S. Bae, Review of progress in solvent-extraction techniques for the determination of polyaromatic hydrocarbons as airborne pollutants, TrAC, Trends Anal. Chem. 61 (2014) 40-48. https://doi.org/10.1016/j.trac.2014.07.001

[135] Q. Zhou, Y. Wang, J. Xiao, H. Fan, C. Chen, Preparation and characterization of magnetic nanomaterial and its application for removal of polycyclic aromatic hydrocarbons, J. Hazard. Mater. 371 (2019) 323-331. https://doi.org/10.1016/j.jhazmat.2019.03.027

[136] L. Talavat, A. Güner, Thermodynamic computational calculations for preparation 5-fluorouracilmagnetic moleculary imprinted polymers and their application in controlled drug release, Inorg. Chem. Commun. 103 (2019) 119-127. https://doi.org/10.1016/j.inoche.2019.02.009

[137] M.D. Álvarez, E. Turiel, A.M. Esteban, Molecularly imprinted polymer monolith containing magnetic nanoparticles for the stir-bar sorptive extraction of thiabendazole and carbendazim from orange samples, Anal. Chim. Acta 1045 (2019) 117-122. https://doi.org/10.1016/j.aca.2018.09.001

[138] Y.C. Lu, M.H. Guo, J. Hao M.X.H. Xiong, Y.J. Liu, Y. Li, Preparation of core-shell magnetic molecularly imprinted polymer nanoparticle for the rapid and selective enrichment of trace diuron from complicated matrices, Ecotoxicol. Environ Saf. 177 (2019) 66-76. https://doi.org/10.1016/j.ecoenv.2019.03.117

[139] M. Hashemi, Z. Nazari, Preparation of molecularly imprinted polymer based on the magnetic multiwalled carbon nanotubes for selective separation and spectrophotometric determination of melamine in milk samples, J. Food Comp. Anal. 69 (2018) 98-106. https://doi.org/10.1016/j.jfca.2018.02.010

[140] W. Xu, Y. Wang, X. Wei, J. Chen, P. Xu, R. Ni, J. Meng, Y. Zhou, Fabrication of magnetic polymers based on deep eutectic solvent for separation of bovine hemoglobin via molecular imprinting technology, Anal. Chim. Acta 1048 (2019) 1-11. https://doi.org/10.1016/j.aca.2018.10.044

[141] P. Tang, H. Zhang, J. Huo, X. Lin, An electrochemical sensor based on iron (II,III)@ graphene oxide@ molecularly imprinted polymer nanoparticles for interleukin-8 detection in saliva, Anal. Methods 7 (2015) 7784-7791. https://doi.org/10.1039/C5AY01361E

[142] N.B. Messaoud, A.A. Lahcen, C. Dridi, A. Amine, Ultrasound assisted magnetic imprinted polymer combined sensor based on carbon black and gold nanoparticles for selective and sensitive electrochemical detection of bisphenol A, Sens. Actuators B 276 (2018) 304-312. https://doi.org/10.1016/j.snb.2018.08.092

[143] M. Li, X. Meng, X. Liang, J. Yuan, X. Hu, Z. Wu, X. Yuan, A novel In(III) ion-imprinted polymer (IIP) for selective extraction of In (III) ions from aqueous solutions, Hydrometallurgy 176 (2018) 243-252. https://doi.org/10.1016/j.hydromet.2018.02.006

[144] M. Li, C. Feng, M. Li, Q. Zeng, Q. Gan, H. Yang, Synthesis and characterization of a surface-grafted Cd (II) ion-imprinted polymer for selective separation of Cd(II) ion from aqueous solution, Appl. Surf. Sci. 332 (2015) 463-472. https://doi.org/10.1016/j.apsusc.2015.01.201

[145] M. Hassanzadeh, M. Ghaemy, S. M. Amininasab, Z. Shami, An effective approach for fast selective separation of Cr(VI) from water by ion-imprinted polymer grafted on the electro-spun nanofibrous mat of functionalized polyacrylonitrile, React. Funct. Polym. 130 (2018) 70-80. https://doi.org/10.1016/j.reactfunctpolym.2018.05.013

[146] E. Najafi, F. Aboufazeli, H.R.L.Z. Zhad, O. Sadeghi, V. Amani, A novel magnetic ion imprinted nano-polymer for selective separation and determination of low levels of mercury(II) ions in fish samples, Food Chem. 141 (2013) 4040-4045. https://doi.org/10.1016/j.foodchem.2013.06.118

[147] J. Fu, X. Wang, J. Li, Y. Ding, L. Chen, Synthesis of multi-ion imprinted polymers based on dithizone chelation for simultaneous removal of Hg^{2+}, Cd^{2+}, Ni^{2+} and Cu^{2+} from aqueous solutions, RSC Adv. 6 (2016) 44087-44095. https://doi.org/10.1039/C6RA07785D

[148] W.R. Zhao, T.F. Kang, L.P. Lu, S.Y. Cheng, Electrochemical magnetic imprinted sensor based on MWCNTs@CS/CTABr surfactant composites for sensitive sensing of diethylstilbestrol, J. Electroanal. Chem. 818 (2018) 181-190. https://doi.org/10.1016/j.jelechem.2018.04.036

[149] H.P. Peng, R.P. Liang, J.D. Qiu, Facile synthesis of $Fe_3O_4@Al_2O_3$ core-shell nanoparticles and their application to the highly specific capture of heme proteins for direct electrochemistry, Biosens. Bioelectron. 26 (2011) 3005-3011. https://doi.org/10.1016/j.bios.2010.12.003

[150] X.T. Peng, L. Jiang, Y. Gong, X.Z. Hu, L.J. Peng, Y.Q. Feng, Preparation of mesoporous ZrO_2-coated magnetic microsphere and its application in the multi-residue analysis of pesticides and PCBs in fish by GC-MS/MS, Talanta 132 (2015) 118-125. https://doi.org/10.1016/j.talanta.2014.08.069

[151] C. Singhal, A. Dubey, A. Mathur, C.S. Pundir, J. Narang, Paper based DNA biosensor for detection of chikungunya virus using gold shells coated magnetic nanocubes, Process Biochem. 74 (2018) 35-42. https://doi.org/10.1016/j.procbio.2018.08.020

[152] H. Niu, S. Zhang, X. Zhang, Y. Cai, Alginate-polymer-caged, C18-functionalized magnetic titanate nanotubes for fast and efficient extraction of phthalate esters from water samples with complex matrix, ACS Appl. Mater. Interfaces 2 (2010) 1157-1163. https://doi.org/10.1021/am100010x

[153] Y. Zhang, S. Ni, X. Wang, W. Zhang, L. Lagerquist, M. Qin, S. Willför, C. Xu, P. Fatehi, Ultrafast adsorption of heavy metal ions onto functionalized lignin-based hybrid magnetic nanoparticles, Chem. Eng. J. 372 (2019) 82-91. https://doi.org/10.1016/j.cej.2019.04.111

[154] E. Najafi, F. Aboufazeli, H.R.L.Z. Zhad, O. Sadeghi, V. Amani, A novel magnetic ion imprinted nano-polymer for selective separation and determination of low levels of mercury (II) ions in fish samples, Food Chem. 141 (2013) 4040-4045. https://doi.org/10.1016/j.foodchem.2013.06.118

Magnetochemistry - Materials and Applications
Materials Research Foundations **66** (2020) 130-172

Materials Research Forum LLC
https://doi.org/10.21741/9781644900611-4

Chapter 4

State of the Art, Challenges and Future Prospects in Magnetochemistry

Fulya Gulbagca[1], Burak Yildiz[1], Fatima Elmusa[1], Mohd Imran Ahamed[2], Fatih Sen[1]*

[1]Sen Research Group, Department of Biochemistry, Faculty of Arts and Science, Dumlupınar University, Evliya Çelebi Campus, 43100 Kütahya, Turkey

[2]Department of Chemistry, Faculty of Science, Aligarh Muslim University, Aligarh-202 002, India

*fatih.sen@dpu.edu.tr

Abstract

Nanoparticles are very suitable for many application areas such as catalysts, sensors, fuel cells etc. Magnetic nanoparticles have attracted a great deal of attention from researchers due to their applications in large number of areas. Recent studies have made significant progress in the use and development of new catalytic systems immobilized on magnetic nanoparticles. Their biocompatible catalytic activity and low toxicity indicate that they are suitable for use in many areas of nanotechnology. The main feature of magnetic nanoparticles is their size-specific properties. Additionally, particle size and specific surface area are other effective features of magnetic nanoparticles. They are very popular in life sciences and biomedical fields because of their advantages. In this study, we have gathered literature information against the problems that may occur in the application areas of magnetochemistry.

Keywords

Magnetochemistry, Magnetic Nanoparticles, Biomedical Applications, Synthesis Strategies, Blood-Brain Barrier, Hyperthermia, Cancer

Contents

1. Introduction

Magnetic nanoparticles (MNPs) are classified according to their differences in size. Those in the range of 10-100 nm are classified as NP and between 100-5000 nm as larger magnetic particles [1]. The most common use of magnetic particles today; the immobilization of proteins, antibodies, enzymes, and drugs, and the areas of use are becoming widespread and diversified. Magnetic immobilization of biologically active molecules used in different fields of biomedicine, nanotechnology, and biotechnology has great importance and effect. So far, the immobilization of enzymes and proteins on biomaterials encapsulated MNPs, and copolymers containing MNPs have been extensively studied, and so many efficient results have been revealed. In this article, studies on magnetochemistry will be discussed, and various methods will be emphasized, and future expectations and challenges will be discussed.

2.   Magnetochemistry and magnetic nanoparticles

Recent studies have shown that the immobilization of materials on MNPs have made significant progress for the applications and usage of new catalytic systems. MNPs are among the most important groups of nanoparticles. MNPs have various properties, such as super paramagnetism, large surface area, surface area, and surface volume, besides they can be easily separated from the environment when exposed to a magnetic field and reduce the limitation of diffusion [2]. In addition, enzymes immobilized on MNP have high activity and advantageous properties as shown in Table 1.

Table 1. Various MNPs used in enzyme immobilization and their biotechnological applications.

Enzyme	Used NP	Application	Ref
Cholesterol oxidase	Fe3O4 NP	Analysis of total cholesterol in serum	[3]
Haloalkane dehalogenase	Silica-Coated Iron Oxide NP	Dehalogenase-containing fusion proteins production	[4]
laccase	CMNP	Environmental pollution and bioremediation studies	[5]
α-amylase	Cellulose-coated Fe3O4 NP	Degradation of starch	[6]
keratinase	Fe3O4 NP	Synthesis of keratin	[7]
β-galactosidase	ZnO NP	Hydrolysis of lactose	[8]
lipase	Fe3O4 NP	p-Nitro phenyl propionate hydrolysis	[9]

Among the magnetic nanomaterials, magnetites were the first magnetic structures discovered in the 1500s before Christ. Even though there are many magnetic materials since then, it still remains the most studied structure. Magnetite has a cubic structure formed on an inverted spinel. For example, the oxygen atoms that form this structure are aligned to the center, while the Fe cations are surrounded by 4 oxygen atoms and 6 oxygen atoms in some regions [10]. At room temperature, there are the electron transition between Fe^{+2} and Fe^{+3} ions. Due to this feature, magnetite is in the class of semi-metallic materials. Magnetite; can be synthesized in many different ways. However, the synthesis of metal salts by binary precipitation is one of the most commonly used methods. The reason for this is that it is an efficient method that can be easily applied [11]. The size, form, and content of the synthesized nanoparticles may vary depending on the type of salt used (e.g. chloride, sulfate, nitrate, perchlorate), metal content, pH and ionic strength of the medium [12]. In the double precipitation method, the reaction takes place by adding base (1:2 molar) to the aqueous solution of metal salts [10,13]. Besides, one of the advantages of the use of nanoparticles in pharmaceutical formulations is the potential to pass the blood-brain barrier (BBB). However, this may also be a major disadvantage for the systemic administration of nanoparticles in terms of potential brain toxicity. In order for drug delivery to occur in the brain, the physical relationship of the drug to the nanoparticles is required. Considering the nanoparticles with different surface properties, neutral nanoparticles, and low concentrations of anionic nanoparticles have no effect on the integrity of BBB, whereas high concentrations of anionic nanoparticles and cationic nanoparticles are toxic to BBB. The brain uptake rates of anionic nanoparticles at lower concentrations are superior to the neutral or cationic formulations of the same concentrations. The surface load of the nanoparticle is considered in the toxicity and brain distribution profiles.

Due to its minimum toxicity and biolytic compatibility, Fe_3O_4 is a widely used magnetic particle. There have been many studies on Fe_3O_4 MNPs, and positive results have been obtained as shown in Table 2.

Table 2. MNPs (Fe_3O_4) Blood-Brain Barrier applications.

MNPs	Magnetic	Mechanism	Model	Effect	Preparation	Toxicity	Ref
PEGylated fluorescent liposomes + Transferrin Diameter = 130 nm	Static 0.08 T 24 h	Transferrin (RMT) + magnetic force promote crossing	In vitro BBB human endothelium + astrocytes	+50–100% transmigration 2 pg Fe/cell uptake	Coprecipitation aqueous 2 h stability	TEER and cell viability unchanged at 48 h	[14][15][16][17][18]
Polysorbate 80 Diameter = 11 nm (hydro 29 nm) ζ = 19 mV	Static 0.3 T 2 h	Poly adsorb protein (RMT) + magnetic force promote crossing	In vivo rat BBB	Accumulation in near cortex 0.6 mg Fe/g tissue uptake 9-fold increase	Mixing 0.2 g Tween 80 with 0.4 PEG IONs	Cell viability unchanged at 72 h	
Silica-coated nanocapsule Diameter = 100–150 nm Ms = 5-fold **SPION**	Static 1000 Oe 1 week RF (100 MHz)	Cell membrane translocation	In vivo mice BBB	25-fold increased concentration	Emulsion polymerization	Slight reversible astrogliosis No immunotoxicity	
Gold coated Diameter = 25–40 nm Ms = 30 emu/g	Static 0.01 T 6 h	Accumulation by magnetic force	In vitro human CSFBB + in vivo rat CSFBB	Up to 50% MRI signal difference confirmed local histological accumulation	Coprecipitation aqueous	High cell viability	

BDNF binded Diameter = 60 nm	Weak static magnet exposition	Magnetic force	In vitro BBB human endothelium + astrocytes	73% BDNF cross 3.5-fold increase Suppress apoptosis Spine loss reversed	Coprecipitation	High cell viability and TEER unchanged
Oleic acid coated Diameter = 220–250 nm $\zeta = -4$ to -17 mV	Static 8000 Gauss	Passive diffusion or RMT + magnetic force	In vivo rats	Indocyanine green load increased brain concentration 5% tot dose	Thermal decomposition	-
Aminosilane or EDT coating Diameter = hydro 25 or 29 nm ζ = 21 or -39 mV	Static 0.06– 0.1 T 24 h	Improve concentration after mannitol opening BBB	In vitro mice endothelium	Flux increase to 44% for EDT after osmotic opening	Aqueous phase reduction/hydrolysis	No change in permeability
Lipophilic fluorescence dye covered by α-D-glucose units Ms = 350 kA/m Diameter = hydro 117 nm ζ = -17 mV	Static 5 h	Magnetic force	In vitro BBB human endothelium + rat astroglia	11, 8, and 29 fold uptake increase of 35, 70, and 140 µg/mL	-	TEER and cell viability unchanged at 29 h
Amphotericin B magnetic liposomes Diameter = 240 nm Ms = 32 memu/g	Static	Magnetic force	In vivo rats	Histological accumulation 400 ng/g brain (after 30 min)	Film dispersion–ultrasonication	Reduced death with magnetic field
Cationic polymeric liposome Diameter = 20 nm	Static 0.5 T	Magnetic force	In vivo rats	Paclitaxel concentration 3-fold histological accumulation	Thin-layer evaporation	-
SiO2 - coated+Amino Tat peptide Diameter = 100 nm ζ = 42 mV Ms = 19 emu/g	Static 2 h	Magnetic force + transport Tat	In vitro BBB human endothelium + glioma	Cell internalization 2.6-fold increase Permeability 2.3-fold increase	Alkaline co-precipitation	TEER moderate decrease High cell viability
Cross-linked poly(ethylene glycol)-poly(aspartate) or citrate-coated Diameter = 25 nm or 90 nm	Alternate 33.4 kA/m at 300 kHz	Temperature opening BBB	In vitro mice or dogs	2–3 -fold flux increase 3-fold cell uptake increase	Co-precipitation	No cell death
Poly(maleic acid-co-olefin) coated Diameter = 11–13 nm	Alternate 7.6 kA/m at 150 kHz	Temperature opening BBB	In vivo rats	Histological accumulation only after RF	-	No cell death

Numerous materials have been prepared with improvements in nanotechnology and hybrid technology for the development of the field of enzyme immobilization, and as a result, nanoparticle-based materials based on various combinations of many organic and inorganic species have become of great interest and importance as a support material for immobilization. It is known that they have many physicochemical properties, such as pore size and distinctive large surface area, hydrophilic/hydrophobic balance, hydrophilicity, and surface chemistry. MNP carriers consisting of 3 separate functional parts are classified as magnetic core for surface protection and functional exterior coating. The outer cover used for connecting the various catalytic species is conveniently functionalized. The materials are divided into 5 groups according to their response to the applied magnetic field. These are diamagnetism, paramagnetism, ferromagnetism, antiferromagnetism, ferrimagnetism as shown in Figure 1.

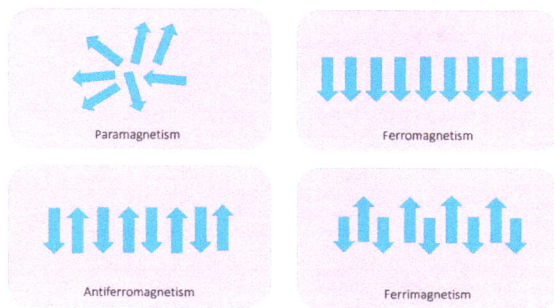

Figure 1. The order of the magnetic moment of a single iron atom.

Magnetic nanoparticles are widely used catalytic materials and are readily used for magnetic removal of immobilized enzymes. With common use, the binding of the catalysts to the magnetic nanoparticles allows this material to be re-used at the end of the reaction. Therefore, a second purification step is not required to separate the catalyst from the environment. Thus, it was found suitable to be called green (green) catalyst compared to the methods reported earlier [2].

2.1 Applications of magnetic nanoparticles

Thanks to the recent advances in nanotechnology; various routes of synthesis, characterization, functionalization of nanoparticles have been identified, and hence nanotechnology has been used for the developments of new applications containing nanoparticles (Table 3) [28–34]. Nano-sized magnetic nanoparticles have been used in *in-vitro* diagnostic studies for nearly 40 years [35–50]. Many iron oxide particles, especially maghemite (γ-Fe_2O_3) and magnetite (Fe_3O_4), have been included in the studies [10,51–53]

Table 3. MNPs and their biomedical applications.

MNPs	Magnetic	Mechanism	Model	Results	Toxicity	Ref
Au + Ni80Fe20 (permalloy) 1 μm radius disk-shaped	Dynamic 1 T at 20 Hz rotating	Vortex shaped rotation	In vivo mice glioma	Increased survival	No change in histology No side effects	[54][55][56][57][58]
pEGFP/p53 conjugated	Static	Gene therapy + magnetofectio n	In vitro BBB + glioblasto ma	Increased induced apoptosis	-	
Aptamer conjugated dextran coated	Alternate 9.55 kA/m at 1 Hz	3D Rotating nanosurgeons	In vitro glioblasto ma	Increased induced apoptosis	-	
Octadecyl-quaternized carboxymethyl chitosan	Static 0.5 T	Delivery loaded paclitaxel	In vivo rats glioma	Increased survival Prolonged bioavailability	Reduced side-effects	
Inhibitor of metalloproteinase-1 conjugated	Static 0.8 T	Crossing BBB and regulation of metalloprotei nases	In vitro BBB + HIV infection	Recovery in spine density ROS and HIV infection level decrease	Unchanged TEER and cells viability	
Bilayers: Tenofovir + dextran Sulphate + vorinostat	Static 0.08 T 6 h	Crossing BBB and antiretroviral therapy	In vitro BBB + HIV infection	HIV infection level decrease Prolonged bioavailability	Unchanged TEER and cells viability	[59][60][61][62][63]
Azidothymidine 50 -triphosphate loaded CoFe2O4@BaTiO3	Static 22 Oe/cm 6 h Alternate 66 Oe at 100 Hz 5 min	Crossing BBB and controlled release of antiviral drug	In vitro BBB + HIV infection	Functional and structural integrity of the drug after the release	High cell viability	
Beclin1 siRNA binded CoFe2O4@BaTiO3	Static 0.8 T 3 h	Crossing BBB and regulate autophagy	In vitro BBB + HIV infection	Attenuate HIV-1 replication and viral-induced inflammation	Unchanged TEER and occludin expression	
Morphine antagonist, CTOP conjugated	Static 0.5 T	Crossing BBB and drug delivery	In vitro BBB + HIV infection	Recovery in spine density Prevention of morphine induced apoptosis	High cell viability Unchanged TEER	
PEG shell	Alternate 15 kA/m 500 KHz	Heat-sensitive capsaicin receptor TRPV1 activated by magnetother mal genetic stimulation	In vitro neurons + in vivo mice	On demand activation of neurons in deep nuclei (VTA)	Lower glial activation and macrophage accumulation compared to implant	

CoFe2O4 -BaTiO3 GMO coated	Static 3000 Oe/cm Alternate 100 Oe at 0–20 Hz	Crossing and concentrate in brain then modulate neural activity	In vitro + in vivo mice	EEG detectable modulation activity of 1 mV	No toxicity for astrocytes and blood cells in vitro
Co-ferrite core and Mn-ferrite shell Polymer PMA coated	Alternate 7–30 kA/m at 412–570 KHz	Heat-sensitive capsaicin receptor TRPV1 activated by magnetothermal genetic stimulation	In vitro neurons + in vivo mice	On demand evoked motor ambulation, striatum rotation or freezing	- [64][65][66][67]
GFP-tagged ferritin	Alternate 23–31 mT at 465 kHz	Heat-sensitive capsaicin receptor TRPV1 activated by magnetothermal genetic stimulation	In vitro neurons + in vivo mice	Glucose-sensing hypothalamus neurons modulate feed behavior	-
Starch-coated	Static 150 mT	Magnetic force open neuron channels	In vitro neuron	Mechanical opening of N-type mechanosensitive Ca2+ channels	Reversibility of opening
Starch and chitosan coated	Static 150 mT	Magnetic force open neuron channels	In vitro neuron	Mechanical opening of N-type mechanosensitive Ca2+ channels	Unchanged cell viability, reversibility of opening [68][69][70][71][72]
Ferumoxide-labeled human neural stem cells	Static 0.32 T	Magnetic targeting	In vivo stroke rats	Better targeting and recovery in a stroke model	Unchanged differentiation into neurons or astrocytes
Dextran-coated	Alternate 1–6 A at 0.25–2 Hz 10 min	Osmotin load targeting in hippocampus and delivery	In vitro + in vivo AD rats	Memory improvement Reduced protein accumulation and synaptotoxicity	Unchanged viability No apoptosis No BBB leakage
Oleic acid-coated	3 days	Gene therapy delivery Alpha-Synuclein RNAi Plasmid	In vitro + in vivo PD mice	Motor improvement Reduced neurodegeneration	No organ damage Normal blood test 12 days [73][74]
Uncoated Fe3O4	Alternate 2 h/d for 1 week	Synergic effect of magnetic stimulation and MNPs	In vivo PD rats	Motor improvement/recover feeding behavior Reduced ROS and lesion volume	Normal mitochondrial activity

MNPs in life sciences and biomedical fields can be listed as follows due to their advantages and superior qualities [75,76]:

- *In-vivo* diagnostic purposes
- Magnetic resonance imaging [77];
- *In-vivo* treatment

Materials Research Forum LLC
https://doi.org/10.21741/9781644900611-4

- Controlled drug applications [78];

- Carrier vector or orientation and triggering agent in gene therapy [79]

- Hyperthermia (destruction of undesirable tissues with high temperature) (Table 4) [80]

- *In-vitro* diagnosis and supply

- Isolation and purification of various biological molecules [81,82].

Biocompatible catalytic activity and low toxicity show that nanoparticles are suitable for use in the biosensor field [92]. In addition, it has been reported that enzyme-substrate interaction occurs faster because of the increase in activity of immobilized enzymes on magnetic nanoparticles due to increased surface area and there is no limitation of mass transfer [10,93]. The use of enzymes in organic synthesis is encouraging and rapidly growing. Because enzymes generally exhibit high chemo-, regio-, and stereoselectivity under mild reaction conditions. However, the use of enzymes is often prevented by denaturation and the deactivation of the biocatalyst under the reaction conditions, and recycling. Enzyme immobilization to the solid matrix has been seen as one of the most effective methods to solve these problems [94–96]. Magnetic nanoparticles have become the focus of attention due to their importance in biomedicine and biotechnology [10], as well as in catalysis studies [97,98]. Various enzymes are able to bind to magnetic particles covalently and to exhibit higher stability than their free conditions, but the enzyme activity has generally been reduced due to the chemical bond formation between the support protein [94,99]. The use of nanoparticles in tumor treatment is also on the agenda. In the studies conducted, it is possible to direct the nanoparticles with a magnetic center to the tumor area with the help of a magnetic field. Thus, only a treatment specific to the tumor site can be applied, the dose of the drug used and the resulting side effects will be reduced [100]. Super paramagnetic nanoparticles offer some impressive possibilities in biomedicine [98,101,102]. Studies on super paramagnetic iron oxide nanoparticles (SPIONs) and surface modifications for their improved applications in cancer theranostics are described in Table 5. Super paramagnetic behavior states that the nanoparticles respond to the magnet more easily and faster than the sample and that the magnetism of the magnetic nanoparticles disappeared as soon as the external magnetic field was removed (Figure 2).

Table 4. This table describes many of the nanoparticle hyperthermia studies which have been done in biological settings.

Particles used	Experimental setting	Effect seen	Ref
Dextran or aminosilane-coated magnetite. Superparamagnetic. 313 nm crystal diameters.	In vitro	Similar cytotoxic effcacy of water ba hyperthermia when used to heat ce	
Magnetite nanoparticles of several different shapes, aspect ratios and hydrodynamic diameters	Ex vivo (human breast tissue) and in vivo (tumors in mice)	Significant average temperature increase and in vivo treated tissues. During experi little increase in mouse core body te	[83][84][85]
Coated magnetite. Superparamagnetic. 10 nm and 200 nm hydrodynamic diameter.	In vivo (tumors in mice)	Heterogenous tumor heating. 1273C temp seen within tumors.	
Magnetite nanoparticles coated with lipid membrane. Administered with Interleukin-2 (IL-2).	In vivo (tumors in mice)	Significant improvement in mouse sur regrowth delay.	
Aminosilane-coated iron oxide. Superparamagnetic. 15 nm crystal diameter.	In vivo (human brain tumors)	Little toxicity seen in human patients af heated to 42.449.5C and treated with the	[86][87][88]
Dextran- and PEG-coated iron oxide, conjugated to Chimeric L6 antibody. Superparamagnetic. 20 nm hydrodynamic diameter.	In vivo (tumors in mice)	Significant tumor growth d	
Iron-based magnetic nanoparticles (10 nm crystals) loaded into liposomes. Liposomes conjugated to Trastuzumab antibody.	In vivo (tumors in mice)	Significant tumor necrosis and growth complete responses.	
Ferromagnetic, dextran-coated nanoparticles. Average hydrodynamic diameter of approximately 100 nm.	In vivo (tumors in mice)	Significant tumor regrowth d	[89][90][91]
Ferromagnetic, dextran-coated nanoparticles. Average hydrodynamic diameter of approximately 100 nm.	In vitro and in vivo (tumors in mice)	Significant tumor regrowth delay. Addi with chemotherapy and radi	

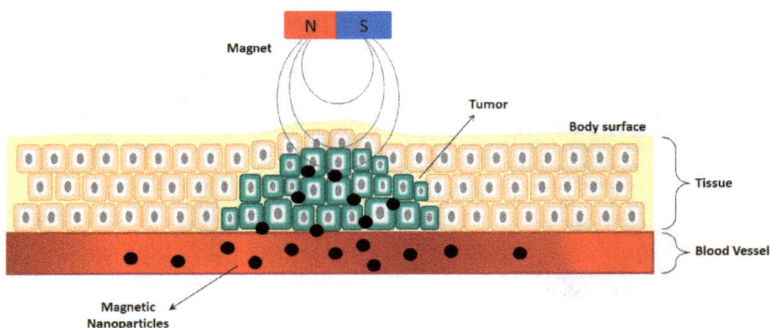

Figure 2. Schematic illustration of MNP treatment for cancer.

Based on these properties, super paramagnetic nanoparticles can be delivered by the vascular system and can be concentrated in certain areas of the body using a magnetic field. These properties of magnetite nanoparticles are therefore used in most areas of application ranging from cancer hypothermia to the target drug release to DNA purification [97,103,104].

Table 5. SPIONs and it's surface modification for their improved application in cancer theranostics.

SPIONs+Surface Modification	Cancer type	Models (in vitro and in vivo)	Result	Ref
DOX@FA-SPIONs	Breast cancer	MCF-7 cell line	Improved MRI, drug delivery and growth inhibition of cancerous cells	[105][106][107][108]
HA-SPIONs	Breast cancer	MDA-MB-231 cell line and Nude Mice	Enhanced imaging and tumor ablation following photothermal therapy	
DOX-HA-SPION	Breast cancer	MDA-MB-231 cell line	Down-regulation of pro-inflammatory cytokines (IL-6), NF-κB and vimentin, enhanced apoptosis, reduction in cell proliferation and angiogenesis	

PEG-GIHN	Breast cancer	MCF-7 cell line	Enhanced imaging and decreased cell viability following photothermal therapy	[109][110][111][112]
DOX@Ps 80-SPIONs	Glioma cancer	C6 cell line and Rats	Enhanced in vitro cytotoxicity and growth inhibition of glioma in vivo through apoptosis	
IOCCP-PEI-PEG	Glioblastoma	SF767 cell line	Enhanced transfection and gene delivery with improved Imaging	
HA-SPIONs	Lung cancer	A459 cell line	Reduced cytotoxicity but improved biocompatibility	
FA-PEG-SPIONs-Cy5.5	Lung cancer	Mice	Enhanced optical imaging	[113][114][115][116]
DOX- PNIPAAm-MAA-SPIONs	Lung cancer	A549 cell line	Increased drug release at lower pH and increase temperature, Time-dependent in vitro cytotoxicity	
SPION^{Hyp}	T cell leukemia, colon cancer	Jurkat and HT-29 cell lines	Increase in vitro cytotoxicity through ROS production under control exposure of light	
SPION$^{CMD-Hyp}$	T cell leukemia,	Jurkat cell line	Increase in vitro cytotoxicity through ROS production under control exposure of light	

SPIONs@FA-PAMAM-CDF	Ovarian and Cervical cancer	SKOV3 and HeLa cell lines	Enhanced MR contrast imaging and apoptosis	
(SPIONs)-(Dox)-(PLGA)- AS1411 aptamer (Apt)	Colon cancer	C26 cell line and Xenograft Mice	Enhance imaging, cytotoxicity and tumor inhibition with prolonging survival of mice.	[117][118][119][119]
MUC1(apt)-Au@SPIONs	Colon cancer	HT-29 cell line	Enhanced MR imaging and photothermal therapy	
SP204-PEG-DOX/NVB-QD@SPIONs	Prostate cancer	PC3, DU145, and LNCaP cell lines and SCID mice	Precise Targeted delivery, enhanced imaging and growth inhibition of xenograft	
YCC-DOX	Liver cancer	Hep3B cell line, Rat, and Rabbit	Enhanced MR imaging, decrease cell viability and tumor volume	[120][121]
siPLK1-StAv-SPIONs	Pancreatic ductal adenocarcinoma	6606PDA cell line and Mice	Enhanced MR imaging and gene silencing of cell cycle specific enzyme PLK1 and inhibition of tumor growth	

3.   Factors affecting the main characteristics of magnetic nanoparticles

3.1   Size effect

Another main feature of MNPs is that they show specific features dependent on size. With these properties, the size and surface area of the particles are highly effective in their physical and chemical characteristics. In order to increase the active areas of the particle, it is necessary to reduce the size to the nanoscale [122]. When the dimensions of the MNPs are below the critical dimension (Dc), the free electron spins are aligned in a single direction. Thus, NPs begin to act as a single domain magnet. When global MNPs reach a single domain limit, the critical dimension (Dc) can be determined by the following equation [123].

$$Dc \approx \frac{36\sqrt{AK}}{\mu^0 Ms^2}$$

A: change constant, K: anisotropy constant, μ^0: vacuum permeability,

Ms: Magnetic saturation [123,124] (Table 6).

Table 6. Typical magnetic materials Ms, K, Dc values.

Materials	Ms (emu cm^{-3})	K (10^3 Jm^{-3})	Dc (nm)
Fe	1745	0.048	15
Co	1400	0.45	70
Ni	490	-0.005	55
Fe$_3$O$_4$	460	-0.011	128
L10-FePt	1140	7	60
SmCo$_5$	910	20	750

For many MNPs, there are Dc values between 10-100 nm. However, a single domain of NPs with anisotropy constant K rom at a very high value is in micrometer (μm) size [123]. The control of the magnetization direction in each domain in control multi-domain NPs with domain wall movements is controlled by the anisotropy energy (KV, V: domain volume). The wall motion of the domain is not found in single-domain NPs, and therefore the magnetization direction is mostly dependent on KV. As a result, single-domain NPs has higher coercivity (Hc) than multi-domain ones.

The lower the size of the nanoparticles, the more thermal energy (kBT, kB; Boltzmann constant and T; temperature) start with KV at that rate (V; the volume of a single domain NP) and leads to a decrease in Hc. When the size of the NP falls below the critical value Ds, kBT exceeds the KV value and causes the magnetization direction to change spontaneously [123]. The performance of NPs below the critical size is at the highest level [125]. The domains in these NPs also decrease until a single domain remains. Thus, super paramagnetic NPs are obtained [122]. Super paramagnetic NPs respond quickly to the applied magnetic field. The external magnetic field to be applied to reach the magnetic saturation (Ms) (to direct all spins in the same direction) is very high. In super paramagnetic structures, no magnetization is observed when the applied external magnetic field is removed. That is, the residual magnetization and coercivity values are zero. They do not show a magnetic loop (hysteresis) [125,126]. These properties reduce the possibility of agglomeration *in vivo* studies by preventing coagulation. They are the focus of many studies due to their biocompatibility and ease of synthesis. Particle size and distribution can be controlled by the synthesis method. For example, when preparing NP by the coprecipitation method, adjusting the size of NPs by playing with pH and ion intensity is often one of the methods used [122].

3.2 Effect of structure and shape

The structures of NPs are interrelated with their magnetic properties. The crystalline structures of the nanoparticles affect the distinctive spin-orbital interactions, namely magneto crystal isotropic. The magneto crystal isotropic constant determines the size of the nanoparticle. The high anisotropy constant generally results in large coercivity [127–129].

3.3 Effect of composition

The magnetic properties of NPs can be adjusted in their multi-component MNP systems according to their composition. Control of the composition of MNPs can be used not only for the magnetization value but also for the adjustment of coercivity. As seen in Figure 3, the structure has a cubic structure. Iron ions in Fe_2O_3s are distributed in (T) and (O) regions. Furthermore, the difference of Fe_2O_3 from Fe_3O_4 is due to the presence of cationic voids in the octahedral region. The order of the gaps is related to the sample preparation method. These gaps can be located in 3 separate positions; it may be completely random, sequential, or partially sequential [130–134].

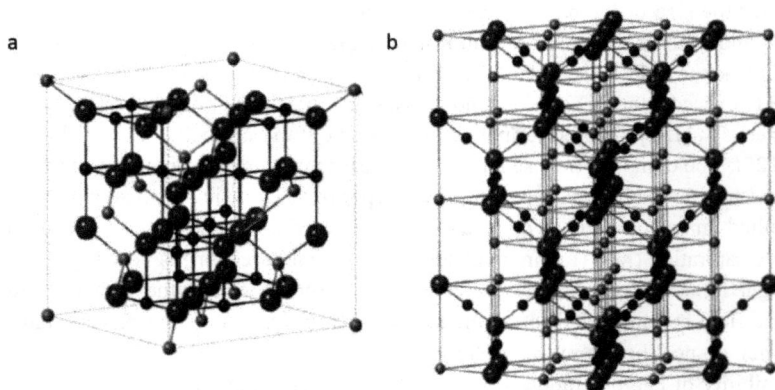

Figure 3. (a) Magnetite and (b) maghemite crystal structure.

4. Synthesis strategies of magnetic nanoparticles

MNP's size, composition, and structure are the most important issues that determine the characteristics. It is known that intensive studies have been carried out by controlling the sizes and shapes of MNPs in order to perform their synthesis. Many chemical methods

Materials Research Forum LLC
https://doi.org/10.21741/9781644900611-4

have been used to synthesize MNPs. The basic synthesis pathways for the synthesis of nanoparticles are given in Table 7.

Table 7. Comparison of MNPs synthesis methods [135,136].

Methods		Benefits	Drawbacks
Microbial methods	High Yield	Microbial incubation, repeatability, lower costs	Time
Chemical methods	Sol-gel synthesis	Precise control of size, internal structure, and ratio	Poor binding and high permeability
	Oxidation method	Uniform and small size synthesis	Small size ferrite colloids
	Chemical precipitation	Simple and efficient	Not suitable for high quality and accuracy synthesis
	Hydrothermal reaction	Easy to control particle size and shape	High temperature and pressure
	Flow injection synthesis	Repeatability and precise control of the process availability	Continuous mixing of reagents along with the bed flow in the capillary reactor
	Electrochemical methods	Particle size control is easy	Repeatability
	Aerosol/vapor phase method	High efficiency	Very high-temperature requirement
	Sonochemical accumulation reactions	Easy to control particle size	The mechanism is not yet fully understood
	Supercritical fluid methods	Organic solvent free and particle size control	Critical temperature and pressure
	Synthesis using nanoreactor	Precise control of particle size	Complex conditions
Physical methods	Gas-phase spooling	Easy	Particle Size difficult to control
	Electron beam lithography	Good interstellar gap control	Expensive and complex device requirement

Super paramagnetic NPs have colloidal structures and are complex due to these structures (Table 8). The first of these difficulties is the difficulty in defining the experimental conditions that provide the acquisition of monodisperse structures of appropriate shape and size. The second difficulty is the choice of the process because it is the choice of an impression of a repeatable process that is not suitable for industrial use, including any ultracentrifugation, size separation chromatography, magnetic filtration, or flow complex gradients. These processes use all of these methods in the preparation of homogeneous composition particles having small size distribution. However, the most commonly used method for the synthesis of MNPs is chemical coprecipitation technique [137–139].

Table 8. Colloidal synthetic strategies for other shape-controlled magnetic composites.

Thermal decomposition
Composition: CoxFe3-xO4
Shape: Nanocubes
Dimensions (nm): 15-27
Metal Precursors: Fe(Acac)3/Co(Acac)2
Reagents involved: Decanoic acid

Thermal decomposition
Composition: MxFe3-xO4 (M2+,Fe2+, Mn2+,Zn2+, Cu2+, Ca2+, Mg2+)
Shape: Nanocubes
Dimensions (nm): 10-20
Metal Precursors: M2+/Fe3+←Oleate
Reagents involved: TOPO, OA

Thermal decomposition/ Galvanic replacement
Composition: MnOx/FeOx
Shape: Hollow nanospheres
Dimensions (nm): 24 (7-10 inner voids)
Metal Precursors: Mn(Oleate)2/Fe(Acac)3
Reagents involved: OA, Oleylamine

Polyol
Composition: MnFe2O4
Shape: Flower-like
Dimensions (nm): 50
Metal Precursors: FeCl3/MnCl2
Reagents involved: PAA

Polyol
Composition: ZnxFe3-xO4
Shape: Nanorings
Dimensions (nm): 13-20 x 100-150 x 70-110
Metal Precursors: Zn(Acac)2/ FeCl3
Reagents involved: EG / H2O / (NH2)2CO

4.1   Electron bunch lithography

The electron bunch lithography method is a physical method. It usually involves oxidizing the particles under electron bombardment. The formation of MNPs in nanoscale takes place by scattering the electron bunch on a surface surrounded by particle films [135].

4.2   Gas-phase deposition

The gas phase condensation method in NPs is of great importance nowadays and the main advantage of the gas phase condensation method, which has an increasing echo, is that it can be easily applied to almost any material. This phenomenon is commonly based on the accumulation of chemical vapor by the help of catalysts on aluminum substrates coated with gold $[Fe(OBut)_3]_2$ or $Sn(OBut)_4$ molecules. Cheap and easily available materials can be used as starting material for synthesis, and non-agglomerated particles can be produced [145–147].

4.3   Coprecipitation

The coprecipitation method, which is one of the most effective methods, is an easily performed chemical effective way to obtain MNPs. With 2 basic methods, synthesis of spherical magnetic particles in solution in nm size is carried out. As the first method, magnetic metals are partially oxidized by different oxidizing agents such as hydroxide. In the other method; The preparation of an aqueous solution and the precipitation of this magnetic solution containing a mixture of metal-hydroxides in a stoichiometric ratio yield, the magnetites of spherical magnetic particles in homogeneous appearance and size [148–151] (Figure 4).

4.4   Microemulsions

Nanoparticles synthesized by coprecipitation strategy have different sizes. There are different methods used and developed to improve the uniform size expression of nanoparticles. When it is found in an environment, two liquids (oil and water), which do not mix, are formed by the isotropic distribution of microemulsion structures in the active substance environment on the surface. The hydrophobic parts of the amphoteric surfactants forming a single layer between these two phases are dissolved in the aqueous phase in the hydrophilic head in the oil phase. With this feature, different types of structures can be formed, from cylindrical micellar to spherical micelle structures [151].

Amphoteric surfactants are present in a restricted environment for the development or formation of apolar solvents in water-soluble reverse micelle structures [151]. Water-in-oil microemulsions are thermodynamically stable in a permeable, isotropic, and liquid medium such as reverse micelle solution. With this strategy, the fine and small microwaves in the aqueous phase in the systems are trapped within the surfactant group dispersed in the oil phase. The micro-cavities in the size of 10 nm stabilized by the surfactant exhibit a trapping effect. Thus, they are semi-limiting to limit particle precipitation, growth, and particle nucleation. Water/oil microemulsion methods are simple, sufficient, and versatile to prepare particles in nanoscale, for *in vivo* and *in vitro* applications [148,152–155].

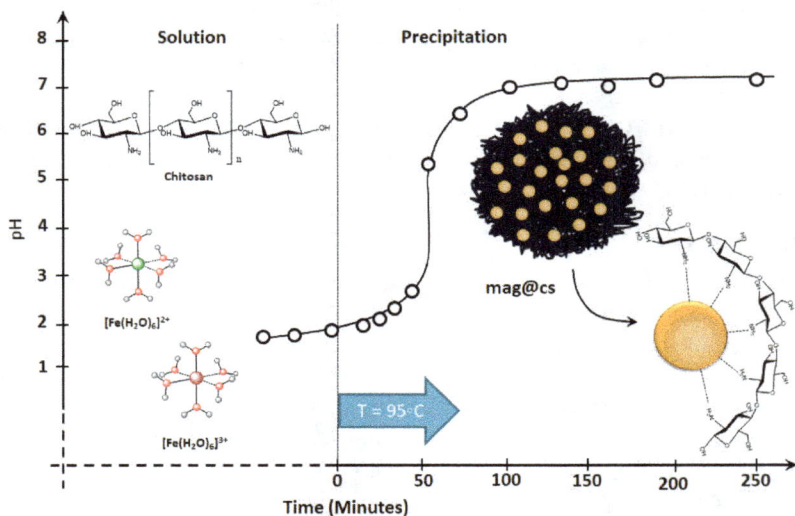

Figure 4. Coprecipitation schematic.

4.5   Hydrothermal synthesis

Hydrothermal synthesis reactions; operates in an autoclave and in an aqueous environment in which the pressure is 2000 psi and the temperature rises above 200 °C. There are two methods; 1) hydrolysis and 2) neutralization or oxidation of the mixture of metal hydroxides. These two reactions have similarities. In this method, there are some effective conditions for the synthesis. These; reaction conditions include solvent, temperature and time, etc. The size of the MNPs increases in direct proportion to the prolonged reaction time and high water content. The reaction rate of these processes is temperature dependent when other conditions are constant [125,156–158].

4.6   Sol-gel reactions

One of the wet chemical synthesis methods of nanostructured metal oxides is the sol-gel method. It is one of the wet chemical synthesis methods of nanostructured metal oxides. This method is based on the condensation and hydroxylation of molecular precursors in the solution. Advanced condensation and inorganic polymerization of the left structure result in the formation of a gel structure consisting of a 3D metal oxide network. These reactions are optimally carried out at room temperature. The application of heat results in the final crystal structure. The properties of the gel structure are very dependent on the structure formed along with the medium of the sol-gel method.

Some advantages of the sol-gel process include;

1. To be able to obtain the desired molecular structure based on working conditions

2. Control of particle size, monodisperse and pure amorphous phase

3. Homogeneity and microstructure control of products synthesized as a result of the reaction

4. Sol-gel matrix in its properties and stability of the molecules that can protect the placement.

As for disadvantages of this method; the system is expensive, requires labor force and has very little efficiency [159–161].

4.7   Polyols

The polyol method is a well-defined multi-faceted method used for the synthesis of nano- and microparticles. Polyols are used as stabilizing agents and are used as reducing agents in order to control the growth of particles and to prevent inter-particle aggregation.

The reduction of the dissolved metallic salts and the direct precipitation of the metal salts from a polyol-containing solution can result in regular metallic particles. In this method, the suspension is stirred for a while after the precursor compound is dispersed in a liquid polyol and heated to a certain temperature (the boiling temperature of the polyol). Metal precursors are dissolved in the diol after forming an interface, and the metal nuclei are reduced to form another nucleus. It then nucleates to form metal particles throughout the nucleate reaction. Particles under the micrometer can be synthesized by increasing the temperature of the reaction or by inducing heterogeneous nucleation due to the process of forming or adding different (foreign) nuclei (nuclei) *in-situ*.

The second method is more efficient because the temperature increase leads to significant thermal degradation of the polyols. The properties and advantages of this method compared to other methods can be summarized as follows: The surface of the prepared metal oxide nanoparticles can be coated *in-situ* by hydrophilic polyols. Nanoparticles can easily spread in aqueous media or other organic solvents. As a result, the size distribution of nanoparticles is smaller than that of nanoparticles produced by basic methods [130,162–166].

4.8   Flow injection synthesis

Alvarez et al., developed the method of flow injection synthesis.[167]. As a working principle, it involves the laminar flow of reagents through a capillary reactor, either continuously or progressively. Inflow injection synthesis technique, the good homogeneous mixing of the reagents is one of the advantages of high reproducibility and the possibility of externally controlling the process [151].

4.9   Electrochemical methods

It has been studied and developed by Pascal and Reetz et al. [168,169]. It was prepared with a metal electrode from an aqueous solution containing dimethylformamide or cationic surfactants. The particle size can be controlled in proportion to the current density. MNPs are synthesized as a result of electrochemical accumulation in the presence of oxidizing agents [170–172].

4.10   Aerosol / steam methods

For direct and continuous production of good MNPs, spray, and laser pyrolysis methods have shown to be the best method under experimental conditions that need to be heavily controlled. Since it has a high production rate in preparation of MNPs for use in *in-vivo* and *in-vitro* applications, it is thought to be a forward-looking method. In the spray pyrolysis, a reducing agent in the solution of the metal salts and the organic solvent is sprayed into a fast reactor. The aerosol reactor comprises a densified, dissolved and evaporated solvent. The final stage contains residual particles. The size is the initial size of the drop formed as a result of the spray. The reaction volume can be reduced in laser pyrolysis. The laser heats the gas cloud containing the metal precursors, and the flow of this gas cloud produces nanoparticles that do not form aggregates in nano dimensions [151]. The main difference between spray and laser pyrolysis is related to the final state of the synthesized nanoparticles. While aggregate formation and larger particles are frequently observed in spray pyrolysis, those observed in laser pyrolysis are less aggregated particles due to shorter reaction time [75].

Magnetochemistry - Materials and Applications   Materials Research Forum LLC
Materials Research Foundations **66** (2020) 130-172   https://doi.org/10.21741/9781644900611-4

4.11 Sonolysis

Polymers, capping agents, and structural hosts are used to limit the size of the nanoparticles. In the presence of sodium dodecyl sulfate, the sonolysis of $Fe(CO)_5$ aqueous solution allows the formation of amorphous iron oxide nanoparticles of stable hydrosol (suspended in water). Super paramagnetic nanoparticles that are highly magnetized and crystalline are produced by sonochemical methods [173].

5.   Stabilization / protection of magnetic nanoparticles

Although there are many important developments in the synthesis of MNPs, long-term agglomeration, and the development of their stability without collapse are also seen as an important situations. The properties of MNPs are changing due to a thin oxidation layer formed on the surface of the oxidation with nanoparticles at the optimum conditions. The stabilites required for almost all applications of MNPs, especially of pure metals such as Fe, Ni and Co, and alloys of these metals. NPs are highly sensitive to air. Their susceptibility to oxidation increases with the reduction in particle size. Therefore, the development of effective strategies is important in order to improve the chemical stability of MNPs. For this purpose, it is best practice to protect the surface of the MNP with a layer that does not react with oxygen. MNP has a layer on its surface and protects it from reaching oxygen. MNPs contain a core-shell structure in all protective strategies. With a clear expression, a core is used as a core and covered with a protective shell called core-shell structure. Nanoparticle, called the core, is protected against the environment. Coating strategies that are widely applied are divided into two basic groups (Figure 5).

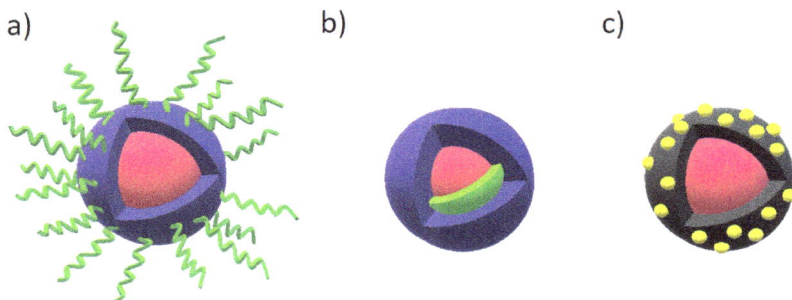

Figure 5. Schematic representation of silica plated NPs by three different methods. a) water/oil microemulsion method (on the outer surface of silica, surfactants are shown.) b) silica coating by polymer or surfactants, c) self-assembly through nanomaterial surface.

Magnetochemistry - Materials and Applications   Materials Research Forum LLC
Materials Research Foundations 66 (2020) 130-172   https://doi.org/10.21741/9781644900611-4

The organics (such as surfactants and polymers) may be coated with a shell or inorganic components (such as silica, carbon, sensitive metals (such as Ag, Au) or oxides) (Figure 6). In another method, MNPs can be placed or dispersed in a dense matrix, typically silica, carbon or polymer, to form a composite. In this way, agglomeration and oxidation are inhibited or reduced to a low level [125,174,175].

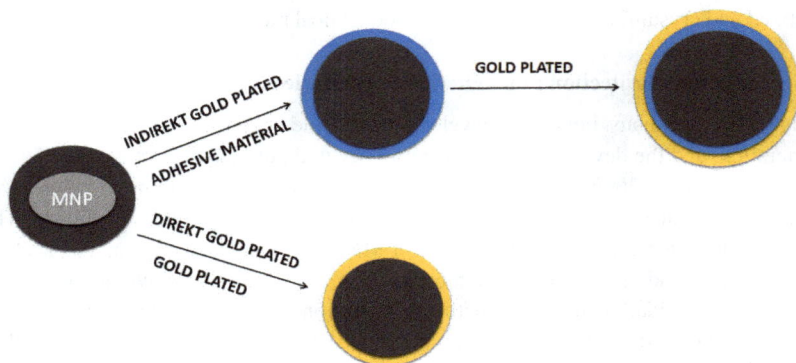

Figure 6. Gold plating pathways of MNPs.

Conclusions and Recommendations

In recent years, the use of magnetic nanoparticles (MNPs) has increased considerably, especially in biological applications. Much research has been done on magnetic nanoparticles intended for use in biomedical applications so far, but more research is needed to make magnetic nanoparticles more widely available. MNPs, which can be used in many fields such as nanostructured systems, have the advantage of being able to give less soluble active substances to the body in the same carrier due to their reduced side effects and modifiable properties. However, in studies, it has been stated that it is not enough to increase the bioavailability, to improve solubility, to reduce side effects and toxicity, to formulate the active substances only with nanoparticles. It has been found that it is possible to target nanotubes using properties which can be varied in a variety of effects, agents, or structures, which are difficult to reach in the body or to transport drugs to specific areas. Nanostructuring systems can be targeted either by external or internal stimulation with changes in the desired site in the body or by agents conjugated to nanoparticles. Nanoparticles can be activated by ultrasound or IR laser by passive or external magnetic methods through the pH, temperature, and enzymatic changes in the diseased areas by conjugating various agents to their structures.

In this chapter, the synthesis processes and the difficulties encountered in the experiments are presented. The synthesis techniques used in the early studies were insufficient for the synthesis of biocompatible MNPs. For this purpose, new synthesis techniques should be developed to obtain more biocompatible and less cytotoxicity MNPs. Second, more efficient methods should be used to prepare magnetic nanoparticle-based composites with fine microstructure and high-efficiency performance for biomedical studies. Third, it is necessary to establish the possible mechanism of action of these two properties for the detection and analysis of the cytotoxicity and biocompatibility of the magnetic nanoparticles to be synthesized and to investigate more recent and appropriate experiment techniques both *in-vitro* and *in-vivo*. Fourth, since the composites based on magnetic nanoparticles with the core-shell structure described in this chapter have unique biological and physicochemical properties, which enable them to have great potential for biomedical applications, the various diagnoses and treatments of these MNPs need to be further investigated. Finally, in order to present new and different methods of biomedical applications, the efficient synthesis of magnetic nanoparticles needs to be better understood and further investigation.

References

[1] I. Koh, L. Josephson, I. Koh, L. Josephson, Magnetic nanoparticle sensors, Sensors 9 (2009) 8130–8145. https://doi.org/10.3390/s91008130

[2] H. Vaghari, H. Jafarizadeh-Malmiri, M. Mohammadlou, A. Berenjian, N. Anarjan, N. Jafari, S. Nasiri, Application of magnetic nanoparticles in smart enzyme immobilization, Biotechnol. Lett. 38 (2016) 223–233. https://doi.org/10.1007/s10529-015-1977-z

[3] G.K. Kouassi, J. Irudayaraj, G. McCarty, Examination of Cholesterol oxidase attachment to magnetic nanoparticles, J. Nanobiotechnol. 3 (2005) 1. https://doi.org/10.1186/1477-3155-3-1

[4] A.K. Johnson, A.M. Zawadzka, L.A. Deobald, R.L. Crawford, A.J. Paszczynski, Novel method for immobilization of enzymes to magnetic nanoparticles, J. Nanoparticle Res. 10 (2008) 1009–1025. https://doi.org/10.1007/s11051-007-9332-5

[5] N.A. Kalkan, S. Aksoy, E.A. Aksoy, N. Hasirci, Preparation of chitosan-coated magnetite nanoparticles and application for immobilization of laccase, J. Appl. Polym. Sci. 123 (2012) 707–716. https://doi.org/10.1002/app.34504

[6] M. Namdeo, S.K. Bajpai, Immobilization of a-amylase onto cellulose-coated magnetite (CCM) nanoparticles and preliminary starch degradation study, J.

Molecular Catal. B, Enzymatic 59 (2009) 134-139.
https://doi.org/10.1016/j.molcatb.2009.02.005

[7] R. Konwarh, N. Karak, S.K. Rai, A.K. Mukherjee, Polymer-assisted iron oxide magnetic nanoparticle immobilized keratinase, Nanotechnology 20 (2009) 225107. https://doi.org/10.1088/0957-4484/20/22/225107

[8] S.A. Ansari, Q. Husain, S. Qayyum, A. Azam, Designing and surface modification of zinc oxide nanoparticles for biomedical applications, Food Chem. Toxicol. 49 (2011) 2107–15. https://doi.org/10.1016/j.fct.2011.05.025

[9] S.H. Huang, M.H. Liao, D.H. Chen, Direct binding and characterization of lipase onto magnetic nanoparticles, Biotechnol. Prog. 19 (2003) 1095–1100. https://doi.org/10.1021/bp025587v

[10] I. Cicha, S. Lyer, C. Janko, R.P. Friedrich, M. Pöttler, C. Alexiou, Magnetic nanoparticles for medical applications, Nanomedicine 12 (2017) 825–829. https://doi.org/10.2217/nnm-2017-0038

[11] P. Tartaj, M.P. Morales, T. González-Carreño, S. Veintemillas-Verdaguer, C.J. Serna, Advances in magnetic nanoparticles for biotechnology applications, J. Magn. Magn. Mater. 290–291 (2005) 28–34. https://doi.org/10.1016/J.JMMM.2004.11.155

[12] C.E. Sjøgren, K. Briley-Saebø, M. Hanson, C. Johansson, Magnetic characterization of iron oxides for magnetic resonance imaging, Magn. Reson. Med. 31 (1994) 268–272. https://doi.org/10.1002/mrm.1910310305

[13] D. Maity, D.C. Agrawal, Synthesis of iron oxide nanoparticles under oxidizing environment and their stabilization in aqueous and non-aqueous media, J. Magn. Magn. Mater. 308 (2007) 46–55. https://doi.org/10.1016/j.jmmm.2006.05.001

[14] H. Ding, V. Sagar, M. Agudelo, S. Pilakka-Kanthikeel, V.S.R. Atluri, A. Raymond, T. Samikkannu, M.P. Nair, Enhanced blood–brain barrier transmigration using a novel transferrin embedded fluorescent magneto-liposome nanoformulation, Nanotechnology 25 (2014) 055101. https://doi.org/10.1088/0957-4484/25/5/055101

[15] Y. Huang, B. Zhang, S. Xie, B. Yang, Q. Xu, J. Tan, Superparamagnetic iron oxide nanoparticles modified with tween 80 pass through the intact blood–brain barrier in rats under magnetic field, ACS Appl. Mater. Interfaces 8 (2016) 11336–11341. https://doi.org/10.1021/acsami.6b02838

[16] S.D. Kong, J. Lee, S. Ramachandran, B.P. Eliceiri, V.I. Shubayev, R. Lal, S. Jin, Magnetic targeting of nanoparticles across the intact blood–brain barrier, J. Control. Release 164 (2012) 49–57. https://doi.org/10.1016/j.jconrel.2012.09.021

[17]   E. Lueshen, I. Venugopal, T. Soni, A. Alaraj, A. Linninger, Implant-assisted intrathecal magnetic drug targeting to aid in therapeutic nanoparticle localization for potential treatment of central nervous system disorders, J. Biomed. Nanotechnol. 11 (2015) 253–261.   https://doi.org/10.1166/jbn.2015.1907

[18]   I. Venugopal, N. Habib, A. Linninger, Intrathecal magnetic drug targeting for localized delivery of therapeutics in the CNS, Nanomedicine 12 (2017) 865–877. https://doi.org/10.2217/nnm-2016-0418

[19]   S. Pilakka-Kanthikeel, V.S.R. Atluri, V. Sagar, S.K. Saxena, M. Nair, Targeted brain derived neurotropic Factors (BDNF) delivery across the blood-brain barrier for neuro-protection using magnetic nano carriers: An In-Vitro Study, PLoS One 8 (2013) e62241.   https://doi.org/10.1371/journal.pone.0062241

[20]   S.L. Raut, B. Kirthivasan, M.M. Bommana, E. Squillante. M. Sadoqi, The formulation, characterization and in vivo evaluation of a magnetic carrier for brain delivery of NIR dye, Nanotechnology 21 (2010) 395102. https://doi.org/10.1088/0957-4484/21/39/395102

[21]   Z. Sun, M. Worden, Y. Wroczynskyj, V. Yathindranath, J. van Lierop, T. Hegmann, D. Miller, Magnetic field enhanced convective diffusion of iron oxide nanoparticles in an osmotically disrupted cell culture model of the blood-brain barrier, Int. J. Nanomedicine 9 (2014) 3013-3026.   https://doi.org/10.2147/IJN.S62260

[22]   L.B. Thomsen, T. Linemann, K.M. Pondman, J. Lichota, K.S. Kim, R.J. Pieters, G.M. Visser, T. Moos, Uptake and transport of superparamagnetic iron oxide nanoparticles through human brain capillary endothelial cells, ACS Chem. Neurosci. 4 (2013) 1352–1360.   https://doi.org/10.1021/cn400093z

[23]   M. Zhao, J. Hu, L. Zhang, L. Zhang, Y. Sun, N. Ma, X. Chen, Z. Gao, Study of amphotericin B magnetic liposomes for brain targeting, Int. J. Pharm. 475 (2014) 9–16.   https://doi.org/10.1016/j.ijpharm.2014.08.035

[24]   M. Zhao, J. Chang, X. Fu, C. Liang, S. Liang, R. Yan, A. Li, Nano-sized cationic polymeric magnetic liposomes significantly improves drug delivery to the brain in rats, J. Drug Target. 20 (2012) 416–421. https://doi.org/10.3109/1061186X.2011.651726

[25]   X. Zhao, T. Shang, X. Zhang, T. Ye, D. Wang, L. Rei, Passage of magnetic tat-conjugated $Fe_3O_4@SiO_2$ nanoparticles across in vitro blood-brain barrier, Nanoscale Res. Lett. 11 (2016) 451.   https://doi.org/10.1186/s11671-016-1676-2

[26]   M. Dan, Y. Bae, T.A. Pittman, R.A. Yokel, Alternating magnetic field-induced

Magnetochemistry - Materials and Applications
Materials Research Foundations 66 (2020) 130-172

Materials Research Forum LLC
https://doi.org/10.21741/9781644900611-4

Hyperthermia increases iron oxide nanoparticle cell association/uptake and flux in blood–brain barrier models, Pharm. Res. 32 (2015) 1615–1625. https://doi.org/10.1007/s11095-014-1561-6

[27]   S.N. Tabatabaei, H. Girouard, A.-S. Carret, S. Martel, Remote control of the permeability of the blood–brain barrier by magnetic heating of nanoparticles: A proof of concept for brain drug delivery, J. Control. Release 206 (2015) 49–57. https://doi.org/10.1016/j.jconrel.2015.02.027

[28]   N. Lolak, E. Kuyuldar, H. Burhan, H. Goksu, S. Akocak, F. Sen, Composites of palladium–nickel alloy nanoparticles and graphene oxide for the knoevenagel condensation of aldehydes with malononitrile, ACS Omega 4 (2019) 6848–6853. https://doi.org/10.1021/acsomega.9b00485

[29]   R. Ayranci, B. Demirkan, B. Sen, A. Şavk, M. Ak, F. Şen, Use of the monodisperse Pt/Ni@rGO nanocomposite synthesized by ultrasonic hydroxide assisted reduction method in electrochemical nonenzymatic glucose detection, Mater. Sci. Eng. C. 99 (2019) 951–956.   https://doi.org/10.1016/J.MSEC.2019.02.040

[30]   B. Sen, A. Şavk, F. Sen, Highly efficient monodisperse pt nanoparticles confined in the carbon black hybrid material for hydrogen liberation, J. Colloid Interface Sci. 520 (2018) 112–118.   https://doi.org/10.1016/j.jcis.2018.03.004

[31]   E. Erken, H. Pamuk, Ö. Karatepe, G. Başkaya, H. Sert, O.M. Kalfa, F. Şen, New Pt(0) nanoparticles as highly active and reusable catalysts in the C1–C3 Alcohol Oxidation and the room temperature dehydrocoupling of dimethylamine-borane, J. Clust. Sci. 27 (2016) 9–23.   https://doi.org/10.1007/s10876-015-0892-8

[32]   B. Şen, E.H. Akdere, A. Şavk, E. Gültekin, Ö. Paralı, H. Göksu, F. Şen, A novel thiocarbamide functionalized graphene oxide supported bimetallic monodisperse Rh-Pt nanoparticles (RhPt/TC@GO NPs) for Knoevenagel condensation of aryl aldehydes together with malononitrile, Appl. Catal. B Environ. 225 (2018) 148–153. https://doi.org/10.1016/J.APCATB.2017.11.067

[33]   İ. Esirden, E. Erken, M. Kaya, F. Sen, Monodisperse Pt NPs@rGO as highly efficient and reusable heterogeneous catalysts for the synthesis of 5-substituted 1H-tetrazole derivatives, Catal. Sci. Technol. 5 (2015) 4452–4457. https://doi.org/10.1039/C5CY00864F

[34]   B. Çelik, Y. Yıldız, H. Sert, E. Erken, Y. Koşkun, F. Şen, Monodispersed palladium–cobalt alloy nanoparticles assembled on poly(N-vinyl-pyrrolidone) (PVP) as a highly effective catalyst for dimethylamine borane (DMAB) dehydrocoupling, RSC Adv. 6 (2016) 24097–24102.   https://doi.org/10.1039/C6RA00536E

[35]   A.K. Gupta, M. Gupta, Synthesis and surface engineering of iron oxide nanoparticles for biomedical applications, Biomaterials 26 (2005) 3995–4021. https://doi.org/10.1016/j.biomaterials.2004.10.012

[36]   B. Sen, S. Kuzu, E. Demir, S. Akocak, F. Sen, Monodisperse palladium–nickel alloy nanoparticles assembled on graphene oxide with the high catalytic activity and reusability in the dehydrogenation of dimethylamine–borane, Int. J. Hydrogen Energy 42 (2017) 23276–23283.   https://doi.org/10.1016/j.ijhydene.2017.05.113

[37]   F. Sen, Y. Karatas, M. Gulcan, M. Zahmakiran, Amylamine stabilized platinum(0) nanoparticles: Active and reusable nanocatalyst in the room temperature dehydrogenation of dimethylamine-borane, RSC Adv. 4 (2014) 1526–1531. https://doi.org/10.1039/C3RA43701A

[38]   H. Pamuk, B. Aday, F. Şen, M. Kaya, Pt NPs@GO as a highly efficient and reusable catalyst for one-pot synthesis of acridinedione derivatives, RSC Adv. 5 (2015) 49295–49300.   https://doi.org/10.1039/C5RA06441D

[39]   B. Sen, S. Kuzu, E. Demir, E. Yıldırır, F. Sen, B. Şen, S. Kuzu, E. Demir, E. Yıldırır, F. Şen, B. Sen, S. Kuzu, E. Demir, E. Yıldırır, F. Sen, B. Şen, S. Kuzu, E. Demir, E. Yıldırır, F. Şen, Highly efficient catalytic dehydrogenation of dimethyl ammonia borane via monodisperse palladium–nickel alloy nanoparticles assembled on PEDOT, Int. J. Hydrogen Energy 42 (2017) 23307–23314. https://doi.org/10.1016/j.ijhydene.2017.05.115

[40]   Z. Ozturk, F. Sen, S. Sen, G. Gokagac, The preparation and characterization of nano-sized Pt-Pd/C catalysts and comparison of their superior catalytic activities for methanol and ethanol oxidation, J. Mater. Sci. 47 (2012) 8134–8144. https://doi.org/10.1007/s10853-012-6709-3

[41]   H. Goksu, Y. Yıldız, B. Çelik, M. Yazici, B. Kilbas, F. Sen, Eco-friendly hydrogenation of aromatic aldehyde compounds by tandem dehydrogenation of dimethylamine-borane in the presence of a reduced graphene oxide furnished platinum nanocatalyst, Catal. Sci. Technol. 6 (2016) 2318–2324. https://doi.org/10.1039/C5CY01462J

[42]   B. Şahin, E. Demir, A. Aygün, H. Gündüz, F. Şen, Investigation of the effect of pomegranate extract and monodisperse silver nanoparticle combination on MCF-7 cell line, J. Biotechnol. 260 (2017) 79–83. https://doi.org/10.1016/j.jbiotec.2017.09.012

[43]   B. Şahin, A. Aygün, H. Gündüz, K. Şahin, E. Demir, S. Akocak, F. Şen, Cytotoxic effects of platinum nanoparticles obtained from pomegranate extract by the green

synthesis method on the MCF-7 cell line, Colloids Surf. B Biointerfaces 163 (2018) 119–124. https://doi.org/10.1016/j.colsurfb.2017.12.042

[44] B. Aday, Y. Yıldız, R. Ulus, S. Eris, F. Sen, M. Kaya, One-Pot, Efficient and green synthesis of acridinedione derivatives using highly monodisperse platinum nanoparticles supported with reduced graphene oxide, New J. Chem. 40 (2016) 748–754. https://doi.org/10.1039/C5NJ02098K

[45] R. Ulus, Y. Yıldız, S. Eriş, B. Aday, F. Şen, M. Kaya, Functionalized multi-walled carbon nanotubes (f-MWCNT) as highly efficient and reusable heterogeneous catalysts for the synthesis of acridinedione derivatives, ChemistrySelect 1 (2016) 3861–3865. https://doi.org/10.1002/slct.201600719

[46] F. Şen, G. Gökağaç, Pt nanoparticles synthesized with new surfactants: Improvement in C1-C3 alcohol oxidation catalytic activity, J. Appl. Electrochem. 44 (2014) 199–207. https://doi.org/10.1007/s10800-013-0631-5

[47] B. Şen, A. Aygün, A. Şavk, S. Akocak, F. Şen, Bimetallic palladium–iridium alloy nanoparticles as highly efficient and stable catalyst for the hydrogen evolution reaction, Int. J. Hydrogen Energy 43 (2018) 20183–20191. https://doi.org/10.1016/J.IJHYDENE.2018.07.081

[48] E. Demir, A. Savk, B. Sen, F. Sen, A novel monodisperse metal nanoparticles anchored graphene oxide as counter electrode for dye-sensitized solar cells, Nano-Structures and Nano-Objects 12 (2017) 41–45. https://doi.org/10.1016/j.nanoso.2017.08.018

[49] G. Li, Q. Yi, X. Yang, Y. Chen, X. Zhou, G. Xie, Ni-Co-N doped honeycomb carbon nano-composites as cathodic catalysts of membrane-less direct alcohol fuel cell, Carbon 140 (2018) 557–568. https://doi.org/10.1016/j.carbon.2018.08.037

[50] B. Sen, S. Kuzu, E. Demir, S. Akocak, F. Sen, Highly monodisperse RuCo nanoparticles decorated on functionalized multiwalled carbon nanotube with the highest observed catalytic activity in the dehydrogenation of dimethylamine–borane, Int. J. Hydrogen Energy 42 (2017) 23292–23298. https://doi.org/10.1016/j.ijhydene.2017.06.032

[51] S.M. Moghimi, A.C. Hunter, J.C. Murray, Long-circulating and target-specific nanoparticles: Theory to practice, Pharmacol. Rev. 53 (2001) 283–318.

[52] A. Curtis, C. Wilkinson, Nantotechniques and approaches in biotechnology, Trends Biotechnol. 19 (2001) 97–101. https://doi.org/10.1016/S0167-7799(00)01536-5

[53] J. Panyam, V. Labhasetwar, Biodegradable nanoparticles for drug and gene delivery to cells and tissue, Adv. Drug Deliv. Rev. 55 (2003) 329–47.

[54] Y. Cheng, M.E. Muroski, D.C.M.C. Petit, R. Mansell, T. Vemulkar, R.A. Morshed, Y. Han, I. V. Balyasnikova, C.M. Horbinski, X. Huang, L. Zhang, R.P. Cowburn, M.S. Lesniak, Rotating magnetic field induced oscillation of magnetic particles for in vivo mechanical destruction of malignant glioma, J. Control. Release 223 (2016) 75–84. https://doi.org/10.1016/j.jconrel.2015.12.028

[55] T. Eslaminejad, S.N. Nematollahi-Mahani, M. Ansari, Glioblastoma targeted gene therapy based on pEGFP/p53-loaded superparamagnetic ıron oxide Nanoparticles, Curr. Gene Ther. 17 (2017) 59–69.
https://doi.org/10.2174/1566523217666170605115829

[56] B.G. Nair, Y. Nagaoka, H. Morimoto, Y. Yoshida, T. Maekawa, D. Sakthi Kumar, Aptamer conjugated magnetic nanoparticles as nanosurgeons, Nanotechnology 21 (2010) 455102. https://doi.org/10.1088/0957-4484/21/45/455102

[57] M. Zhao, C. Liang, A. Li, J. Chang, H. Wang, R. Yan, J. Zhang, J. Tai, Magnetic paclitaxel nanoparticles inhibit glioma growth and improve the survival of rats bearing glioma xenografts, Anticancer Res. 30 (2010) 2217–23.

[58] V. Atluri, R. Jayant, S. Pilakka-Kanthikeel, G. Garcia, S. Thangavel, A. Yndart, A. Kaushik, M. Nair, Development of TIMP1 magnetic nanoformulation for regulation of synaptic plasticity in HIV-1 infection, Int. J. Nanomedicine 11 (2016) 4287–4298. https://doi.org/10.2147/IJN.S108329

[59] R. Jayant, V. Atluri, M. Agudelo, V. Sagar, A. Kaushik, M. Nair, Sustained-release nanoART formulation for the treatment of neuroAIDS, Int. J. Nanomedicine 10 (2015) 1077. https://doi.org/10.2147/IJN.S76517

[60] M. Nair, R. Guduru, P. Liang, J. Hong, V. Sagar, S. Khizroev, Externally controlled on-demand release of anti-HIV drug using magneto-electric nanoparticles as carriers, Nat. Commun. 4 (2013) 1707. https://doi.org/10.1038/ncomms2717

[61] M. Rodriguez, A. Kaushik, J. Lapierre, S.M. Dever, N. El-Hage, M. Nair, Electro-magnetic nano-particle bound Beclin1 siRNA crosses the blood–brain barrier to attenuate the ınflammatory effects of HIV-1 ınfection in vitro, J. Neuroimmune Pharmacol. 12 (2017) 120–132. https://doi.org/10.1007/s11481-016-9688-3

[62] V. Sagar, S. Pilakka-Kanthikeel, V.S.R. Atluri, H. Ding, A.Y. Arias, R.D. Jayant, A. Kaushik, M. Nair, Therapeutical neurotargeting via magnetic nanocarrier: Implications to opiate-ınduced neuropathogenesis and neuroAIDS, J. Biomed.

Nanotechnol. 11 (2015) 1722–33.   https://doi.org/10.1166/jbn.2015.2108

[63]   R. Chen, G. Romero, M.G. Christiansen, A. Mohr, P. Anikeeva, Wireless magnetothermal deep brain stimulation, Science 347 (2015) 1477–1480. https://doi.org/10.1126/science.1261821

[64]   R. Guduru, P. Liang, J. Hong, A. Rodzinski, A. Hadjikhani, J. Horstmyer, E. Levister, S. Khizroev, Magnetoelectric 'spin' on stimulating the brain, Nanomedicine 10 (2015) 2051–2061.   https://doi.org/10.2217/nnm.15.52

[65]   R. Munshi, S.M. Qadri, Q. Zhang, I. Castellanos Rubio, P. del Pino, A. Pralle, Magnetothermal genetic deep brain stimulation of motor behaviors in awake, freely moving mice, Elife 6 (2017).   https://doi.org/10.7554/eLife.27069

[66]   S.A. Stanley, L. Kelly, K.N. Latcha, S.F. Schmidt, X. Yu, A.R. Nectow, J. Sauer, J.P. Dyke, J.S. Dordick, J.M. Friedman, Bidirectional electromagnetic control of the hypothalamus regulates feeding and metabolism, Nature 531 (2016) 647–650. https://doi.org/10.1038/nature17183

[67]   A. Tay, D. Di Carlo, Magnetic Nanoparticle-Based Mechanical Stimulation for Restoration of Mechano-Sensitive Ion Channel Equilibrium in Neural Networks, Nano Lett. 17 (2017) 886–892.   https://doi.org/10.1021/acs.nanolett.6b04200

[68]   A. Tay, A. Kunze, C. Murray, D. Di Carlo, Induction of Calcium Influx in Cortical Neural Networks by Nanomagnetic Forces, ACS Nano 10 (2016) 2331–2341. https://doi.org/10.1021/acsnano.5b07118

[69]   M. Song, Y.-J. Kim, Y.-H. Kim, J. Roh, E.-C. Kim, H.J. Lee, S.U. Kim, B.-W. Yoon, Long-term effects of magnetically targeted ferumoxide-labeled human neural stem cells in focal cerebral ischemia, Cell Transplant 24 (2015) 183–190. https://doi.org/10.3727/096368913X675755

[70]   M. Song, Y.-J. Kim, Y. Kim, J. Roh, S.U. Kim, B.-W. Yoon, Using a Neodymium magnet to target delivery of ferumoxide-labeled human neural stem cells in a rat model of focal cerebral ischemia, Hum. Gene Ther. 21 (2010) 603–610. https://doi.org/10.1089/hum.2009.144

[71]   F.U. Amin, A.K. Hoshiar, T.D. Do, Y. Noh, S.A. Shah, M.S. Khan, J. Yoon, M.O. Kim, Osmotin-loaded magnetic nanoparticles with electromagnetic guidance for the treatment of Alzheimer's disease, Nanoscale 9 (2017) 10619–10632. https://doi.org/10.1039/c7nr00772h

[72]   T.D. Do, F. Ul Amin, Y. Noh, M.O. Kim, J. Yoon, Guidance of magnetic nanocontainers for treating alzheimer's disease using an electromagnetic, targeted

drug-delivery actuator, J. Biomed. Nanotechnol. 12 (2016) 569–74.
https://doi.org/10.1166/jbn.2016.2193

[73] S. Niu, L.-K. Zhang, L. Zhang, S. Zhuang, X. Zhan, W.-Y. Chen, S. Du, L. Yin, R. You, C.-H. Li, Y.-Q. Guan, Inhibition by multifunctional magnetic nanoparticles loaded with alpha-synuclein RNAi plasmid in a Parkinson's disease model, Theranostics 7 (2017) 344–356. https://doi.org/10.7150/thno.16562

[74] P. Umarao, S. Bose, S. Bhattacharyya, A. Kumar, S. Jain, Neuroprotective potential of superparamagnetic iron oxide nanoparticles along with exposure to electromagnetic field in 6-OHDA rat model of Parkinson's disease, J. Nanosci. Nanotechnol. 16 (2016) 261–9.

[75] P. Tartaj, M. a del P. Morales, S. Veintemillas-Verdaguer, T. Gonz lez-Carre o, C.J. Serna, The preparation of magnetic nanoparticles for applications in biomedicine, J. Phys. D. Appl. Phys. 36 (2003) R182–R197. https://doi.org/10.1088/0022-3727/36/13/202

[76] C. Scherer, A.M. Figueiredo Neto, Ferrofluids: Properties and applications, Brazilian J. Phys. 35 (2005) 718–727. https://doi.org/10.1590/S0103-97332005000400018

[77] N. Yanase, H. Noguchi, H. Asakura, T. Suzuta, Preparation of magnetic latex particles by emulsion polymerization of styrene in the presence of a ferrofluid, J. Appl. Polym. Sci. 50 (1993) 765–776. https://doi.org/10.1002/app.1993.070500504

[78] V. Veiga, D.H. Ryan, E. Sourty, F. Llanes, R.H. Marchessault, Formation and characterization of superparamagnetic cross-linked high amylose starch, Carbohydr. Polym. 42 (2000) 353–357. https://doi.org/10.1016/S0144-8617(99)00166-6

[79] M. Muthana, S.D. Scott, N. Farrow, F. Morrow, C. Murdoch, S. Grubb, N. Brown, J. Dobson, C.E. Lewis, A novel magnetic approach to enhance the efficacy of cell-based gene therapies, Gene Ther. 15 (2008) 902–910. https://doi.org/10.1038/gt.2008.57

[80] D. Müller-Schulte, H. Brunner, Novel magnetic microspheres on the basis of poly(vinyl alcohol) as affinity medium for quantitative detection of glycated haemoglobin, J. Chromatogr. A. 711 (1995) 53–60. https://doi.org/10.1016/0021-9673(95)00114-3

[81] M. Safarikova, I. Safarik, The application of magnetic techniques in biosciences, Magn. Electr. Sep. 10 (2001) 223–252. https://doi.org/10.1155/2001/57434

[82] S. Berensmeier, Magnetic particles for the separation and purification of nucleic

acids, Appl. Microbiol. Biotechnol. 73 (2006) 495–504.
https://doi.org/10.1007/s00253-006-0675-0

[83] A. Jordan, R. Scholz, P. Wust, H. Fähling, Roland Felix, Magnetic fluid
hyperthermia (MFH): Cancer treatment with AC magnetic field induced excitation of
biocompatible superparamagnetic nanoparticles, J. Magn. Magn. Mater. 201 (1999)
413–419. https://doi.org/10.1016/S0304-8853(99)00088-8

[84] I. Hilger, W.A. Kaiser, Iron oxide-based nanostructures for MRI and magnetic
hyperthermia, Nanomedicine 7 (2012) 1443–1459.
https://doi.org/10.2217/nnm.12.112

[85] I. Hilger, R. Hiergeist, W.A. Winnefeld, K. Schubert, H. Kaiser, Thermal ablation
of tumors using magnetic nanoparticles: An in vivo feasibility study, Investigative
Radiology 10 (2002) 580–586.
https://doi.org/10.1097/01.RLI.0000028491.19254.EE

[86] A. Ito, K. Tanaka, K. Kondo, M. Shinkai, H. Honda, K. Matsumoto, T. Saida, T.
Kobayashi, Tumor regression by combined immunotherapy and hyperthermia using
magnetic nanoparticles in an experimental subcutaneous murine melanoma, Cancer
Sci. 94 (2003) 308–313. https://doi.org/10.1111/j.1349-7006.2003.tb01438.x

[87] K. Maier-Hauff, R. Rothe, R. Scholz, U. Gneveckow, P. Wust, B. Thiesen, A.
Feussner, A. von Deimling, N. Waldoefner, R. Felix, A. Jordan, Intracranial
thermotherapy using magnetic nanoparticles combined with external beam
radiotherapy: Results of a feasibility study on patients with glioblastoma multiforme,
J. Neurooncol. 81 (2007) 53–60. https://doi.org/10.1007/s11060-006-9195-0

[88] S.J. DeNardo, G.L. DeNardo, A. Natarajan, L.A. Miers, A.R. Foreman, C.
Gruettner, G.N. Adamson, R. Ivkov, Thermal dosimetry predictive of efficacy of
111In-ChL6 nanoparticle AMF-induced thermoablative therapy for human breast
cancer in mice, J. Nucl. Med. 48 (2007) 437–44.

[89] T. Kikumori, T. Kobayashi, M. Sawaki, T. Imai, Anti-cancer effect of
hyperthermia on breast cancer by magnetite nanoparticle-loaded anti-HER2
immunoliposomes, Breast Cancer Res. Treat. 113 (2009) 435–441.
https://doi.org/10.1007/s10549-008-9948-x

[90] C.L. Dennis, A.J. Jackson, J.A. Borchers, P.J. Hoopes, R. Strawbridge, A.R.
Foreman, J. van Lierop, C. Grüttner, R. Ivkov, Nearly complete regression of tumors
via collective behavior of magnetic nanoparticles in hyperthermia, Nanotechnology
20 (2009) 395103. https://doi.org/10.1088/0957-4484/20/39/395103

[91] P.J. Hoopes, J.A. Tate, J.A. Ogden, R.R. Strawbridge, S.N. Fiering, A.A. Petryk, S.M. Cassim, A.J. Giustini, E. Demidenko, R. Ivkov, S. Barry, P. Chinn, A. Foreman, Assessment of intratumor non-antibody directed iron oxide nanoparticle hyperthermia cancer therapy and antibody directed IONP uptake in murine and human cells, in: T.P. Ryan (Ed.), Proc. SPIE--the Int. Soc. Opt. Eng., 2009: p. 71810P. https://doi.org/10.1117/12.812056

[92] D.-H. Chen, M.-H. Liao, Preparation and characterization of YADH-bound magnetic nanoparticles, J. Molecular Catal. B, Enzymatic 16 (2002) 283-291. https://doi.org/10.1016/S1381-1177(01)00074-1

[93] L.M. Rossi, A.D. Quach, Z. Rosenzweig, Glucose oxidase?magnetite nanoparticle bioconjugate for glucose sensing, Anal. Bioanal. Chem. 380 (2004) 606–613. https://doi.org/10.1007/s00216-004-2770-3

[94] M. Bilal, Y. Zhao, T. Rasheed, H.M.N. Iqbal, Magnetic nanoparticles as versatile carriers for enzymes immobilization: A review, Int. J. Biol. Macromol. 120 (2018) 2530–2544. https://doi.org/10.1016/J.IJBIOMAC.2018.09.025

[95] L. Cao, Carrier-bound Immobilized Enzymes, Wiley, 2005. https://doi.org/10.1002/3527607668

[96] U.T. Bornscheuer, Immobilizing Enzymes: How to create more suitable biocatalysts, Angew. Chemie Int. Ed. 42 (2003) 3336–3337. https://doi.org/10.1002/anie.200301664

[97] R.V. Mehta, Synthesis of magnetic nanoparticles and their dispersions with special reference to applications in biomedicine and biotechnology, Mater. Sci. Eng. C. 79 (2017) 901–916. https://doi.org/10.1016/j.msec.2017.05.135

[98] X. Li, J. Wei, K.E. Aifantis, Y. Fan, Q. Feng, F.-Z. Cui, F. Watari, Current investigations into magnetic nanoparticles for biomedical applications, J. Biomed. Mater. Res. Part A. 104 (2016) 1285–1296. https://doi.org/10.1002/jbm.a.35654

[99] K. Khoshnevisan, F. Vakhshiteh, M. Barkhi, H. Baharifar, E. Poor-Akbar, N. Zari, H. Stamatis, A.-K. Bordbar, Immobilization of cellulase enzyme onto magnetic nanoparticles: Applications and recent advances, Mol. Catal. 442 (2017) 66–73. https://doi.org/10.1016/J.MCAT.2017.09.006

[100] T. Kang, F. Li, S. Baik, W. Shao, D. Ling, T. Hyeon, Surface design of magnetic nanoparticles for stimuli-responsive cancer imaging and therapy, Biomaterials 136 (2017) 98–114. https://doi.org/10.1016/J.BIOMATERIALS.2017.05.013

[101] L. Maldonado-Camargo, M. Unni, C. Rinaldi, Magnetic Characterization of Iron

Oxide Nanoparticles for Biomedical Applications, in: Methods Mol. Biol., 2017: pp. 47–71. https://doi.org/10.1007/978-1-4939-6840-4_4

[102] K.W. Huang, J.J. Chieh, C.K. Yeh, S.H. Liao, Y.Y. Lee, P.-Y. Hsiao, W.C. Wei, H.C. Yang, H.E. Horng, Ultrasound-induced magnetic imaging of tumors targeted by biofunctional magnetic nanoparticles, ACS Nano 11 (2017) 3030–3037. https://doi.org/10.1021/acsnano.6b08730

[103] A. Rafati, A. Zarrabi, P. Gill, Fabrication of DNA nanotubes with an array of exterior magnetic nanoparticles, Mater. Sci. Eng. C. 79 (2017) 216–220. https://doi.org/10.1016/j.msec.2017.05.044

[104] P. Das, M. Colombo, D. Prosperi, Recent advances in magnetic fluid hyperthermia for cancer therapy, Colloids Surf. B Biointerfaces. 174 (2019) 42–55. https://doi.org/10.1016/J.COLSURFB.2018.10.051

[105] Y. Huang, K. Mao, B. Zhang, Y. Zhao, Superparamagnetic iron oxide nanoparticles conjugated with folic acid for dual target-specific drug delivery and MRI in cancer theranostics, Mater. Sci. Eng. C. 70 (2017) 763–771. https://doi.org/10.1016/J.MSEC.2016.09.052

[106] A. Zarrin, S. Sadighian, K. Rostamizadeh, O. Firuzi, M. Hamidi, S. Mohammadi-Samani, R. Miri, Design, preparation, and in vitro characterization of a trimodally-targeted nanomagnetic onco-theranostic system for cancer diagnosis and therapy, Int. J. Pharm. 500 (2016) 62–76. https://doi.org/10.1016/J.IJPHARM.2015.12.051

[107] R.-M. Yang, C. Fu, J. Fang, X. Xu, X. Wei, W. Tang, X. Jiang, L. Zhang, Hyaluronan-modified superparamagnetic iron oxide nanoparticles for bimodal breast cancer imaging and photothermal therapy, Int. J. Nanomedicine 12 (2016) 197–206. https://doi.org/10.2147/IJN.S121249

[108] D. Vyas, N. Lopez-Hisijos, S. Gandhi, M. El-Dakdouki, M.D. Basson, M.F. Walsh, X. Huang, A.K. Vyas, L.S. Chaturvedi, Doxorubicin-hyaluronan conjugated super-paramagnetic iron oxide nanoparticles (DOX-HA-SPION) Enhanced cytoplasmic uptake of doxorubicin and modulated apoptosis, il-6 release and nf-kappab activity in human MDA-MB-231 breast cancer cells, J. Nanosci. Nanotechnol. 15 (2015) 6413–6422. https://doi.org/10.1166/jnn.2015.10834

[109] M. Khafaji, M. Vossoughi, M.R. Hormozi-Nezhad, R. Dinarvand, F. Börrnert, A. Irajizad, A new bifunctional hybrid nanostructure as an active platform for photothermal therapy and MR imaging, Sci. Rep. 6 (2016) 27847. https://doi.org/10.1038/srep27847

[110] H.L. Xu, K.L. Mao, Y.P. Huang, J.J. Yang, J. Xu, P.-P. Chen, Z.L. Fan, S. Zou, Z.-Z. Gao, J.-Y. Yin, J. Xiao, C.-T. Lu, B.-L. Zhang, Y.-Z. Zhao, Glioma-targeted superparamagnetic iron oxide nanoparticles as drug-carrying vehicles for theranostic effects, Nanoscale 8 (2016) 14222–14236. https://doi.org/10.1039/C6NR02448C

[111] Z.R. Stephen, C.J. Dayringer, J.J. Lim, R.A. Revia, M. V. Halbert, M. Jeon, A. Bakthavatsalam, R.G. Ellenbogen, M. Zhang, Approach to rapid synthesis and functionalization of iron oxide nanoparticles for high gene transfection, ACS Appl. Mater. Interfaces 8 (2016) 6320–6328. https://doi.org/10.1021/acsami.5b10883

[112] K. Yu, M. Lin, H.-J. Lee, K.-S. Tae, B.-S. Kang, J. Lee, N. Lee, Y. Jeong, S.-Y. Han, D. Kim, K.S. Yu, M.M. Lin, H.-J. Lee, K.-S. Tae, B.-S. Kang, J.H. Lee, N.S. Lee, Y.G. Jeong, S.-Y. Han, D.K. Kim, Receptor-Mediated Endocytosis by Hyaluronic Acid@Superparamagnetic Nanovetor for Targeting of CD44-Overexpressing Tumor Cells, Nanomaterials 6 (2016) 149. https://doi.org/10.3390/nano6080149

[113] M.-K. Yoo, I.-K. Park, H.-T. Lim, S.-J. Lee, H.-L. Jiang, Y.-K. Kim, Y.-J. Choi, M.-H. Cho, C.-S. Cho, Folate–PEG–superparamagnetic iron oxide nanoparticles for lung cancer imaging, Acta Biomater. 8 (2012) 3005–3013. https://doi.org/10.1016/J.ACTBIO.2012.04.029

[114] A. Akbarzadeh, M. Samiei, S.W. Joo, M. Anzaby, Y. Hanifehpour, H.T. Nasrabadi, S. Davaran, Synthesis, characterization and in vitro studies of doxorubicin-loaded magnetic nanoparticles grafted to smart copolymers on A549 lung cancer cell line, J. Nanobiotechnology 10 (2012) 46. https://doi.org/10.1186/1477-3155-10-46

[115] L. Mühleisen, M. Alev, H. Unterweger, D. Subatzus, M. Pöttler, R. Friedrich, C. Alexiou, C. Janko, L. Mühleisen, M. Alev, H. Unterweger, D. Subatzus, M. Pöttler, R.P. Friedrich, C. Alexiou, C. Janko, Analysis of hypericin-mediated effects and implications for targeted photodynamic therapy, Int. J. Mol. Sci. 18 (2017) 1388. https://doi.org/10.3390/ijms18071388

[116] H. Unterweger, D. Subatzus, R. Tietze, C. Janko, M. Poettler, A. Stiegelschmitt, M. Schuster, C. Maake, A. Boccaccini, C. Alexiou, Hypericin-bearing magnetic iron oxide nanoparticles for selective drug delivery in photodynamic therapy, Int. J. Nanomedicine. 10 (2015) 6985. https://doi.org/10.2147/IJN.S92336

[117] D. Luong, S. Sau, P. Kesharwani, A.K. Iyer, Polyvalent folate-dendrimer-coated iron oxide theranostic nanoparticles for simultaneous magnetic resonance imaging and precise cancer cell targeting, Biomacromolecules 18 (2017) 1197–1209. https://doi.org/10.1021/acs.biomac.6b01885

[118] J. Mosafer, K. Abnous, M. Tafaghodi, A. Mokhtarzadeh, M. Ramezani, In vitro and in vivo evaluation of anti-nucleolin-targeted magnetic PLGA nanoparticles loaded with doxorubicin as a theranostic agent for enhanced targeted cancer imaging and therapy, Eur. J. Pharm. Biopharm. 113 (2017) 60–74. https://doi.org/10.1016/J.EJPB.2016.12.009

[119] M. Azhdarzadeh, F. Atyabi, A.A. Saei, B.S. Varnamkhasti, Y. Omidi, M. Fateh, M. Ghavami, S. Shanehsazzadeh, R. Dinarvand, Theranostic MUC-1 aptamer targeted gold coated superparamagnetic iron oxide nanoparticles for magnetic resonance imaging and photothermal therapy of colon cancer, Colloids Surf. B Biointerfaces. 143 (2016) 224–232. https://doi.org/10.1016/J.COLSURFB.2016.02.058

[120] J.H. Maeng, D.-H. Lee, K.H. Jung, Y.-H. Bae, I.-S. Park, S. Jeong, Y.-S. Jeon, C.-K. Shim, W. Kim, J. Kim, J. Lee, Y.-M. Lee, J.-H. Kim, W.-H. Kim, S.-S. Hong, Multifunctional doxorubicin loaded superparamagnetic iron oxide nanoparticles for chemotherapy and magnetic resonance imaging in liver cancer, Biomaterials 31 (2010) 4995–5006. https://doi.org/10.1016/J.BIOMATERIALS.2010.02.068

[121] U.M. Mahajan, S. Teller, M. Sendler, R. Palankar, C. van den Brandt, T. Schwaiger, J.-P. Kühn, S. Ribback, G. Glöckl, M. Evert, W. Weitschies, N. Hosten, F. Dombrowski, M. Delcea, F.U. Weiss, M.M. Lerch, J. Mayerle, Tumour-specific delivery of siRNA-coupled superparamagnetic iron oxide nanoparticles, targeted against PLK1, stops progression of pancreatic cancer, Gut. 65 (2016) 1838–1849. https://doi.org/10.1136/gutjnl-2016-311393

[122] L. Mohammed, H.G. Gomaa, D. Ragab, J. Zhu, Magnetic nanoparticles for environmental and biomedical applications: A review, Particuology 30 (2017) 1–14. https://doi.org/10.1016/j.partic.2016.06.001

[123] L. Wu, A. Mendoza-Garcia, Q. Li, S. Sun, Organic phase syntheses of magnetic nanoparticles and their applications, Chem. Rev. 116 (2016) 10473–10512. https://doi.org/10.1021/acs.chemrev.5b00687

[124] G.C. Papaefthymiou, Nanoparticle magnetism, Nano Today 4 (2009) 438–447. https://doi.org/10.1016/J.NANTOD.2009.08.006

[125] A.-H. Lu, E.L. Salabas, F. Schüth, Magnetic nanoparticles: Synthesis, Protection, functionalization, and application, Angew. Chemie Int. Ed. 46 (2007) 1222–1244. https://doi.org/10.1002/anie.200602866

[126] Z. Hedayatnasab, F. Abnisa, W. Mohd Ashri Wan Daud, Review on magnetic nanoparticles for magnetic nanofluid hyperthermia application, Mater. Design 123 (2017) 174-196. https://doi.org/10.1016/j.matdes.2017.03.036

[127] A.K. Singh, O.N. Srivastava, K. Singh, Shape and size-dependent magnetic properties of Fe_3O_4 nanoparticles synthesized using piperidine, Nanoscale Res. Lett. 12 (2017) 298. https://doi.org/10.1186/s11671-017-2039-3

[128] D.P. Joshi, G. Pant, N. Arora, S. Nainwal, Effect of solvents on morphology, magnetic and dielectric properties of (α-Fe_2O_3@SiO_2) core-shell nanoparticles, Heliyon 3 (2017) e00253. https://doi.org/10.1016/j.heliyon.2017.e00253

[129] Y. Jun, J. Seo, J. Cheon, Nanoscaling laws of magnetic nanoparticles and their applicabilities in biomedical sciences, Acc. Chem. Res. 41 (2008) 179–189. https://doi.org/10.1021/ar700121f

[130] J.-H. Park, S.-H. Shin, S.-H. Kim, J.-K. Park, J.-W. Lee, J.-H. Shin, J.-H. Park, S.-W. Kim, H.-J. Choi, K.-S. Lee, J.-C. Ro, C. Park, S.-J. Suh, Effect of Synthesis Time and Composition on Magnetic Properties of FeCo Nanoparticles by Polyol Method, J. Nanosci. Nanotechnol. 18 (2018) 7115–7119. https://doi.org/10.1166/jnn.2018.15477

[131] N. Manuchehrabadi, Z. Gao, J. Zhang, H.L. Ring, Q. Shao, F. Liu, M. McDermott, A. Fok, Y. Rabin, K.G.M. Brockbank, M. Garwood, C.L. Haynes, J.C. Bischof, Improved tissue cryopreservation using inductive heating of magnetic nanoparticles, Sci. Transl. Med. 9 (2017) eaah4586. https://doi.org/10.1126/scitranslmed.aah4586

[132] M. Virumbrales-Del Olmo, A. Delgado-Cabello, A. Andrada-Chacón, J. Sánchez-Benítez, E. Urones-Garrote, V. Blanco-Gutiérrez, M.J. Torralvo, R. Sáez-Puche, Effect of composition and coating on the interparticle interactions and magnetic hardness of MFe_2O_4 (M = Fe, Co, Zn) nanoparticles, Phys. Chem. Chem. Phys. 19 (2017) 8363–8372. https://doi.org/10.1039/c6cp08743d

[133] S. Khoee, Y. Bagheri, A. Hashemi, Composition controlled synthesis of PCL-PEG Janus nanoparticles: magnetite nanoparticles prepared from one-pot photo-click reaction, Nanoscale 7 (2015) 4134–48. https://doi.org/10.1039/c4nr06590e

[134] J. Nowak, F. Wiekhorst, L. Trahms, S. Odenbach, The influence of hydrodynamic diameter and core composition on the magnetoviscous effect of biocompatible ferrofluids, J. Phys. Condens. Matter. 26 (2014) 176004. https://doi.org/10.1088/0953-8984/26/17/176004

[135] L.H. Reddy, J.L. Arias, J. Nicolas, P. Couvreur, Magnetic nanoparticles: Design and characterization, toxicity and biocompatibility, pharmaceutical and biomedical applications, Chem. Rev. 112 (2012) 5818–5878. https://doi.org/10.1021/cr300068p

[136] J. Xu, J. Sun, Y. Wang, J. Sheng, F. Wang, M. Sun, Application of iron magnetic

nanoparticles in protein immobilization, Molecules 19 (2014) 11465–11486. https://doi.org/10.3390/molecules190811465

[137] D. Ni, W. Bu, E.B. Ehlerding, W. Cai, J. Shi, Engineering of inorganic nanoparticles as magnetic resonance imaging contrast agents, Chem. Soc. Rev. 46 (2017) 7438–7468. https://doi.org/10.1039/c7cs00316a

[138] W. Wu, C.Z. Jiang, V.A.L. Roy, Designed synthesis and surface engineering strategies of magnetic iron oxide nanoparticles for biomedical applications, Nanoscale 8 (2016) 19421–19474. https://doi.org/10.1039/c6nr07542h

[139] L. Rao, B. Cai, L.-L. Bu, Q.-Q. Liao, S.-S. Guo, X.-Z. Zhao, W.-F. Dong, W. Liu, Microfluidic electroporation-facilitated synthesis of erythrocyte membrane-coated magnetic nanoparticles for enhanced imaging-guided cancer therapy, ACS Nano 11 (2017) 3496–3505. https://doi.org/10.1021/acsnano.7b00133

[140] Y. Xu, J. Sherwood, Y. Qin, R.A. Holler, Y. Bao, A general approach to the synthesis and detailed characterization of magnetic ferrite nanocubes, Nanoscale 7 (2015) 12641–12649. https://doi.org/10.1039/C5NR03096J

[141] A. Sathya, P. Guardia, R. Brescia, N. Silvestri, G. Pugliese, S. Nitti, L. Manna, T. Pellegrino, $Co_x Fe_{3-x}O_4$ Nanocubes for theranostic applications: effect of cobalt content and particle size, Chem. Mater. 28 (2016) 1769–1780. https://doi.org/10.1021/acs.chemmater.5b04780

[142] A. López-Ortega, A.G. Roca, P. Torruella, M. Petrecca, S. Estradé, F. Peiró, V. Puntes, J. Nogués, Galvanic replacement onto complex metal-oxide nanoparticles: Impact of water or other oxidizers in the formation of either fully dense onion-like or multicomponent hollow MnO_x/FeO_x structures, Chem. Mater. 28 (2016) 8025–8031. https://doi.org/10.1021/acs.chemmater.6b03765

[143] M.F. Casula, E. Conca, I. Bakaimi, A. Sathya, M.E. Materia, A. Casu, A. Falqui, E. Sogne, T. Pellegrino, A.G. Kanaras, Manganese doped-iron oxide nanoparticle clusters and their potential as agents for magnetic resonance imaging and hyperthermia, Phys. Chem. Chem. Phys. 18 (2016) 16848–16855. https://doi.org/10.1039/C6CP02094A

[144] S. Yang, J.-T. Jiang, C.-Y. Xu, Y. Wang, Y.-Y. Xu, L. Cao, L. Zhen, Synthesis of Zn(II)-Doped Magnetite Leaf-Like Nanorings for Efficient Electromagnetic Wave Absorption, Sci. Rep. 7 (2017) 45480. https://doi.org/10.1038/srep45480

[145] M. Jang, G. Cao, Deposition of magnetic nanoparticles suspended in the gas phase on a specific target area, Environ. Sci. Technol. 40 (2006) 6730–7

https://doi.org/10.1021/es060018v

[146] S.H. Baker, M.S. Kurt, M. Roy, M.R. Lees, C. Binns, Structure and magnetism in Cr-embedded Co nanoparticles, J. Phys. Condens. Matter. 28 (2016) 046003. https://doi.org/10.1088/0953-8984/28/4/046003

[147] S. Bartling, C. Yin, I. Barke, K. Oldenburg, H. Hartmann, V. von Oeynhausen, M.-M. Pohl, K. Houben, E.C. Tyo, S. Seifert, P. Lievens, K.-H. Meiwes-Broer, S. Vajda, Pronounced size dependence in structure and morphology of gas-phase produced, partially oxidized cobalt nanoparticles under catalytic reaction conditions, ACS Nano 9 (2015) 5984–5998. https://doi.org/10.1021/acsnano.5b00791

[148] S. Majidi, F. Zeinali Sehrig, S.M. Farkhani, M. Soleymani Goloujeh, A. Akbarzadeh, Current methods for synthesis of magnetic nanoparticles, Artif. Cells, Nanomedicine, Biotechnol. 44 (2014) 1–13. https://doi.org/10.3109/21691401.2014.982802

[149] S. Jeon, R. Subbiah, T. Bonaedy, S. Van, K. Park, K. Yun, Surface functionalized magnetic nanoparticles shift cell behavior with on/off magnetic fields, J. Cell. Physiol. 233 (2018) 1168–1178. https://doi.org/10.1002/jcp.25980

[150] M. Ul-Islam, M.W. Ullah, S. Khan, S. Manan, W.A. Khattak, W. Ahmad, N. Shah, J.K. Park, Current advancements of magnetic nanoparticles in adsorption and degradation of organic pollutants, Environ. Sci. Pollut. Res. 24 (2017) 12713–12722. https://doi.org/10.1007/s11356-017-8765-3

[151] S. Laurent, D. Forge, M. Port, A. Roch, C. Robic, L. Vander Elst, R.N. Muller, Magnetic iron oxide nanoparticles: Synthesis, stabilization, vectorization, physicochemical characterizations, and biological applications, Chem. Rev. 108 (2008) 2064–2110. https://doi.org/10.1021/cr068445e

[152] T.K. Indira, P.K. Lakshmi, Magnetic nanoparticles-A review, Int. J. Pharmaceutical Nanotechnol. 3 (2010) 1035-1042. https://ijpsnonline.com/Issues/1035_full.pdf

[153] R. López, M. Pineda, G. Hurtado, R. León, S. Fernández, H. Saade, D. Bueno, Chitosan-Coated Magnetic Nanoparticles Prepared in One Step by Reverse Microemulsion Precipitation, Int. J. Mol. Sci. 14 (2013) 19636–19650. https://doi.org/10.3390/ijms141019636

[154] M. Pineda, S. Torres, L. López, F. Enríquez-Medrano, R. de León, S. Fernández, H. Saade, R. López, Chitosan-coated magnetic nanoparticles prepared in one-step by precipitation in a high-aqueous phase content reverse microemulsion, Molecules 19

(2014) 9273–9287. https://doi.org/10.3390/molecules19079273

[155] C. Okoli, M. Sanchez-Dominguez, M. Boutonnet, S. Järås, C. Civera, C. Solans, G.R. Kuttuva, Comparison and functionalization study of microemulsion-prepared magnetic iron oxide nanoparticles, Langmuir 28 (2012) 8479–8485. https://doi.org/10.1021/la300599q

[156] Y. Liu, X. Shen, Synthesis and application of surface-modified NiFe nanoparticles as a new magnetic nano adsorbent for the removal of nickel ions from aqueous solution, Water Sci. Technol. 76 (2017) 2851–2857. https://doi.org/10.2166/wst.2017.453

[157] J. Li, S. Wang, X. Shi, M. Shen, Aqueous-phase synthesis of iron oxide nanoparticles and composites for cancer diagnosis and therapy, Adv. Colloid Interface Sci. 249 (2017) 374–385. https://doi.org/10.1016/j.cis.2017.02.009

[158] E. Kılınç, Fullerene C_{60} functionalized γ-Fe_2O_3 magnetic nanoparticle: Synthesis, characterization, and biomedical applications, Artif. Cells, Nanomedicine, Biotechnol. 44 (2014) 298–304. https://doi.org/10.3109/21691401.2014.948182

[159] J. Sánchez, D.A. Cortés-Hernández, J.C. Escobedo-Bocardo, R.A. Jasso-Terán, A. Zugasti-Cruz, Bioactive magnetic nanoparticles of Fe–Ga synthesized by sol–gel for their potential use in hyperthermia treatment, J. Mater. Sci. Mater. Med. 25 (2014) 2237–2242. https://doi.org/10.1007/s10856-014-5197-1

[160] E. Ozyilmaz, S. Sayin, M. Arslan, M. Yilmaz, Improving catalytic hydrolysis reaction efficiency of sol–gel-encapsulated Candida rugosa lipase with magnetic β-cyclodextrin nanoparticles, Colloids Surf. B Biointerfaces. 113 (2014) 182–189. https://doi.org/10.1016/j.colsurfb.2013.08.019

[161] W. Wang, Y. Zhang, Q. Yang, M. Sun, X. Fei, Y. Song, Y. Zhang, Y. Li, Fluorescent and colorimetric magnetic microspheres as nanosensors for Hg^{2+} in aqueous solution prepared by a sol–gel grafting reaction and host–guest interaction, Nanoscale. 5 (2013) 4958. https://doi.org/10.1039/c3nr00580a

[162] G. Hemery, A.C. Keyes, E. Garaio, I. Rodrigo, J.A. Garcia, F. Plazaola, E. Garanger, O. Sandre, Tuning sizes, morphologies, and magnetic properties of monocore versus multicore iron oxide nanoparticles through the controlled addition of water in the polyol synthesis, Inorg. Chem. 56 (2017) 8232–8243. https://doi.org/10.1021/acs.inorgchem.7b00956

[163] G. Zhang, Q. Zhang, T. Cheng, X. Zhan, F. Chen, Polyols-infused slippery surfaces based on magnetic Fe_3O_4-functionalized polymer hybrids for enhanced

multifunctional anti-icing and deicing properties, Langmuir 34 (2018) 4052–4058. https://doi.org/10.1021/acs.langmuir.8b00286

[164] G. Martínez, A. Malumbres, R. Mallada, J.L. Hueso, S. Irusta, O. Bomatí-Miguel, J. Santamaría, Use of a polyol liquid collection medium to obtain ultrasmall magnetic nanoparticles by laser pyrolysis, Nanotechnology 23 (2012) 425605. https://doi.org/10.1088/0957-4484/23/42/425605

[165] R. Hachani, M. Lowdell, M. Birchall, A. Hervault, D. Mertz, S. Begin-Colin, N.T.K. Thanh, Polyol synthesis, functionalisation, and biocompatibility studies of superparamagnetic iron oxide nanoparticles as potential MRI contrast agents, Nanoscale 8 (2016) 3278–87. https://doi.org/10.1039/c5nr03867g

[166] A. Mardinoglu, P.J. Cregg, Modelling the effect of SPION size in a stent assisted magnetic drug targeting system with interparticle interactions, Sci. World J. 2015 (2015) 1–7. https://doi.org/10.1155/2015/618658

[167] M. Jiao, J. Zeng, L. Jing, C. Liu, M. Gao, Flow synthesis of biocompatible Fe_3O_4 nanoparticles: Insight into the effects of residence time, fluid velocity, and tube reactor dimension on particle size distribution, Chem. Mater. 27 (2015) 1299–1305. https://doi.org/10.1021/cm504313c

[168] C. Pascal, and J. L. Pascal, F. Favier, M.L.E. Moubtassim, C. Payen, Electrochemical synthesis for the control of γ-Fe_2O_3 nanoparticle size, morphology, microstructure, and magnetic behavior, Chem. Mater. 11 (1998) 141-147. https://doi.org/10.1021/CM980742F

[169] M.S. Balula, J.A. Gamelas, H.M. Carapuça, A.M.V. Cavaleiro, W. Schlindwein, Electrochemical behaviour of first row transition metal substituted polyoxotungstates: A comparative study in acetonitrile, Eur. J. Inorg. Chem. 2004 (2004) 619–628. https://doi.org/10.1002/ejic.200300292

[170] E. Mazarío, P. Herrasti, M.P. Morales, N. Menéndez, Synthesis and characterization of $CoFe_2O_4$ ferrite nanoparticles obtained by an electrochemical method, Nanotechnology 23 (2012) 355708. https://doi.org/10.1088/0957-4484/23/35/355708

[171] J.G. Ovejero, A. Mayoral, M. Cañete, M. García, A. Hernando, P. Herrasti, Electrochemical synthesis and magnetic properties of MFe_2O_4 (M = Fe, Mn, Co, Ni) nanoparticles for potential biomedical applications, J. Nanosci. Nanotechnol. 19 (2019) 2008–2015. https://doi.org/10.1166/jnn.2019.15313

[172] D.-H. Kim, J.-S. Park, M.-S. Kang, Continuous Preparation of Water-Dispersible

Magnetite Nanoparticles by Electrochemical Synthesis, J. Nanosci. Nanotechnol. 18 (2018) 5721–5725. https://doi.org/10.1166/jnn.2018.15398

[173] J. Pinkas, V. Reichlova, R. Zboril, Z. Moravec, P. Bezdicka, J. Matejkova, Sonochemical synthesis of amorphous nanoscopic iron(III) oxide from $Fe(acac)_3$, Ultrason. Sonochem. 15 (2008) 257–264. https://doi.org/10.1016/j.ultsonch.2007.03.009

[174] E. Tombácz, R. Turcu, V. Socoliuc, L. Vékás, Magnetic iron oxide nanoparticles: Recent trends in design and synthesis of magnetoresponsive nanosystems, Biochem. Biophys. Res. Commun. 468 (2015) 442–453. https://doi.org/10.1016/j.bbrc.2015.08.030

[175] R. Cassano, S. Mellace, M. Marrelli, F. Conforti, S. Trombino, α-Tocopheryl linolenate solid lipid nanoparticles for the encapsulation, protection, and release of the omega-3 polyunsaturated fatty acid: in vitro anti-melanoma activity evaluation, Colloids Surf. B Biointerfaces 151 (2017) 128–133. https://doi.org/10.1016/j.colsurfb.2016.11.043

Magnetochemistry - Materials and Applications
Materials Research Foundations **66** (2020) 173-216

Materials Research Forum LLC
https://doi.org/10.21741/9781644900611-5

Chapter 5

Magnetic Nanoparticles in Analytical Chemistry

D.M.A. Neto[1], J.S. Rocha[1], P.B.A. Fechine[1], Leonardo Vivas[2], Dinesh Pratap Singh[2*] and R.M. Freire[3*]

[1]Grupo de Química de Materiais Avançados (GQMAT)- Departamento de Química Analítica e Físico-Química, Universidade Federal do Ceará – UFC, Campus do Pici, Fortaleza, CE, Brazil

[2]Millennium Institute for Research in Optics (MIRO), Department of Physics, University of Santiago Chile (USACH), Avenida Ecuador 3493, Estacion Central, Santiago, Chile

[3]Department of Physics/CEDENNA, University of Santiago Chile (USACH), Avenida Ecuador 3493, Estacion Central, Santiago, Chile

*rafael.m.freire@gmail.com/dineshpsingh@gmail.com

Abstract

Recently magnetic nanoparticles (MNPs) have attracted significant attention for various applications in the analytical chemistry. In this regard this chapter first gives a brief overview of the use of these kinds of nanomaterials in the relevant research areas and subsequently the readers can find the different methodologies to produce MNPs having distinct structural, morphological and surface characteristics which are important to produce for corresponding applications. Various analytical chemistry-based applications of MNPs such as magnetic separation, capture and pre-concentration etc. are described besides the sensor application.

Keywords

Magnetic Nanoparticles, Characteristics, Pre-Concentration, Magnetic Capture, Sensors

Contents

1.   Introduction

Materials produced by using of nanotechnology can be made more durable, better electrical conductors, lighter and offers the guarantee of developing multifunctional materials. Nanotechnology holds a great promise for industry applications. For example, automotive [1] and aerospace [2] researches have been focused in design and manufacturing of vehicles using nanotechnology: weight reduction, energy efficiency, physical sensors, nanocoatings to enhance the durability of metals and catalysts for emissions cleaning. Many applications are associated with several different types of nanomaterials, such as carbon-based (graphene, carbon nanotubes and nanofibers), nanocrystals, nanocomposite, polymeric micelles, nanoparticles (NPs) etc. Among them, the use of NPs must be highlighted since it shows size-dependent properties, which emerge due to their high surface-to-volume ratio and quantum effect. Surface plasmon resonance in metallic NPs (Au and Ag), quantum confinement effect in fluorescent quantum dots (CdSe and ZnS) or the superparamagnetism in NPs (Fe_3O_4 and FeCo) are some examples of properties arising from smaller sizes [3–7]. Therefore, it is easy to see that a great control over the properties of the NPs allows tuning the properties of the final desired product, which is highly desired.

By definition, to be considered a NP, a material must have, at least, one dimension (or a quasi-zero-dimensional, 0D) that measures is in the size range between 1 and 100 nanometres (nm). In this class, the magnetic nanoparticles (MNPs) can be placed as one of the most studied nanomaterials and with the highest potential of application in the industry. Actually, MNPs are present everywhere and inside us. For instance, there are MNPs in migratory birds [8], ants (head, thorax and abdomen) [9], magnetotactic bacteria [10], human brain [11] and extra-terrestrial dust [12]. These type of NPs can exhibit different morphologies, such as nanoplatelets, nanorods, hollow nanospheres, nanocubes, peony-like and coral-like nanostructures [13,14]. Further, there is a wide range of applications in many fields. Currently, the literature displays publication reporting the use of MNPs for energy storage, target drug delivery, cancer hyperthermia therapy, magnetic resonance imaging (MRI), spintronics, fluorescent-magnetic nanocomposites for biological imaging and synthetic pigments in ceramics [15–17]. A large number of researches have studied MNPs as an potential material for environmental remediation: capture and degradation of pollutants [18]. However, one of the most important applications of these materials is the magnetic decantation, when the analyte can be harvested and separated from the rest of the solution by simply applying an external magnetic field. In addition, this kind of material can be used for chemical sensors [19].

Taking into consideration all the exposed, which is the basis for the application of MNPs in analytical chemistry [20], it is important to note the demand of surface engineering in order to trap the target molecules. Since each one can have different structural characteristics. The superficial modification of NPs is commonly called functionalization [16,20]. This process is important because the material acquires stability in the colloidal suspension, for example. Regarding analytical applications, a greater superficial area is exposed. As a consequence, the efficiency in the adsorption of the analytes may increase allowing multifunctionalities, such as fluorescence for use in chemosensors [20]. Functionalized NPs can also prevent their capture by the immune system [21]. Moreover, it is possible to obtain hydrophobic, conductive and anticorrosive systems for other applications. There are many coating agents and strategies of surface functionalization, that can leave reactive functional groups ($-NH_2$, $-COOH$, $-SH$ etc.) for a specific application. Also, it can be used for small molecules, polymers, amino acids, dendrimers, surfactants, biomolecules, porphyrins etc. [21–26].

There is another method very similar to the functionalization, which is called encapsulation. However, many authors use both terms for the same purpose. In general, the encapsulation is performed with inorganic or polymeric materials [17,24,27,28]. This method is commonly used for MNPs that are not air stable (easily oxidized). For example, the literature reports core-shell-based nanosystem like $Fe_3O_4@SiO_2$ [17,29], $FeCo@Al_2O_3$ [30] and $Fe@Au$ [31]. At this point it is worth to mention that magnetic bioseparation using core-shell MNPs is a powerful analytical tool. Since allows to purify biomolecules to further analyze their functionalities in diagnostics [32]. Therefore, the choice of the synthesis methodology and their parameters control to obtain MNPs is an important step to obtain homogeneity distribution, particle size and morphology. These properties will determine the chemical and physical properties of the material, as well as the final product manufactured using the synthesized NPs.

Besides the protocol to produce the MNPs with desired characteristics, the surface of the material is a critical point as already mentioned. For this purpose, the protective shell must also provide electrostatic properties, which is frequently obtained performing a surface functionalization. Some researchers have used antibody immobilized on MNPs and use for antigen capture to obtain selectivity [33], for example. Although, functionalized MNPs can also be used for extraction of oil and oil droplets dispersed in oilfield produced water from aqueous environments by magnetic decantation [34–36]. These nanodemulsifiers can also be operated in crude oil to promote eco-friendly process for oil absorption. Recent studies have showed that metal-organic frameworks (MOFs) may be used for removal of heavy metal ions (As(III), As(IV), Pb(II), Hg(II), Cd(II), Ni(II) and Zn(II)) and radionuclides (U(VI), Se(VI), Cs(I) and Th(IV)) from aquatic environment [37,38]. Some examples of composite systems with these synthetic porous materials applied to heavy metals removal include Fe_3O_4@TAR, Fe_3O_4@ZIF-8, Fe_3O_4@SiO_2@HKUST-1 and $CoFe_2O_4$@MIL-100(Fe) [37–41]. In general, MNPs cores were used to grow MOFs shell *in situ* or using templates. The advantage in using MOFs as absorbent is due to high potential in separation and selective adsorption of analytes from water, besides their capacities ranging anywhere from hundreds to a few mg g^{-1} [37].

2.   Synthesis of MNPs

The production of MNPs is one of the most important steps to develop a successful nanostructure regarding any application. Since their characteristics may be modulated by picking one methodology or another. In this sense, the reader must know that each application demands special requirements. For instance, the literature displays the use of MNPs in the biomedical field for over 100 years. In this case, the intrinsic magnetism is a special characteristic, which allows a drug delivery into a very specific site. However, after removal of the external magnetic field, the MNPs should not aggregate in the biological environment. So, the superparamagnetism at room temperature is a characteristic highly desired once the NPs did not present remnant magnetization. Obviously, this is not the only factor to obtain the aqueous stability and MNPs surface engineering also needs to be performed. On the other hand, the nanomaterials should not have the same aqueous stability regarding analytical applications, like magnetic separation, for example. Therefore, the produced magnetic structure used for this kind of application must not be as stable as the nanosystems utilized in the biomedical field. Since this would lead to larger period of times to remove the target analyte from an investigated mixture. Additionally, MNPs characteristics, such as size, morphology and surface area, must also be observed.

Magnetochemistry - Materials and Applications   Materials Research Forum LLC
Materials Research Foundations **66** (2020) 173-216   https://doi.org/10.21741/9781644900611-5

As it can be seen, many characteristics must be observed in order to obtain an appropriate magnetic nanomaterial for analytical applications. Therefore, several methodologies to produce MNPs with distinct features can be found in the literature. Some examples include solid state, co-precipitation, solvo/hydrothermal, microwave, sonochemical, microemulsion, thermal decomposition, sputtering gas condensation and others. Each one may generate different nanomaterials regarding size, morphology, dispersity and surface area. So, to choose the right methodology is not an easy task and the reader has to take into account some additional factors. For instance, the solid-state route uses high temperatures to produce the MNPs, which raises the cost of the process. Therefore, this type of protocol may not be suitable to synthesize the required nanomaterials for analytical applications. Further, the nanomaterials obtained through the solid-state route used to have bigger sizes. As a consequence, the surface area is smaller, which is undesired. Therefore, to make the right choice, the reader must know the methodologies. So, this section will focus on the most established, as well as emergent methods to synthesize MNPs.

2.1 Co-precipitation

The co-precipitation protocol is probably one of the simplest and most common pathways used to obtain MNPs. In this sense, it is worth to mention that iron-based MNPs like magnetite (Fe_3O_4), the most used magnetic nanomaterial in the biomedical field, is usually prepared through this method. The typical synthesis is based on the chemical equation displayed below:

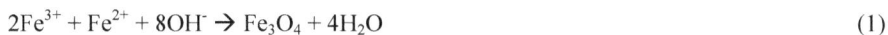

$$2Fe^{3+} + Fe^{2+} + 8OH^- \rightarrow Fe_3O_4 + 4H_2O \tag{1}$$

Firstly, a stoichiometric mixture of the ferric and ferrous salts is prepared in aqueous medium. In the second procedure, the resultant solution is heated up and a base (NaOH, KOH [42] or NH_4OH [43,44], for example) is added in order to obtain Fe_3O_4. The reaction is fast and a pH value between 8 and 14 is expected to carry out a complete precipitation of this MNP. No further thermal treatment is necessary, which is specifically for this case. Since taking into consideration other similar nanostructures, such as $CoFe_2O_4$, $MnFe_2O_4$ and $NiFe_2O_4$, a calcination process at 400°C is demanded to produce the oxide [45]. Therefore, the co-precipitation protocol results in stoichiometric hydroxides, which are subsequently converted into the corresponding oxide. The procedure may also be performed in a reverse way, where the stoichiometric mixture of the metallic salts is added into the hydroxide solution to rapidly precipitate the desired product [46].

As it can be noted, the protocol is very simple, cost-effective and easy to scale up if a large amount of MNPs is desired. However, the control over the distribution size of the MNPs is hardly reached [47]. Since the crystal growth mechanism may not be well tuned by using the co-precipitation process as described. Therefore, researchers need to study different parameters like pH, concentration and nature of the metallic precursors reaction temperature and/or ionic strength, for example [48]. Further, from a general point of view, the NPs synthesis in solution include two steps, which are nucleation and growth [49]. Taking this into consideration, it is worth to infer the nuclei is firstly formed then grow to become the desired NPs. However, if the size is very small, high surface area to ratio volume is found and agglomerate is produced in order to minimize the energy. To prevent this process, one common strategy is to add chelating agents into the mixture before starting the reaction. In this case, the chemical entity added interacts with the nuclei limiting its growth. Thus, it helps to modulate the size of the crystal formed. Small molecules like citric [50] and oleic [51] acid, dopamine [52] and others [53] are good examples of chelating agents. The use of polymers to stabilize the MNPs size is also reported in the literature [54,55].

2.2   Hydro-solvothermal

The solvothermal reactions are carried out in a reactor or autoclave under pressure and temperature values higher than 2000 psi and 200ºC, respectively. However, the term hydrothermal can be applied when the reaction is performed in aqueous media. In this regard, the MNPs are usually synthesized through oxidation of the mixed metal hydroxides, which were previously precipitated. Obviously, the conditions utilized during the procedure have a huge influence towards the desired nanomaterial. Therefore, parameters like temperature, salts nature and concentration, reaction time and solvent must be strictly controlled. Since these factors are found to affect the equilibrium between nucleation and growth processes. For example, Freire and co-authors [14] recently investigated the nanocrystal growth of $Ni_{0.5}Zn_{0.5}Fe_2O_4$ MNPs under different hydrothermal conditions (base concentration and reaction time, while the temperature was held constant at 250ºC). Under high NaOH concentration (5.2 M), the growth was found to be the predominant step, which leads to a small number of large particles. On the other hand, no significant nanocrystal growth was observed by increasing the hydrothermal reaction time from 6 to 24 h. In another work [56], the change of the solvent composition was studied. In this case, different concentrations of ethylene glycol were added in water and the mixture was used to perform the reaction. $Mn_{0.5}Zn_{0.5}Fe_2O_4$ MNPs were produced and their average diameter was found to drop along the concentration of ethylene glycol.

Magnetochemistry - Materials and Applications   Materials Research Forum LLC
Materials Research Foundations **66** (2020) 173-216   https://doi.org/10.21741/9781644900611-5

The term hydrothermal may be unprecise for some studies. Since it may lead the reader to infer that an aqueous medium was applied to carry out the reaction. For instance, Fe_3O_4 nanocrystals were synthesized in a procedure using toluene as a solvent [57]. The reaction was performed in a typical stainless-steel reaction kettle with a Teflon-liner. These MNPs were subsequently used as a seed to produce Fe_3O_4/Mn_3O_4 hybrid nanocrystals with defined anisotropic magnetic domains. In another study [58], Fe_3O_4 uniform nanorods (Fig. 1) were produced using n-octanol to carry out the reaction under pressure. Therefore, in these cases, the term solvothermal seems to be more appropriate. In general, this kind of procedure did not imply a previous precipitation of the corresponding hydroxides as observed for hydrothermal protocols. Basically, the precursors are dissolved into the organic solvent and the mixture is enclosed inside the reactor. Afterward, it is conducted to a muffle at a given temperature and time, where the solvothermal reaction is going to happen to obtain the desired nanomaterial. As it can be seen, the solvothermal protocol is easy and versatile, since different morphologies can be synthesized. Furthermore, the nucleation occurs directly in solution, which leads to high purity and crystallinity [59]. Also, the liquid phase allows the modulation of parameters like diffusion, adsorption and reaction rate. As a consequence, the control over the size and morphology is reached and the MNPs aggregation is reduced [60].

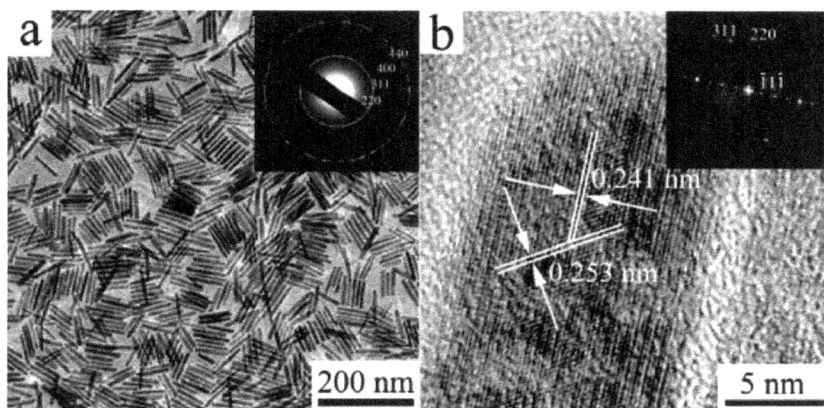

Figure 1. TEM micrograph (a) and high-resolution TEM image of the Fe_3O_4 nanorods. The inset in (a) and (b) denotes the selected area electron diffraction and fast Fourier transform pattern, respectively. Reproduced with permission from Ref. 58.

2.3   High-temperature reactions

Probably, this is one of the simplest methods to synthesize MNPs with a high monodispersity and size control. Many publications can be easily found in the literature. For instance, different monodisperse bimetallic MNPs were produced through an facile route using oleylamine as solvent, surfactant and reducing agent [61]. In this case, the acetylacetonate-based metallic precursors are dissolved in the organic medium and their reduction was carried out at 300°C for 1 h. Subsequently, the dispersion was cooled down and the product was separated, purified and re-dispersed in hexane for further characterization. TEM images displayed monodisperse $Co_{47}Pt_{53}$ MNPs of 9.5 nm were produced (Fig. 2), but the synthesis can also be performed to obtain FePt, NiPt, CuPt, and ZnPt NPs. The authors presented the methodology as a general approach to get Pt-based bimetallic NPs. However, other NPs using the same chemical elements may be obtained. Furthermore, it is well-known that monodisperse ferrite-based nanocrystals can be achieved in the presence of oleylamine, 1,2-hexadecanediol, oleic acid and phenol ether [62,63].

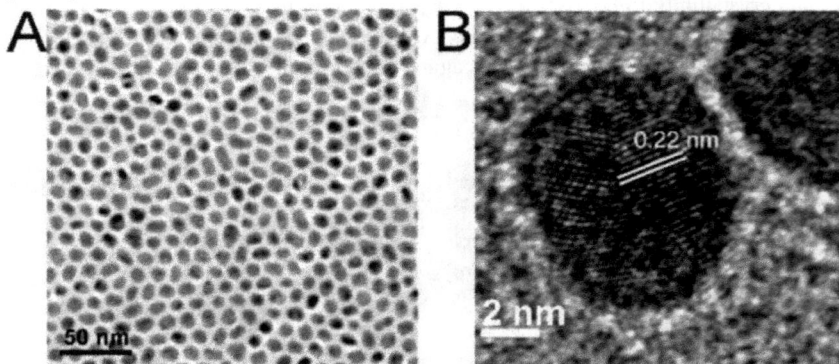

Figure 2. TEM micrograph (a) and high-resolution TEM image of the 9.5 nm $Co_{47}Pt_{53}$ MNPs. The white lines denote a spaced lattice fringes with an interplanar distance of 0.22 nm, which can be attributed to the (111) plane in the face centered cubic CoPt alloy structure. Reproduced with permission from Ref. 61.

Another approach used to obtain monodisperse MNPs is the thermal decomposition, which is a reaction frequently performed utilizing organometallic complexes as the metal precursors. For example, FePt bimetallic MNPs are normally synthesized through thermal decomposition of iron pentacarbonyl ($Fe(CO)_5$), as well as the reduction of $Pt(acac)_2$

[64,65]. In another work [66], Co@Fe core-shell MNPs were produced through sequential decomposition of $Co_2(CO)_8$ and $Fe(CO)_5$, respectively. Further, the mentioned iron-based complex may also play as a precursor to produce iron oleate, which is another chemical applied to synthesized MNPs at high temperatures [67,68]. However, it is worth to mention that metallic MNPs are very oxygen-reactive species, which leads to a fast oxidation. Therefore, an efficient coating must be used to prevent this chemical process, since it may cause the loss of the magnetic properties. In this sense, the synthesis is usually performed using an organic compound to cover the surface of the MNPs. Also, this surface passivation process also helps to control the size and monodispersity level [69].

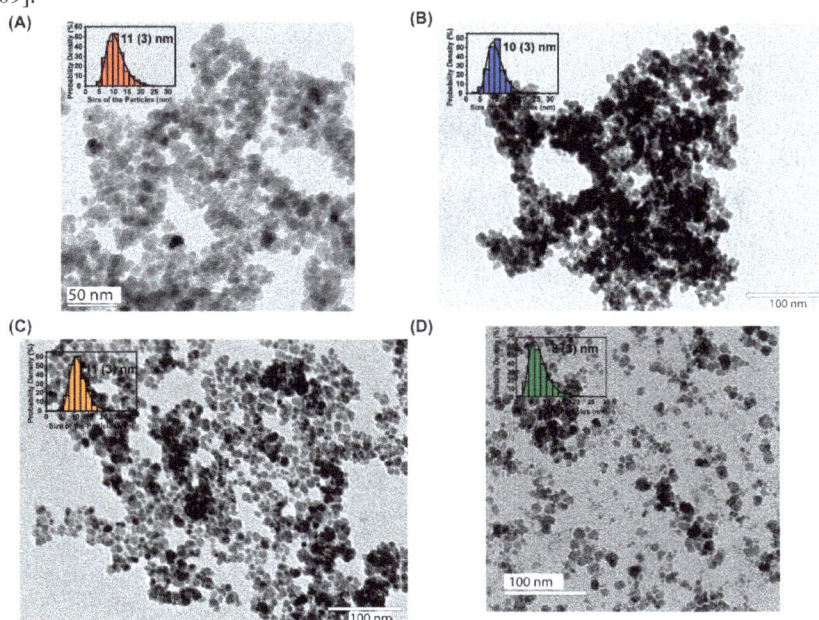

Figure 3. TEM micrographs and size-distribution graph of the Fe_3O_4 MNPs synthesized and passivated through fast sonochemical one step approach. Each picture displayed a MNP functionalized with a different coating agent: (A) sodium poly acrylate, (B) branched-polyethylenimine, (C) trisodium citrate, and (D) sodium oleate. Reproduced with permission from Ref. 70.

2.4 Sonochemical

A sonochemical approach can also be used to synthesize MNPs. For instance, in a work recently published, D.M.A. Neto and co-authors [70] performed a facile and fast methodology to produce functionalized Fe_3O_4 NPs. Basically, the iron precursors were dissolved in deionized water and ammonium hydroxide was added into the solution under sonication. A black precipitate is rapidly observed, and the system was further sonicated for more 4 minutes. Subsequently, the coating agent (polysodium acrylate, trisodium citrate, branched polyethyleneimine and sodium oleate) is added into the slurry and the dispersion was kept under sonication for another 4 minutes. The whole ultrasound-assisted process took 12 minutes 9 – 11 nm Fe_3O_4 NPs. In this case, the rapid process is possible due to the extremely high pressure and temperature spot generated by the collapse of the ultrasound originated cavities [71]. This allows the conversion of the corresponding hydroxides into the MNPs. In another study [72], a nanostructured system of general formula Fe_3O_4@NH_2-mesoporous silica@Polypyrrole/Pd (core/double shell nanocomposite) was synthesized through versatile three-step sonochemical method. Therefore, besides low period of time to produce the desired MNP, this type approach also allows the user to produce complex systems. This clearly demonstrates the enormous potential of the sonochemistry towards the production of MNPs, as well as magnetic nanostructures with complex architecture.

3. Structure and functionalization of MNPs

MNPs have brought unique improvements in analytical chemistry areas. These features come from their interactions with magnetic fields and field gradients as well as magnetically induced motion, enhance of signals and switch behaviors [73]. There are many types of MNPs available in the literature with potential to be applied in analytical chemistry. Included in these groups are oxide nanoparticles, mainly in the form of inverse spinel ferrite oxides, and metallic nanoparticles, which can be single metallic nanoparticles or metallic alloy nanoparticles [74]. Additionally, MNPs must be combined with other non-magnetic organic or inorganic compounds, in order to form multifunctional materials. These include Ag, Au, CdS, SiO_2 NPs as well as polymers, small organic molecules, biomolecules.

Among all available MNPs, magnetite (Fe_3O_4) nanoparticles are the most applied, because they are easy and cheap to prepare, and possess a high magnetic moment amidst ferrite MNPs. Additionally, Fe_3O_4 is considered biocompatible to the human body, as consequence, is the only magnetic nanoparticles approved by the U.S. Food and Drug Administration [73,75]. This fact makes Fe_3O_4 NPs the best option for nanosensors that act in real biological environments.

Materials Research Forum LLC
https://doi.org/10.21741/9781644900611-5

In terms of magnetic properties, the main improvement that the MNPs have brought is their superparamagnetism (*SPM*) behavior. In the sequence, it is explained briefly how a material becomes superparamagnetic. However, first it is important to define some terms: a) superconducting quantum interference device (SQUID) magnetometer is the machine that is used to characterize magnetic materials, and it is the record of the magnetization generated by the particles when a certain magnetic field is applied. If the nanoparticles are superparamagnetic at room temperature, the curve reveals a saturation of the magnetization and no hysteresis around the origin [73]; b) coercivity (H_C) is the field that must be applied in the negative direction to bring the magnetization of the sample back to zero [76]; c) remnant magnetization (M_r) is the residual magnetization left behind after an external magnetic field is applied. Complete information about magnetic characterization and properties is presented in these references [71,77].

The magnetic properties of a magnetic material are greatly affected when the size of the particles achieves nanometer scale. When a certain magnetic material possesses particles in a micro- or macroscale, the particle is separated by boundaries (domain walls, Fig. 4 A) of uniform magnetization within these boundaries. If this material reduces its size of the particles below so-called critical size to single-domain (D_{SD}), it is more energetically favorable to become single-domain, which means that the magnetic moment of the particle will point to the same direction and it is determined by magnetocrystalline anisotropy. Materials in the *SD* regime have greater H_C values than those ones in multi-domain regime, once the magnetization/demagnetization process are dominated by domain-wall motion [71]. As a consequence, *SD* materials have larger M_r and H_C values (Fig. 4 B). A continuous decrease in the particle size leads to decrease in the thermal energy barrier for the demagnetization process, and consequently, a decrease in the M_r and H_C values (Fig. 4 A and B). A further decrease in the size of the nanoparticles leads to a formation of *SPM* material. For this case, thermal energy is higher enough to overcome the anisotropy forces and the relaxation of the spins happens in the absence of a magnetic field. Thus M_r and H_C values are equal to zero [71] (Fig, 4 A and B). SPM behavior, which is exclusive of nanoparticles (NPs), is a breakthrough in the research field of MNPs, once the aggregation of SPM MNPs through magnetic forces is absent or very small. This fact made possible the application of these MNPs in biological environment of humans [78]. Additionally, SPM behavior can be easier dispersed in liquid solvents, and the absent of M_r is important for magnetic sensors [73].

Considering analytical chemistry as the intended application of MNPs, the surface of the MNPs is the place in which the interactions between the MNPs and the analyte occurs. Originally, the surface of magnetic oxides is composed of hydroxyl groups [79] and quite often MNPs must interact with the target in a very specific manner. For instance, the magnetic capture of a biological biomarker in complex environment such as the blood [80]. Obviously, hydroxyl groups, in the original surface of the nanoparticles, do not fulfill this task. Furthermore, bare MNPs tend to agglomerate due to physical interactions between small NPs, and loss of magnetism due to oxidation of eventually Fe^{2+} present in the lattice structure of the NPs. Functionalization plays the important role of prevent/retard the occurring of these events [81]. Therefore, functionalization regulates hydrodynamic diameter, solvent dispersability, surface charge, colloidal stability or the speed of response to an external magnetic field [82]. In this sense, functionalization of the MNPs is crucial step in the preparation of the MNPs in order to generate nanomaterials with designed properties, and this step should be carefully planned considering the final application of the nanomaterial.

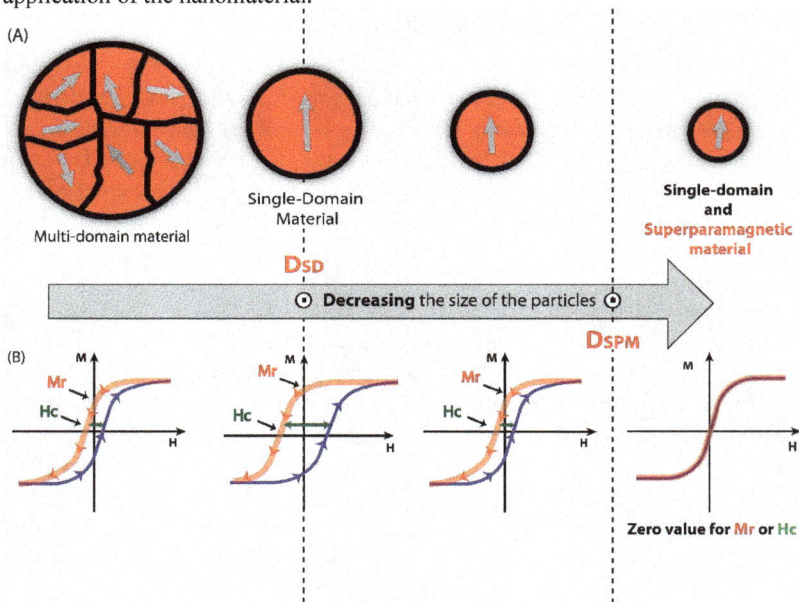

Figure 4. Schematic illustration of the particle size effect on: (A) on the domain state and (B) in Hysteresis loops of spinel ferrite materials. Blue and red lines are the first and the second runs, respectively. Reproduced with permission from Ref. 71.

There are many strategies to prepare functionalized MNPs, which produce different types functionalized materials are: core-shell structure, matrix dispersed structure, Janus type heterostructures and shell–core-shell structure [81], as shown in Figure 5. For all cited cases, the functionalized agent (FA) can be composed of an organic, inorganic, or both materials. Regardless of the type of functionalized material, it is important have occur a covalent interaction between the FA and the surface of MNPs, in order to increase the chemical stability of the hybrid materials.

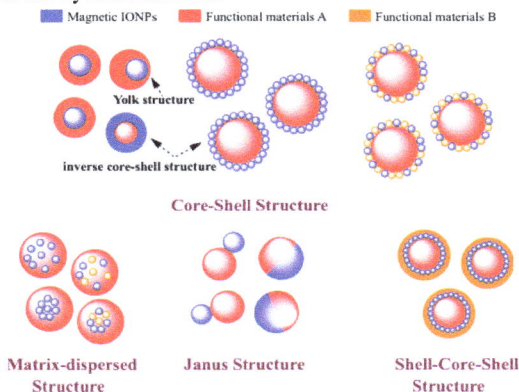

Figure 5. Schematic representation of the types of functionalized MNPs. Blue spheres represent MNPs, whereas nonmagnetic materials and matrix materials are represented in other colors. Reproduced with permission from Ref. 81.

With respect to the application of MNPs in magnetic separation and sensors, the most common type of functionalized material is core-shell structure [74]. Therefore, much of the following discussion focuses on core-shell nanoparticles. Functionalization of MNPs can be performed in several steps, in which each step represents a layer of organic/inorganic component. First layer is composed by the FA that has functional groups with proper electronic density to coordinate to iron atoms at the surface of MNPs (Figure 6 a). They are typically amine groups [70,83], carboxylic acid/carboxylate groups [70], catechol groups [84], ethoxysilane groups [85], phosphonate groups [86]. This organic FA should also have a three-dimensional structure that allows functional groups to be projected out to the working environment. As a matter of fact, the groups that are projected out to the surface of MNPs dictate the chemical features of this nanomaterial, such as colloidal stability, surface charge, hydrodynamic size and so forth. This group is usually amine groups [85], thiol groups [87], carboxylic acid/carboxylate groups [70] and phosphonate groups [86] due to facility in the chemical coupling with amine or

Magnetochemistry - Materials and Applications   Materials Research Forum LLC
Materials Research Foundations **66** (2020) 173-216   https://doi.org/10.21741/9781644900611-5

carboxylic acid chemical functionalities of bio proteins, enzymes, antibodies and targeting agents.

For the second step in the design of functionalized MNPs, most of the time there are two options. The first option is to attach directly the end group to the bio proteins, enzymes, antibodies or targeting agents (Figure 6 b). Wang and co-authors coated Fe_3O_4 NPs with a carboxylic acid-polyethylene glycol (PEG)-derived that projected out carboxylic acid groups [92]. These groups were attached to aptamers through the formation of amide bonds using as activating reactants EDC (1-ethyl-3-(3-dimethylaminopropyl)carbodiimide) and NHS (Nhydroxosuccinimide). Another example of the direct conjugation using EDC and NHS is presented in the publication Kohler and co-authors, in which folic acid were coupled to MNPs functionalized with amine PEG-derived molecules [93].

The second option for the second step in the preparation of functionalized MNPs is to modify the groups that are projected out at the surface of the MNPs for a subsequently conjugation with bio proteins, enzymes, antibodies and targeting agents. The modification must be performed with a molecule that has the capability to react with bio proteins, enzymes, antibodies and targeting agents, and form a bond that is chemically stable in its working environment (Figure 6 c). The most used approach is to attach bifunctional aldehydes (glutaradehyde), mainly due to its facility and simplicity (Figure 6 c). These molecules can form imine bonds to the amine groups at the surface of the MNPs and those ones at the molecule of interest [85,94]. However, if the functional molecule possess a high amount of amine groups, this conjugation technique may be not suitable, due to its poor selectivity [82]. There are strategies to increase selectivity using aldehyde chemistry conjugation, for instance Pereira and Lai treated an antibody with sodium periodate to oxidize minor sugar units of the antibody into the aldehyde functional group. The modified antibody were conjugated with hydrazide coated-MNPs [95].

In order to increase selectivity, the conjugation with bio proteins, enzymes, antibodies and targeting agents can be performed though thiol functional groups (Figure 6 d). The use of this strategy reduces the probability of the aggregation, once the biomolecules have smaller amount of thiol groups, compared to residual amine and carboxylic acid groups [82]. The interaction with thiol groups generally is performed with maleimide, pyridyl disulfide and iodoacetate groups present at the surface of the MNPs [89,90,96]. The insertion these functional groups is generally performed by the conjugation of amine (or carboxylic acid) of the MNPs with the carboxylic acid (or amine) of the molecule containing maleimide, pyridyl disulfide and iodoacetate groups, through EDC and NHS chemistry.

The most advanced approach in surface nano-engineering is called "click" chemistry. The key advantages are the high yields, selectivity and good physiological stability of the bound formed (Figure 6 d). One of the most used examples is azide-alkyne ligation. For this case, it is necessary to insert one group at the surface of the MNPs and another in the molecule of interest, and then perform the conjugation catalyzed by Cu (I) [87].

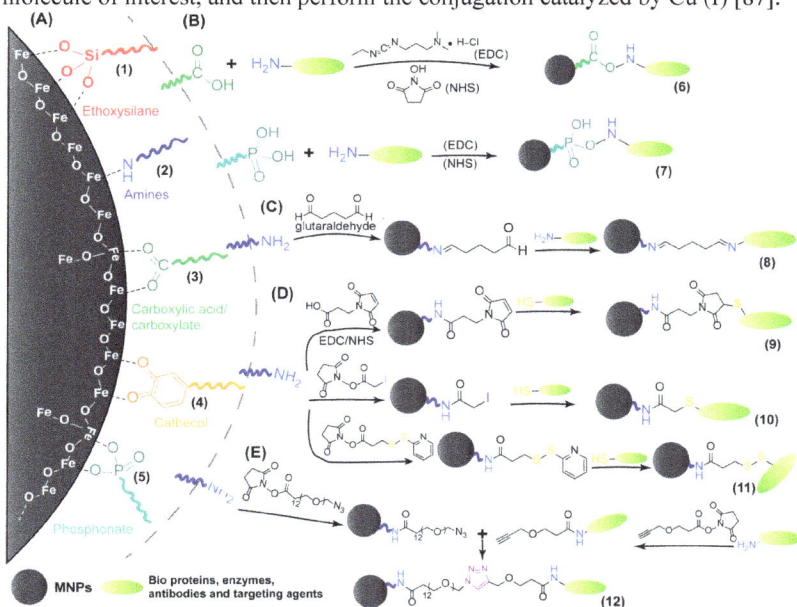

Figure 6. schematic representation of possible forms of functionalization of MNPs surface. (A) Chemical groups that coordinate to iron atoms of the MNPs surface. References for each group: 1 - [85]; 2 - [70]; 3 - [70]; 4 - [88]; 5 - [86]. (B) Direct conjugation of bio proteins, enzymes, antibodies or targeting agents through EDC/NHS chemistry. Groups 6 and 7 - [87]; (C) Conjugation of the molecule of interest though glutaraldehyde. Group 8 - [85]. (D) Conjugation of bio proteins, enzymes, antibodies or targeting agents using residues of thiol groups. Groups 9 - [89]; 10 - [90]; 11 - [89]. (E) Conjugation of the molecules of interest through azide/alkyne click chemistry. Group 12 - [91]. Reproduced with permission from each reference mentioned in this caption.

Finally, it is important to mention that there are others alternatives to the examples shown herein [97]. We recommend McCarthy and co-authors publication for any researcher interested in the developments of an efficient chemical route for the modification of MNPs, where the authors explain in detail the protocols for several chemical modifications [87].

4. Use of MNPs in analytical chemistry

4.1 Pre-concentration, capture and separation

A sample preparation process in analytical chemistry usually involves the separation of the analyte from the sample matrix, removal of interfering elements and pre-concentration of the analytes [98]. This procedure is of fundamental importance for a successfully detection and quantification of the analytes in complex matrices, and/or those ones with smaller concentrations. Complex matrices and low abundance analytes are likely present in the analysis biological samples [99]. For this reason, our discussion will be mainly focused on the magnetic separation of biomolecules. However, it is worth to mention that heavy metals are also difficult to detect. For instance, according to the World Health Organization (WHO), the allowable value of cadmium for public water supply is set at less than $3ng.mL^{-1}$. This means the methodology to sense this metallic cation should be able to perform the detection of smaller concentrations. Hence, it demands a pre-concentration step, which may be carried out through magnetic solid phase extraction [100]. In this sense, B. Zargar and co-author [101] produced a sol-gel magnetic coated using carbon nanotubes to perform a selective pre-concentration step of Cd^{2+} ions in different water samples. Fe_3O_4 NPs were used as the magnetic component of the nanocomposite. A similar nanosystem was also applied by Jinhua Li and co-authors [102] to remove chromium from aqueous solutions. In another work, a pre-concentration step of cobalt and mercury was carried out by using a magnetized fungal solid phase extractor [103]. The reuse of this system was tested 35 times and the biosorption capacities were found to be 25.4 and 30.3 mg/g for Co^{2+} and Hg^{2+}, respectively.

Taking into consideration the publications already mentioned in this section, a common component is easily identified: the MNPs. The advantages of MNPs in analytical methods come from their large surface-to-volume ratio, comparable size to many analytes of interest, easy functionalization and, mainly, to the convenient manipulation with an external magnet [73,98]. Usually, the application of MNPs in magnetic separation is performed as follows: functionalized MNPs, which exhibit affinity toward the isolated structure, are mixed with a sample containing the target analyte. Within a certain period of incubation, the target is linked to the MNPs, through specific or non-specific interactions. Subsequently, MNPs with the compound of interest are separated from the sample using an external magnetic field. Then, the isolated analyte can be directly analyzed or eluted and detected by a certain analytical method [98]. After magnetic separation of the analyte, an analytical technique must be employed to perform the detection and quantification, such as liquid chromatography (LC), liquid chromatography-mass spectroscopy (LC/MS), liquid chromatography–tandem mass spectroscopy (LC/MS/MS), matrix-assisted laser desorption ionization mass spectrometry (MALDI-MS), near-infrared spectroscopy (NIRS), surface-enhanced Raman spectroscopy (SERS), polymerase chain reaction (PCR), flow cytometry (FCM), electron and fluorescence microscopy, etc. [98].

Regarding the analysis of biomolecules, most used techniques for sample preparation for biological analysis are: extraction techniques [104], electrophoresis [105], ultrafiltration [106], precipitation [107]. Although these mentioned techniques have their advantages, literature is consolidated to conclude that their disadvantages outweigh the advantages, with respect to analysis of analytes in complex sample matrix [99]. Therefore, functionalized MNPs have demonstrated to be perfect candidate to replace traditional sample preparation techniques, mainly to their agility, convenience, elegance and efficiency in the sample preparation for biological analysis [99]. Given this, it can be mentioned the use of MNPs for the enrichment of biological macromolecules in proteome [108,109], separation of nucleic acids [110], cell separation [111], enzyme immobilization [112] and extraction of bioactive compounds of low molecular weight, such as pharmaceutical drugs [113]. One of the advances achieved by the use of MNPs in analytical chemistry is the nanoprobe-based affinity mass spectrometry (NBAMS), which comprises in the application of functionalized MNPs to isolate the targeted analyte and the analysis by mass spectrometry (MS) [114–116]. Furthermore, if the mass spectrometer has a matrix-assisted laser desorption/ionization as source of ions, MNPs can act also as a solid support for direct analysis of the analytes, without the step of elution of the analyte from the MNPs [114]. Recently, NBAMS were used for the quantifications of protein biomarkers and their proteoforms related to inflammation and renal dysfunction, in which antibody-bounded MNPs were employed as protein-separation and enrichment of the analyte [115] (Figure 7). The methodology developed by the authors could simultaneously quantify the inflammatory markers C-reactive protein (CRP), serum amyloid A (SAA), and calprotectin (S100A8/9) and the kidney function marker cystatin C (CysC), with low limits of detection (0.01−0.06 μg/mL), low coefficients of variation (3.8−9.4%), low volume of sample (20 μL) the high sample throughput of 384 samples per day [115].

MNPs were also applied in the AIDS diagnosis, through CD4+ T lymphocytes quantification, using instrumentation widely available in low-resource settings laboratories [111]. For this case, CD4 were separated and preconcentrated from whole blood in magnetic nanoparticles, and labeled with an enzyme for the optical readout performed with a standard microplate reader [111] (Figure 8). The attachment of the CD4+ T lymphocytes to the magnetic beads was confirmed and by scanning electron microscopy (Figure 9 a), and flow cytometry (Figure 9 b). The analytical method proposed by the authors has shown to be suitable for clinical analysis, due to its low limit of detection (50 CD4 cells μL^{-1}), which covers the medical interest range from 63 to 979 CD4 CD4 cells μL^{-1}. This method is a highly suitable alternative diagnostic tool for the expensive flow cytometry, providing a sensitive method but using instrumentation widely available in low-resource settings laboratories and requiring low-maintenance.

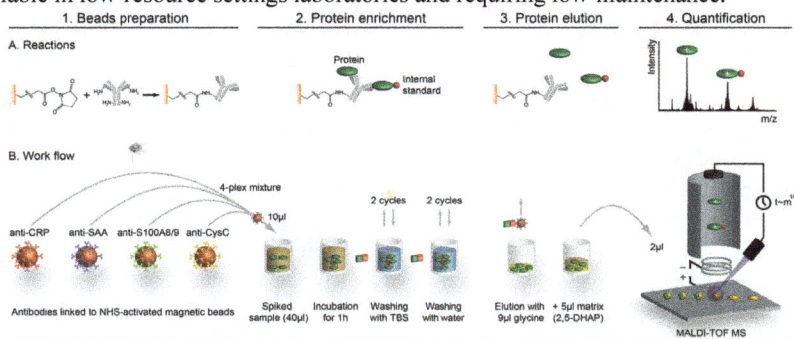

Figure 7. Description of the immuno-MALDI-TOF MS assay. The method consists of four steps starting with preparation of magnetic beads, followed by protein enrichment, protein elution, and finally quantification by MALDI MS. The fundamental reactions for the functionalization of MNPs (A) and the work flow of the analysis for each step (B). Reproduced with permission from Ref. 115.

MNPs have also brought advances in the analysis of environmental-related species. For instance, analytical samples containing low-molecular weight pharmaceuticals and personal care products could also be prepared by using magnetic separation [117]. Tang and co-authors employed Fe_3O_4-layered double hydroxide (Fe_3O_4-LDH) core-shell microspheres as sorbent to enable automation of the integrative extraction and analytical quantification of acetylsalicylic acid (ASA), 2,5-dihydroxybenzoic acid (DBA), 2-phenylphenol (PP), and fenoprofen (FP) in aqueous samples. Through magnetic force, the sorbent, after extraction, was isolated from the sample and then dissolved by acid to release the analytes and inject the sample in a HPLC system with photodiode array

(Figure 10). Thus the customary analyte elution step in conventional SPE was unnecessary.

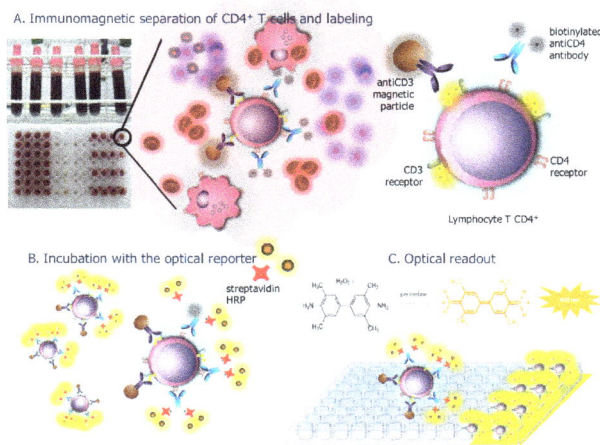

Figure 8. Schematic representation of the CD4 magneto ELISA. (A) The CD4+ T lymphocytes were captured from whole blood by the antiCD3-MPs and labeled in one step with antiCD4-biotin. (B) The incubation with the optical reporter streptavidin-HRP was then performed. (C) Finally, the optical readout was achieved with H2O2 and TMB. The enzymatic reaction was stopped by adding H2SO4 and the supernatant was measured at 450 nm. Reproduced with permission from Ref. 111.

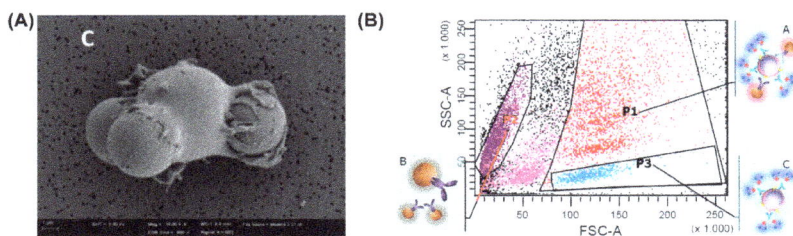

Figure 9. Evaluation of the binding of CD4+ T lymphocytes to antiCD3 magnetic particles by (A) SEM and flow cytometry (B); gate P1 represents the antiCD3 magnetic particles-CD4 cells complex, gate P2 the antiCD3 magnetic particles and gate P3 CD4 lymphocytes. Reproduced with permission from Ref. 111.

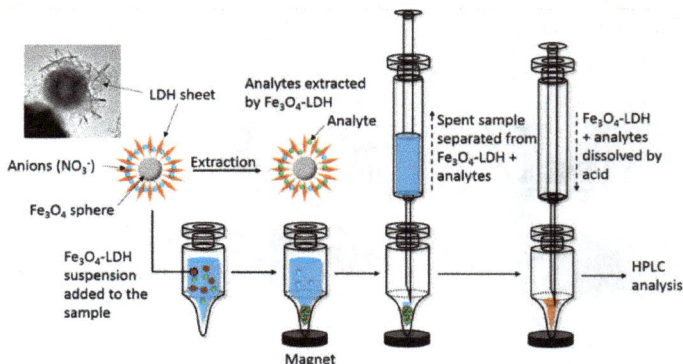

Figure 10. Work flow of automated SPE. Collection of the final extract and injection into the HPLC system are also automated. Reproduced with permission from Ref. 117.

Very interesting MNPs were prepared by Zhang and co-authors for simultaneous dispersive solid phase extraction and determination of trace multitarget analytes [118]. For this case, the authors prepared a novel tetraazacalix[2]arene[2]triazine (TCT) coated MNPs for simultaneous extraction and stepwise elution for five PAHs (phenanthrene, anthracene, pyrene, chrysene, and benzo(a)pyrene), six nitroaromatics (4-nitrotoluene, 2,4-dinitrotoluene, 2,4,6-trinitrotoluene, 4-nitrophenol, 2,4-dinitrophenol, and 2,4,6-trinitrophenol), and four metal ions (Cu, Zn, Mn, Cd) in aqueous samples (Figure 11) . This task was possible to accomplish due to unique structural framework and multiple recognition features of TCT, which includes π electron stacking, charge transfer, hydrogen bonding, and ion-exchange. Analyses of the analytes were performed by HPLC-fluorescence detection, HPLC-UV detection, and atomic absorption spectrometry in ideal and real matrix (tap water, river water, lake water and urine).

The mentioned examples of the application of MNPs in magnetic separation science for analytical chemistry demonstrate the great advance accomplished by the development of proper functionalized MNPs. There are also several other examples of the use of MNPs in sample preparation that are summarized in other publications [98,119–122].

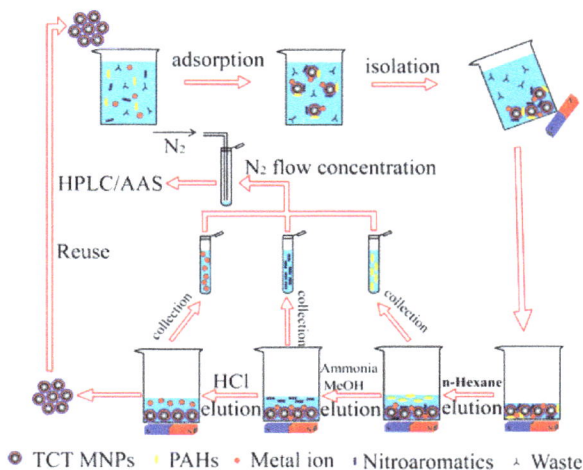

Figure 11. Description of sample preparation of multiple analytes through TCT coated MNPs. Reproduced with permission from Ref. 118.

4.2 Sensors

Sensor devices are of fundamental importance for monitoring environmental, biological and food systems. For this reason, researches always seek manners to improve the features of sensor, mainly through the enhancement of the sensitivity, and decreasing the time of analysis through miniaturized devices. The incorporation of nanomaterials in such devices has brought possibility to perform these improvements. Among the nanomaterials available for sensors applications, MNPs deserve attention for the following reasons:

a) Biocompatibility – this property enables the design of biosensors [123];

b) Manipulation of the NPs by a magnet – besides the extraction of the analyte, as discussed in the previous section, manipulation with a magnet of the MNPs can concentrate the analyte closer to the transducer of the sensor, which enhances sensitivity of the sensor [124];

c) Easy functionalization – as discussed in the previous sections, MNPs have a great capability to be functionalized;

d) Composites of MNPs with carbon nanostructures – these composites has become a hot topic in science, once its properties cannot be achieved by the component alone [124]. It

Magnetochemistry - Materials and Applications   Materials Research Forum LLC
Materials Research Foundations **66** (2020) 173-216   https://doi.org/10.21741/9781644900611-5

is highlighted the performance of these composites as components for electrochemical sensors;

e) Low cost synthesis.

The types of sensors that MNPs have been applied include electrochemical, microfluidic, magnetic resonance imaging (MRI) techniques and others. In this sense, we presented in the next sections a brief explanation of the principles for each kind of sensor and examples of publications that show the advances achieved due to the use of MNPs.

4.2.1 Electrochemical

Among the types of sensors, electroanalytical methods have consolidated as the most rapidly expanding class of chemical sensors, once such devices are sensitive, selective, fast, inexpensive, easy miniaturization and operation [74,124]. The distinction of different electrochemical sensors, regarding the type of signal measured, are potentiometric –measurement charge difference between working and reference electrodes, conductometric – measurement of the conductive property of a medium and amperometric – measurement of the current generated by an oxidation/reduction reaction. Another type of electrochemical sensor is impedimetric sensor, which acquire information of the analyte by measuring the impedance of certain system. Amperometric sensors are undoubtedly the most widely used class of sensors, mainly through the use of voltammetry techniques [124]. More information concerning the fundamental of electrochemical techniques applies in analytical chemistry are found in the literature [125].

Electrochemical sensors based on MNPs can be divided in two categories: non-enzymatic, which functionalized or bare MNPs are incorporated to the surface of the electrode, and enzymatic, in which MNPs act as a mediator [124] and a enzymatic reaction participates of the sensing mechanism. Although enzymatic sensors are selective due to specific enzyme-analyte interaction, enzymes can be easily denatured and they gave relatively high costs to the sensor. In this regard, there is an effort to develop non-enzymatic electrochemical sensors with the same capability of an enzymatic sensor. One of the candidates is for sure MNPs functionalized with molecular imprinted polymers (MIPs). The synthesis of MIPs involves a polymerization of functional monomers and cross-linker around a template, which has the same structure/or similar to the analyte of interest [126]. MIPs a developed in a way that the interaction between the template and monomer/cross-linker is weak enough for a simple extraction remove the template, leaving a specific site for recognition of the analyte [127] (Figure 12 a). The combination of MIPs and MNPs (MMIPs) could produce nanomaterials with superior features once they facilitate pre-concentration, separation, manipulation of the analyte, and, at the same time, they have great selectivity [126]. For instance, Han and co-authors prepared core-shell $Fe_3O_4@Au$ functionalized glutathione MIP for electrochemical determination of estradiol (E2) [128]. The NPs were assembled and immobilized on the surface of the electrode by simply installing a magnet into magnetic glassy carbon electrode (Figure 12 b). This strategy could increase the concentration of the analyte close to the interface of the electrolyte/electrode, which increased the sensitivity, and refresh the surface of the electrode by simply removing the magnet. The MMIP synthesized in this work had a stable adsorption for imprinting molecule E2 which could not be interfered by its analogs. Finally, this electrochemical sensor was successfully used for detecting E2 in milk powder with good sensitivity, selectivity, reproducibility and efficiency.

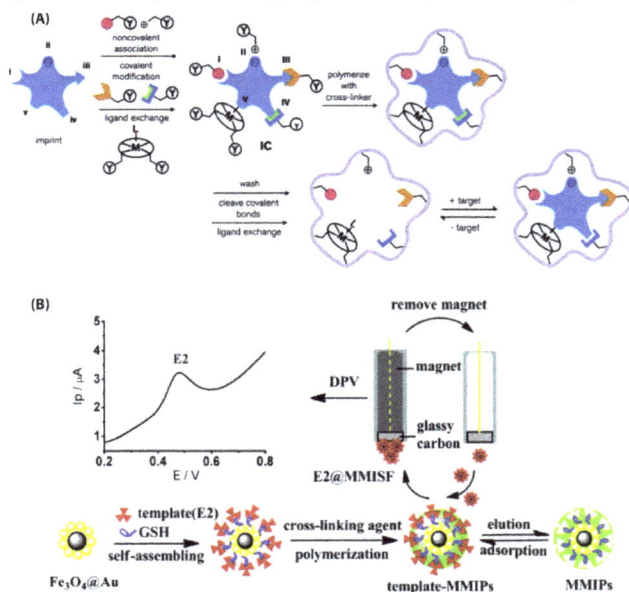

Figure 12. (A) Schematic representation of the synthesis, recognition, and separation of a MIP. The most used types of interaction between the matrix and the template: (i) non-covalent; (ii) electrostatic/ionic; (iii) covalent; (iv) semi-covalent; and (v) metal center coordination. Template molecules and functional monomers, containing a functional group, Y, combine via cross-linkers. Then, the template molecules remove using washing, ligand exchange or cleaving covalent bonds, leaving cavities similar to template molecules which are capable of specific recognition of the template [127,129] (B) Schematic representation of the preparation of MMIPs and recognition and measurement of E2. Reproduced with permission from Ref. 128.

Enzymatic electrochemical sensors are classified as electrochemical biosensors, in which enzymes are ideally in direct spatial contact with transduction devices, such as electrodes. In this type of sensors, the electrochemical signal transduction can be generated by two methods:

a) label-free methods – the signal obtained from oxidation/reduction of the analyte that is electrocative. In such cases the interaction of the analyte with specific biorecognition element can alter the intensity of the signal, improving the quality of the sensor;

b) labeled methods – it is used when the analyte do not have an electrochemical activity or its activity is not suitable for electrochemical quantification. The electrochemical signal comes from a redox probe, which its redox reaction is affected by the presence of the analyte [130].

In this context, MNPs are very important to the research area of biosensors, including enzymatic sensors. One of the representative examples of the use of MNPs in enzymatic sensors are the work performed by Sun and co-authors. For this case, the authors produced a sensitive electrochemical immunoassay for chlorpyrifos by using flake-like Fe_3O_4 modified carbon nanotubes (CNTs) as the enhanced multienzyme label [131]. Schematic representation of the preparation of the electrochemical immunoassay is shown in Fig. 13. In this work, the authors employed a labeled approach with Horseradish peroxidase (HRP), hydroquinone, H_2O_2 system as a redox probe reactions. The quantification of the chlorpyrifos was possible due to competition between chlorpyrifos and Ab_1 for the antigen biding sites. Therefore, more chlorpyrifos in the sample, fewer Ab_1, which caused fewer binding with HRP-Ab_2 label or the multi-HRP-CNTs@f-Fe_3O_4-Ab_2 label. Under optimal conditions, the sensor presented a great performance, with detection limit of 6.3 pg/mL, high specificity, specificity, reproducibility and stability. The importance to use of CNTs decorated with Fe_3O_4 was the following reasons: a) the excellent capability to transport electrons of the hybrid material; b) fake-like Fe_3O_4 film can increase the catalytic activity of the multi-enzyme label due to the peroxidase-mimic activities of Fe_3O_4; c) the presence of the flake-like film could provide a larger surface area to load HRP and Ab_2.

Based on these examples, MNPs increase the quality of the electrochemical sensor by: a) carrying closer to the transducer the electroactive molecules, or any other species related to the electrochemical reaction; b) enabling washing strategies and the processing of larger volumes; c) increasing the electroactive area of the transducers; d) acting like a peroxidase through the peroxidase mimic activity [130]. The examples of electrochemical sensors presented here is representative selection of some advanced electrochemical sensing strategies that MNPs play an important role. There are many other examples of electrochemical sensors based on MNPs, they are reviewed in this references [123,124,126,130].

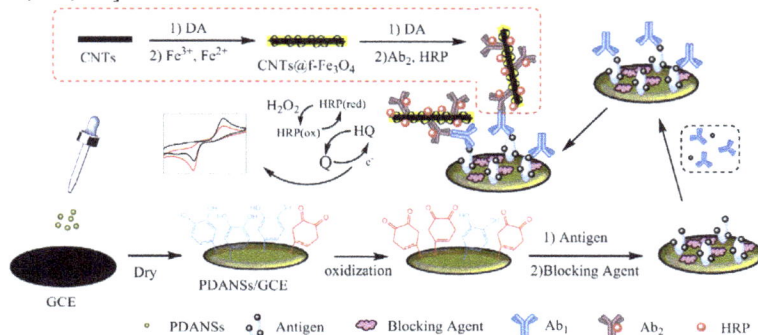

Figure 13. Schematic illustration of electrochemical immunoassay for CPF detection. Reproduced with permission from Ref. 131.

4.2.2 Microfluidic

Microfluidic systems sensors are miniaturized devices that integrate multiple function or steps in one main unit. These functions include sample transfer, mixing, separation and signal output. The main advances accomplished with these systems are smaller sample volume, lower materials consumption, less cost, faster turnaround time and automation of the process [132]. A general scheme of the sections of a microfluidic sensor is shown in the Figure 14 (a). Although the scheme presented in the Figure H (a) are focused in biosensors, microfluidic devices for food and environmental analysis are similar. One of the advantages in microfluidic sensors are the possibility to choose the readout mechanism according to the target. In this context, MNPs gain a special attention to when the transduction of the signal is performed by magnetic readout [133].

Magnetic readout is based commonly in magnetoresistance that is the change of electrical resistance under external applied field [133]. These devices can be used not just for detection and quantification of certain analytes, but also biding affinity characterization, as in the work of Lee and co-authors (Figure 14 (b)) [134]. The authors developed a magneto-nanosensor platform integrated with a microfluidic chip that allows the probing low-affinity protein-protein interactions. MNPs were conjugated with prey-proteins and different bait-proteins were immobilized at magneto-nanosensor. When an interaction between the prey and bait proteins occurs, MNPs attached to the magnetoresistance sensor, an increasing of the signal is recorded. Therefore, the measurement of the magnetoresistance is connected with the biding affinity the pray and bait proteins. Other examples of microfluidic sensors based on MNPs can be found in these reviews [130,133,135].

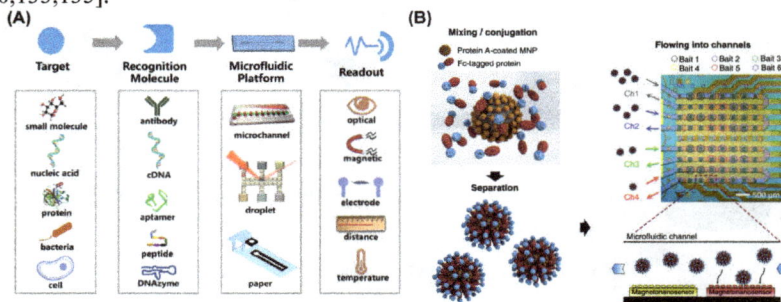

Figure 14. (A) Basics parts of a microfluidic biosensor. REF: [133]; (B) Schematic of kinetic assay with the magneto-nanosensor platform. Serial dilutions of the complexes were flowed into four channels, where each channel contains six immobilized sensors, which act as "bait" for the pray proteins immobilized onto MNPs. Reproduced with permission from Ref. 134.

4.2.3 Magnetic resonance imaging techniques

Magnetic resonance imaging (MRI) is a powerful platform for the noninvasive real-time visualization of cancer-related, cardiovascular, liver, and neurodegenerative diseases. The image is generated by the signal intensity of the nuclear magnetic resonance of hydrogen atoms in the human body. There are two types of signals recorded in MRI, which are related to the time of relaxation process of the nuclear spin of hydrogen after the incidence of radio-frequency radiation: transverse (T_1) and longitudinal (T_2) relaxations process. Each signal has own contrast characteristics and applications. Deeper explanations regarding the mechanisms of relaxations process are in these references [136–138]. As the hydrogen atoms are in different concentrations among human body, intrinsic contrast of the different tissues can be enough for the correct diagnostic. However, very often the contrast produced by intrinsic hydrogen protons is not enough for accurate diagnostic, thus it is necessary to administrate a contrast agent [137]. There are two types of contrast agents: MNPs and paramagnetic metal complexes (most used is gadolinium). As MNPs are usually used as T_2 contrast agent, which means that the presence of MNPs in biological medium induces a darken image compared to tissue without the contrast. On the other hand, complexes of paramagnetic metals that are classified as T_1 contrast agent. Contrary to MNPs, this type of contrast agent produces brighter images [138,139].

Despite the usefulness of MRI with or without contrast agents, there is a seek for contrast agents (or MRI sensors) that can sense pathological changes in local environment [139]. These changes can be pH [140], presence of calcium ions [141,142], proteins, DNA, etc [143]. The presence of these species can induce changes in the aggregation/disaggregation of MNPs or distance between contrast agents, which are closely related to contrast features of the contrast agents [140,143]. Recently, Choi and co-authors introduced distance-dependent magnetic resonance tuning (MRET), which occurs between a paramagnetic (gadolinium chelate) 'enhancer' and superparamagnetic 'quencher' (MNPs) [143]. In this approach, the T_1 signal can be tuned ON or OFF, depending on the separation distance between the two contrast agents (Figure 15 (a)). More interestingly, the chemical stimulation that will lead to an increase/decrease in this distance depends on the species that link the enhancer and quencher. Thus, the sensing principle will be intra-/intermolecular interactions or chemical reactions involving the linker, for example, enzyme-mediated cleavage and ion-mediated conformational change Figure 15 (b). The authors also investigated MRET concept *in vitro* and *in vivo* systems, through the sensing of matrix metalloproteinase-2 (MMP-2), which is a representative cancer biomarker. For this task, the enhancer and quencher were linked by a specific PLGVR sequence, which is cleaved by MMP (Figure 15 (c)). For this hybrid contrast agent, the T_1 signal increased linearly with the MMP-2 concentration, indicating the possibility to quantify MMP-2 in the real microenvironment (Figure 15 (d)). Additionally, it was changed the MMP-2 expression levels in mice using different amounts of MMP inhibitors, and the MRET probe was injected. The plot of the transverse relaxivity with MMP-2 concentration indicates that the MMP-2 expression level is well represented by the T_1 MRI signal (Figure 15 (f)). These results demonstrate the potential of this approach for the quantitative analyses of several biomolecules in its biological microenvironmental, which can help to gain accurate diagnosis with non-invasive procedures.

There are two other types of magnetic resonance sensors to be highlighted: inverse contrast enhancement [140], and magnetic relaxation switching sensors [144]. The first one is based on a stimulus that induces inverse contrast enhancement, for instance, Lu and co-authors coated ultra-small MNPs with i-motif linker, which induced disaggregation of the MNPs under acid medium, such as extracellular cellular matrix of cancer cells, turning a T_2 contrast agent into a T_1 contrast agent [144]. The second is based in a change of the T_2 signal upon aggregation of the MNPs that is motivated by the presence of certain chemical specie. More detail concerning this approach can be found in these references [144,145].

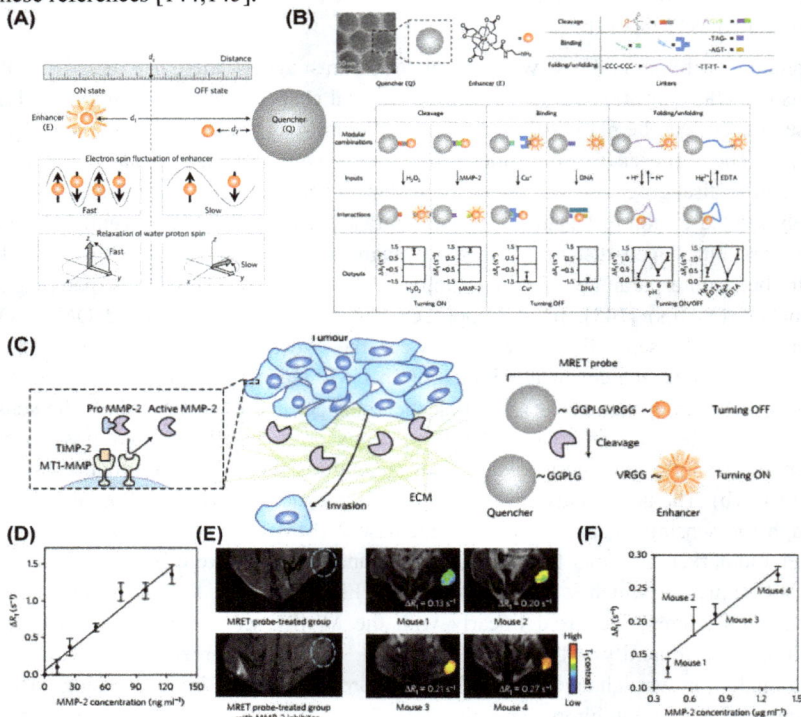

Figure 15. (A) Schematic representation of of distance-dependent magnetic resonance tuning. (B) Different mode of operation of MRET probes. (C) Activation of MRET probe by MMP-2 in living organisms. When the MMP-2 is in contact with MRET probe with specific PLGVR sequence linking the enhancer and quencher, it happens a cleavage of the probe and enhancer is released, resulting in an increase in the T1 MRI signal. Reproduced with permission from Ref. 143.

4.2.4 Others

There are other types of sensors based on MNPs. One of them for which MNPs have a very important role is colorimetric determination involving hydrogen peroxide, due to intrinsic peroxidase mimicking activity [146,147]. The advantages of colorimetric sensors are low cost, portability and absence of complication instrumentation [147]. For instance, Lu and co-authors synthesized Yolk–shell nanostructured $Fe_3O_4@C$ MNPs for label-free detection of glucose and H_2O_2. Commonly, colorimetric sensors are combined with other sensing technique. Shen and co-authors developed a bimodal sensor (colorimetric and MRI) using $Au-Fe_3O_4$ NPs functionalized l-(2-mercaptoethyl)-1,2,3,4,5,6-hexanhydro-s-triazine-2,4,6-trione for the detection of melamine. Interestingly, the presence of melamine induced aggregation of the NPs, which changed the T_2 signal, and caused a red-shift of the absorption peak of the gold NPs attached to Fe_3O_4 [148].

MNPs have also brought advances in surface enhanced Raman scattering (SERS) sensors. SERS is powerful sensing technique for wide range of analytes and applications, once it offers superior fingerprinting, nondestructive data acquisition [29] and single-molecule outstanding sensitivity (enhancement factor up to 10^{14}–10^{15} in total) [149]. The insertion of MNPs in the SERS sensors caused overcome of the SERS disadvantages, mainly those ones related to the concentration/enrichment effect of the analyte, and procedure to fabricating enhancement hot spots [149]. An example of the important role of MNPs in SERS sensors are demonstrated in the work of Pang and co-authors [150]. A Magnetic immunoassay for cancer biomarker (carcinoembryonic antigen) detection based on SERS was developed, and the proposed sensor could detect the analyte with an outstanding low limit of detection of 4.75 fg/mL and a wide dynamic linear range from 10 fg/mL to 100 ng/mL. In this case MNPs were used to extract the carcinoembryonic antigen from the complex matrix and to produce enhancement hotspots.

References

[1]   M.C. Coelho, G. Torrão, N. Emami, J. Gr´cio, Nanotechnology in automotive industry: Research strategy and trends for the future—small objects, big impacts, J. Nanosci. Nanotechnol. 12 (2012) 6621–6630. https://doi.org/10.1166/jnn.2012.4573

[2]   M.A. Meador, Taking nanotechnology to new heights: The potential impact on future aerospace vehicles, MRS Bull. 40 (2015) 815–821. https://doi.org/10.1557/mrs.2015.224

[3]   I. Kamal, Prospects of Some applications of engineered nanomaterials: A review, Open Access J. Biomed. Eng. Biosci. 2 (2018) 245–252. https://doi.org/10.32474/OAJBEB.2018.02.000149

[4]   V. Prakash Sharma, U. Sharma, M. Chattopadhyay, V.N. Shukla, Advance applications of nanomaterials: A Review, Mater. Today Proc. 5 (2018) 6376–6380. https://doi.org/10.1016/j.matpr.2017.12.248

[5]   D. Brabazon, E. Pellicer, F. Zivic, J. Sort, M.D. Baró, N. Grujovic, K.L. Choy, Review of production routes of nanomaterials, Commerial Nanotechnologies-A case study approach. (2018) 15–29. https://doi.org/10.1007/978-3-319-56979-6_2

[6]   S. Singamaneni, V.N. Bliznyuk, C. Binek, E.Y. Tsymbal, Magnetic nanoparticles: recent advances in synthesis, self-assembly and applications, J. Mater. Chem. 21 (2011) 16819–16845. https://doi.org/10.1039/C1JM11845E

[7]   S.M. Ng, M. Koneswaran, R. Narayanaswamy, A review on fluorescent inorganic nanoparticles for optical sensing applications, RSC Adv. 6 (2016) 21624–21661. https://doi.org/10.1039/C5RA24987B

[8]   R. Wiltschko, W. Wiltschko, The magnetite-based receptors in the beak of birds and their role in avian navigation, J. Comp. Physiol. A Neuroethol. Sensory, Neural, Behav. Physiol. 199 (2013) 89–98. https://doi.org/10.1007/s00359-012-0769-3

[9]   D. Acosta-Avalos, E. Wajnberg, P.S. Oliveira, I. Leal, M. Farina, D.M. Esquivel, Isolation of magnetic nanoparticles from Pachycondyla marginata ants, J. Exp. Biol. 202 (1999) 2687–2692. https://doi.org/10.1073/pnas.94.21.11633

[10]   L. Yan, S. Zhang, P. Chen, H. Liu, H. Yin, H. Li, Magnetotactic bacteria, magnetosomes and their application, Microbiol. Res. 167 (2012) 507–519. https://doi.org/10.1016/j.micres.2012.04.002

[11]   S.A. Gilder, M. Wack, L. Kaub, S.C. Roud, N. Petersen, H. Heinsen, P. Hillenbrand, S. Milz, C. Schmitz, Distribution of magnetic remanence carriers in the human brain, Sci. Rep. 8 (2018) 1–9. https://doi.org/10.1038/s41598-018-29766-z

[12]   L. Taylor, H. Schmitt, W. Carrier, M. Nakagawa, Lunar dust problem: From liability to asset, 1st Sp. Explor. Conf. Contin. Voyag. Discov. (2005). https://doi.org/10.2514/6.2005-2510

[13]   A. Chiolerio, A. Chiodoni, P. Allia, P. Martino, Magnetite and other Fe-Oxide nanoparticles: Datasheet from · Volume : "Handbook of Nanomaterials Properties" in SpringerMaterials. https://doi.org/10.1007/978-3-642-31107-9_34), (n.d.)

[14]   R.M. Freire, P.G.C. Freitas, W.S. Galvao, L.S. Costa, T.S. Ribeiro, I.F. Vasconcelos, J.C. Denardin, R.C. de Oliveira, C.P. Sousa, P. de-Lima-Neto, A.N. Correia, P.B.A. Fechine, Nanocrystal growth, magnetic and electrochemical properties of NiZn ferrite, J. Alloys Compd. 738 (2018). https://doi.org/10.1016/j.jallcom.2017.12.088

[15]   A. Akbarzadeh, M. Samiei, S. Davaran, Magnetic nanoparticles: Preparation, physical properties, and applications in biomedicine, Nanoscale Res. Lett. 7 (2012) 144. https://doi.org/10.1186/1556-276x-7-144

[16]   P. Biehl, M. von der Lühe, S. Dutz, F.H. Schacher, Synthesis, characterization, and applications of magnetic nanoparticles featuring polyzwitterionic coatings, Polymers 10 (2018). https://doi.org/10.3390/polym10010091

[17]   R. Serrano, S. Stafford, Recent progress in synthesis and functionalization of multimodal fluorescent-magnetic nanoparticles for biological applications, Appl. Sci. 8 (2018) 172. https://doi.org/10.3390/app8020172

[18]   F.D. Guerra, M.F. Attia, D.C. Whitehead, F. Alexis, Nanotechnology for environmental remediation: Materials and applications, Molecules. 23 (2018) 1–23. https://doi.org/10.3390/molecules23071760

[19]   J.S. Beveridge, J.R. Stephens, M.E. Williams, The use of magnetic nanoparticles in analytical chemistry, Annu. Rev. Anal. Chem. 4 (2011) 251–273. https://doi.org/10.1146/annurev-anchem-061010-114041

[20]   M. Faraji, Recent analytical applications of magnetic nanoparticles, Nanochem Res. 1 (2016) 264–290. https://doi.org/10.7508/ncr.2016.02.014

[21]   R. Mout, D.F. Moyano, S. Rana, V.M. Rotello, Surface functionalization of nanoparticles for nanomedicine, Chem. Soc. Rev. 41 (2012) 2539–2544. https://doi.org/10.1039/c2cs15294k

[22]   A. Ebrahiminezhad, V. Varma, S. Yang, Y. Ghasemi, A. Berenjian, Synthesis and application of amine functionalized iron oxide nanoparticles on Menaquinone-7 fermentation: A step towards process intensification, Nanomaterials 6 (2015) 1. https://doi.org/10.3390/nano6010001

[23]   C.S. Clemente, V.G.P. Ribeiro, J.E.A. Sousa, F.J.N. Maia, A.C.H. Barreto, N.F. Andrade, J.C. Denardin, G. Mele, L. Carbone, S.E. Mazzetto, P.B.A. Fechine, Porphyrin synthesized from cashew nut shell liquid as part of a novel superparamagnetic fluorescence nanosystem, J. Nanoparticle Res. C7 - 1739. 15 (2013) 1–10. https://doi.org/10.1007/s11051-013-1739-6

[24]   T.M. Freire, L.M.U. Dutra, D.C. Queiroz, N.M.P.S. Ricardo, K. Barreto, J.C. Denardin, F.R. Wurm, C.P. Sousa, A.N. Correia, P. De Lima-Neto, P.B.A. Fechine, Fast ultrasound assisted synthesis of chitosan-based magnetite nanocomposites as a modified electrode sensor, Carbohydr. Polym. (2016). https://doi.org/10.1016/j.carbpol.2016.05.095

[25]   D.M.A. Neto, R.M. Freire, J. Gallo, T.M. Freire, D.C. Queiroz, N.M.P.S. Ricardo, I.F. Vasconcelos, G. Mele, L. Carbone, S.E. Mazzetto, M. Bañobre-Lopez, P.B.A. Fechine, Rapid sonochemical approach produces functionalized Fe_3O_4 nanoparticles with excellent magnetic, colloidal, and relaxivity properties for MRI application, J. Phys. Chem. C. 121 (2017). https://doi.org/10.1021/acs.jpcc.7b04941

Materials Research Forum LLC
https://doi.org/10.21741/9781644900611-5

[26] E.E. Nelson, A.E. Guyer, The development of the ventral prefrontal cortex and social flexibility, Dev Cogn Neurosci. 1 (2012) 233–245. https://doi.org/10.1016/j.dcn.2011.01.002

[27] D.M.A. Neto, R.M. Freire, J. Gallo, T.M. Freire, D.C. Queiroz, N.M.P.S. Ricardo, I.F. Vasconcelos, G. Mele, L. Carbone, S.E. Mazzetto, M. Bañobre-López, P.B.A. Fechine, Rapid sonochemical approach produces functionalized Fe_3O_4 nanoparticles with excellent magnetic, colloidal, and relaxivity properties for MRI application, J. Phys. Chem. C. 121 (2017). https://doi.org/10.1021/acs.jpcc.7b04941

[28] P.E. Feuser, L. dos S. Bubniak, M.C. dos S. Silva, A. da C. Viegas, A. Castilho Fernandes, E. Ricci-Junior, M. Nele, A.C. Tedesco, C. Sayer, P.H.H. de Araújo, Encapsulation of magnetic nanoparticles in poly(methyl methacrylate) by miniemulsion and evaluation of hyperthermia in U87MG cells, Eur. Polym. J. 68 (2015) 355–365. https://doi.org/ 10.1016/j.eurpolymj.2015.04.029

[29] T.K.H. Ta, M.T. Trinh, N.V. Long, T.T.M. Nguyen, T.L.T. Nguyen, T.L. Thuoc, B.T. Phan, D. Mott, S. Maenosono, H. Tran-Van, V.H. Le, Synthesis and surface functionalization of Fe_3O_4-SiO_2 core-shell nanoparticles with 3-glycidoxypropyltrimethoxysilane and 1,1′-carbonyldiimidazole for bio-applications, Colloids Surfaces A Physicochem. Eng. Asp. 504 (2016) 376–383. https://doi.org/10.1016/j.colsurfa.2016.05.008

[30] Q. Nguyen, C.N. Chinnasamy, S.D. Yoon, S. Sivasubramanian, T. Sakai, A. Baraskar, S. Mukerjee, C. Vittoria, V.G. Harris, Functionalization of FeCo alloy nanoparticles with highly dielectric amorphous oxide coatings, J. Appl. Phys. 103 (2008) 127–130. https://doi.org/10.1063/1.2839593

[31] T.A. Pham, N.A. Kumar, Y.T. Jeong, Facile preparation of boronic acid functionalized Fe-core/Au-shell magnetic nanoparticles for covalent immobilization of adenosine, Colloids Surfaces A Physicochem. Eng. Asp. 370 (2010) 95–101. https://doi.org/ 10.1016/j.colsurfa.2010.08.053

[32] H. Fatima, K.S. Kim, Magnetic nanoparticles for bioseparation, Korean J. Chem. Eng. 34 (2017) 589–599. https://doi.org/10.1007/s11814-016-0349-2

[33] J. Li, Y. Zhou, M. Li, N. Xia, Q. Huang, H. Do, Y.-N. Liu, F. Zhou, Carboxymethylated dextran-coated magnetic iron oxide nanoparticles for regenerable bioseparation, J. Nanosci. Nanotechnol. 11 (2011) 10187–10192. https://doi.org/10.1166/jnn.2011.5002

[34] G. Simonsen, M. Strand, G. Øye, Potential applications of magnetic nanoparticles within separation in the petroleum industry, J. Pet. Sci. Eng. (2018). https://doi.org/10.1016/j.petrol.2018.02.048

[35]   S. Ko, E.S. Kim, S. Park, H. Daigle, T.E. Milner, C. Huh, M. V. Bennetzen, G.A. Geremia, Amine functionalized magnetic nanoparticles for removal of oil droplets from produced water and accelerated magnetic separation, J. Nanoparticle Res. 19 (2017). https://doi.org/10.1007/s11051-017-3826-6

[36]   M.R. Jafari Nasr, M.R. Rahimpour, M. Arjmand, S.A. Vaziri, Application of a novel magnetic nanoparticle as demulsifier for dewatering in crude oil emulsion AU - Farrokhi, Fatemeh, Sep. Sci. Technol. 53 (2018) 551–558. https://doi.org/10.1080/01496395.2017.1373676

[37]   M. Feng, P. Zhang, H.C. Zhou, V.K. Sharma, Water-stable metal-organic frameworks for aqueous removal of heavy metals and radionuclides: A review, Chemosphere. 209 (2018) 783–800. https://doi.org/10.1016/j.chemosphere.2018.06.114

[38]   S.N. Paulina A. Kobielska, Ashlee J. Howarth, Omar K. Farha, Metal-organic frameworks for heavy metal removal from water, Coord. Chem. Rev. 353 (2018) 92–107. https://doi.org/10.1016/j.ccr.2017.12.010

[39]   Q. Yang, Q. Zhao, S. Ren, Q. Lu, X. Guo, Z. Chen, Fabrication of core-shell $Fe_3O_4@MIL$-100(Fe) magnetic microspheres for the removal of Cr(VI) in aqueous solution, J. Solid State Chem. 244 (2016) 25–30. https://doi.org/10.1016/j.jssc.2016.09.010

[40]   L. Huang, M. He, B. Chen, B. Hu, A designable magnetic MOF composite and facile coordination-based post-synthetic strategy for the enhanced removal of Hg^{2+} from water, J. Mater. Chem. A. 3 (2015) 11587–11595. https://doi.org/10.1039/C5TA01484K

[41]   J.-B. Huo, L. Xu, J.-C.E. Yang, H.-J. Cui, B. Yuan, M.-L. Fu, Magnetic responsive Fe3O4-ZIF-8 core-shell composites for efficient removal of As(III) from water, Colloids Surfaces A Physicochem. Eng. Asp. 539 (2018) 59–68. https://doi.org/10.1016/j.colsurfa.2017.12.010

[42]   M.C.Mascolo, Y. Pei, T.A. Ring, Room Temperature co-precipitation synthesis of magnetite nanoparticles in a large pH window with different bases, Materials 6 (2013) 5549–5567. https://doi.org/10.3390/ma6125549

[43]   A.A.C.H. Barreto, V.V.R. Santiago, R.R.M. Freire, S.S.E. Mazzetto, J.J.C. Denardin, G. Mele, I.M.I. Cavalcante, M.M.E.N.P. Ribeiro, N.N.M.P.S. Ricardo, T. Gonçalves, L. Carbone, T.T.L.G. Lemos, O.D.L.O. Pessoa, P.P.B.A. Fechine, Magnetic nanosystem for cancer therapy using oncocalyxone a, an antitumor secondary metabolite isolated from a Brazilian plant, Int. J. Mol. Sci. 14 (2013) 18269. https://doi.org/10.3390/ijms140918269

[44]   S. Rajput, C.U. Pittman, D. Mohan, Magnetic magnetite (Fe_3O_4) nanoparticle synthesis and applications for lead (Pb^{2+}) and chromium (Cr^{6+}) removal from water, J. Colloid Interface Sci. 468 (2016) 334–346. https://doi.org/ 10.1016/j.jcis.2015.12.008

[45]   A.C.H.C.H. Barreto, V.R.R. Santiago, R.M.M. Freire, S.E.E. Mazzetto, J.M.M. Sasaki, I.F.F. Vasconcelos, J.C.C. Denardin, G. Mele, L. Carbone, P.B.A.B.A. Fechine, grain size control of the magnetic nanoparticles by solid state route modification, J. Mater. Eng. Perform. 22 (2012) 2073–2079. https://doi.org/10.1007/s11665-013-0480-8

[46]   W. Zhang, S. Jia, Q. Wu, J. Ran, S. Wu, Y. Liu, Convenient synthesis of anisotropic Fe3O4 nanorods by reverse co-precipitation method with magnetic field-assisted, Mater. Lett. 65 (2011) 1973–1975. https://doi.org/10.1016/j.matlet.2011.03.101

[47]   S. Laurent, D. Forge, M. Port, A. Roch, C. Robic, L. Vander Elst, R.N. Muller, Magnetic iron oxide nanoparticles: Synthesis, stabilization, vectorization, physicochemical characterizations, and biological applications, Chem. Rev. 108 (2008) 2064–2110. https://doi.org/10.1021/cr068445e

[48]   H.-C. Roth, S.P. Schwaminger, M. Schindler, F.E. Wagner, S. Berensmeier, Influencing factors in the CO-precipitation process of superparamagnetic iron oxide nano particles: A model based study, J. Magn. Magn. Mater. 377 (2015) 81–89. https://doi.org/https://doi.org/10.1016/j.jmmm.2014.10.074

[49]   N.A. Frey, S. Peng, K. Cheng, S. Sun, Magnetic nanoparticles: synthesis, functionalization, and applications in bioimaging and magnetic energy storage, Chem. Soc. Rev. 38 (2009) 2532–2542. https://doi.org/10.1039/b815548h

[50]   A. Mashhadi Malekzadeh, A. Ramazani, S.J. Tabatabaei Rezaei, H. Niknejad, Design and construction of multifunctional hyperbranched polymers coated magnetite nanoparticles for both targeting magnetic resonance imaging and cancer therapy, J. Colloid Interface Sci. 490 (2017) 64–73. https://doi.org/ 10.1016/j.jcis.2016.11.014

[51]   M.S.A. Darwish, Effect of carriers on heating efficiency of oleic acid-stabilized magnetite nanoparticles, J. Mol. Liq. 231 (2017) 80–85. https://doi.org/ 10.1016/j.molliq.2017.01.094

[52]   S. Mumtaz, L.-S. Wang, S.Z. Hussain, M. Abdullah, Z. Huma, Z. Iqbal, B. Creran, V.M. Rotello, I. Hussain, Dopamine coated Fe_3O_4 nanoparticles as enzyme mimics for the sensitive detection of bacteria, Chem. Commun. 53 (2017) 12306–12308. https://doi.org/10.1039/C7CC07149C

[53]   Y.M. Wang, X. Cao, G.H. Liu, R.Y. Hong, Y.M. Chen, X.F. Chen, H.Z. Li, B. Xu, D.G. Wei, Synthesis of Fe_3O_4 magnetic fluid used for magnetic resonance imaging and

hyperthermia, J. Magn. Magn. Mater. 323 (2011) 2953–2959. https://doi.org/10.1016/j.jmmm.2011.05.060

[54]   V.T.A. Nguyen, M. Gauthier, O. Sandre, Templated Synthesis of magnetic nanoparticles through the self-assembly of polymers and surfactants, Nanomaterials 4 (2014) 628–685. https://doi.org/10.3390/nano4030628

[55]   Gozde Unsoy, S. Yalcin, R. Khodadust, G. Gungor, G. Ufuk, Synthesis optimization and characterization of chitosan-coated iron oxide nanoparticles produced for biomedical applications, J. Nanoparticle Res. 14 (2012) 964. https://doi.org/10.1007/s11051-012-0964-8

[56]   R.M.M. Freire, P.G.C.G.C. Freitas, T.S.S. Ribeiro, I.F.F. Vasconcelos, J.C.C. Denardin, G. Mele, L. Carbone, S.E.E. Mazzetto, P.B.A.B.A. Fechine, Effect of solvent composition on the structural and magnetic properties of MnZn ferrite nanoparticles obtained by hydrothermal synthesis, Microfluid. Nanofluidics. 17 (2014) 233–244. https://doi.org/10.1007/s10404-013-1290-x

[57]   M. Jiang, X. Peng, Anisotropic Fe_3O_4/Mn_3O_4 Hybrid Nanocrystals with Unique Magnetic Properties, Nano Lett. 17 (2017) 3570–3575. https://doi.org/10.1021/acs.nanolett.7b00727

[58]   H. Sun, B. Chen, X. Jiao, Z. Jiang, Z. Qin, D. Chen, Solvothermal Synthesis of Tunable Electroactive Magnetite Nanorods by Controlling the Side Reaction, J. Phys. Chem. C. 116 (2012) 5476–5481. https://doi.org/10.1021/jp211986a

[59]   W.L. Suchanek, R.E. Riman, Hydrothermal Synthesis of Advanced Ceramic Powders, Adv. Sci. Technol. 45 (2006) 184–193. https://doi.org/10.4028/www.scientific.net/AST.45.184

[60]   R.E. Riman, W.L. Suchanek, K. Byrappa, C.-W. Chen, P. Shuk, C.S. Oakes, Solution synthesis of hydroxyapatite designer particulates, Solid State Ionics. 151 (2002) 393–402. https://doi.org/10.1016/S0167-2738(02)00545-3

[61]   Y. Yu, W. Yang, X. Sun, W. Zhu, X.-Z. Li, D.J. Sellmyer, S. Sun, Monodisperse MPt (M = Fe, Co, Ni, Cu, Zn) Nanoparticles Prepared from a Facile Oleylamine Reduction of Metal Salts, Nano Lett. 14 (2014) 2778–2782. https://doi.org/10.1021/nl500776e

[62]   S. Sun, H. Zeng, Size-Controlled Synthesis of Magnetite Nanoparticles, J. Am. Chem. Soc. 124 (2002) 8204–8205. https://doi.org/10.1021/ja026501x

[63]   S.H. Sun, H. Zeng, D.B. Robinson, S. Raoux, P.M. Rice, Monodisperse MFe_2O_4 (M = Fe, Co, Mn) nanoparticles, J Am Chem Soc. 126 (2004)

[64]   S.H. Sun, C.B. Murray, D. Weller, L. Folks, A. Moser, Monodisperse FePt nanoparticles and ferromagnetic FePt nanocrystal superlattices, Science (80-.). 287 (2000)

[65]   S. Sun, Recent Advances in Chemical Synthesis, Self-Assembly, and Applications of FePt Nanoparticles, Adv. Mater. 18 (2006) 393–403. https://doi.org/10.1002/adma.200501464

[66]   C. Wang, S. Peng, L.-M. Lacroix, S. Sun, Synthesis of high magnetic moment CoFe nanoparticles via interfacial diffusion in core/shell structured Co/Fe nanoparticles, Nano Res. 2 (2009) 380–385. https://doi.org/10.1007/s12274-009-9037-4

[67]   M. V Kovalenko, M.I. Bodnarchuk, R.T. Lechner, G. Hesser, F. Schäffler, W. Heiss, Fatty acid salts as stabilizers in size- and shape-controlled nanocrystal synthesis: The case of inverse spinel iron oxide, J. Am. Chem. Soc. 129 (2007) 6352–6353. https://doi.org/10.1021/ja0692478

[68]   P. Guardia, A. Riedinger, S. Nitti, G. Pugliese, S. Marras, A. Genovese, M.E. Materia, C. Lefevre, L. Manna, T. Pellegrino, One pot synthesis of monodisperse water soluble iron oxide nanocrystals with high values of the specific absorption rate, J. Mater. Chem. B. 2 (2014) 4426–4434. https://doi.org/10.1039/C4TB00061G

[69]   D. Kim, N. Lee, M. Park, B.H. Kim, K. An, T. Hyeon, Synthesis of uniform ferrimagnetic magnetite nanocubes, J. Am. Chem. Soc. 131 (2009) 454–455. https://doi.org/10.1021/ja8086906

[70]   D.M.A. Neto, R.M. Freire, J. Gallo, T.M. Freire, D.C. Queiroz, N.M.P.S. Ricardo, I.F. Vasconcelos, G. Mele, L. Carbone, S.E. Mazzetto, M. Bañobre-López, P.B.A. Fechine, Rapid sonochemical approach produces functionalized Fe_3O_4 nanoparticles with excellent magnetic, colloidal, and relaxivity properties for MRI application, J. Phys. Chem. C. 121 (2017) 24206–24222. https://doi.org/10.1021/acs.jpcc.7b04941

[71]   W.S. Galvão, D.M.A. Neto, R.M. Freire, P.B.A. Fechine, Superparamagnetic nanoparticles with spinel structure: a review of synthesis and biomedical applications, Solid State Phenom. (2015). https://doi.org/10.4028/www.scientific.net/SSP.241.139

[72]   Y. Snoussi, S. Bastide, M. Abderrabba, M.M. Chehimi, Sonochemical synthesis of Fe_3O_4@NH_2-mesoporous silica@Polypyrrole/Pd: A core/double shell nanocomposite for catalytic applications, Ultrason. Sonochem. 41 (2018) 551–561. https://doi.org/10.1016/j.ultsonch.2017.10.021

[73]   J.S. Beveridge, J.R. Stephens, M.E. Williams, The use of magnetic nanoparticles in analytical chemistry, Annu. Rev. Anal. Chem. 4 (2011) 251–273. https://doi.org/10.1146/annurev-anchem-061010-114041

[74]   T.A.P. Rocha-Santos, Sensors and biosensors based on magnetic nanoparticles, TrAC Trends Anal. Chem. 62 (2014) 28–36. https://doi.org/ 10.1016/j.trac.2014.06.016

[75]   K. El-Boubbou, Magnetic iron oxide nanoparticles as drug carriers: Clinical relevance, Nanomedicine. 13 (2018) 953–971. https://doi.org/10.2217/nnm-2017-0336

[76]   J. Estelrich, E. Escribano, J. Queralt, M.A. Busquets, Iron Oxide Nanoparticles for Magnetically-Guided and Magnetically-Responsive Drug Delivery., Int. J. Mol. Sci. 16 (2015) 8070–8101. https://doi.org/10.3390/ijms16048070

[77]   F. Fiorillo, Characterization and measurement of magnetic materials, A volume in Elsevier Series in Electromagnetism, Academic Press, 2004.

[78]   N.Ž. Knežević, I. Gadjanski, J.-O. Durand, Magnetic nanoarchitectures for cancer sensing, imaging and therapy, J. Mater. Chem. B. 7 (2019) 9–23. https://doi.org/10.1039/C8TB02741B

[79]   S.P. Pujari, L. Scheres, A.T.M. Marcelis, H. Zuilhof, Covalent surface modification of oxide surfaces, Angew. Chemie - Int. Ed. 53 (2014) 6322–6356. https://doi.org/10.1002/anie.201306709

[80]   S. Carinelli, M. Martí, S. Alegret, M.I. Pividori, Biomarker detection of global infectious diseases based on magnetic particles, New Biotechnol. 32 (2015) 521–532. https://doi.org/ 10.1016/j.nbt.2015.04.002

[81]   W. Wu, Z. Wu, T. Yu, C. Jiang, W.-S. Kim, Recent progress on magnetic iron oxide nanoparticles: synthesis, surface functional strategies and biomedical applications, Sci. Technol. Adv. Mater. 16 (2015) 023501. https://doi.org/10.1088/1468-6996/16/2/023501

[82]   K. Hola, Z. Markova, G. Zoppellaro, J. Tucek, R. Zboril, Tailored functionalization of iron oxide nanoparticles for MRI, drug delivery, magnetic separation and immobilization of biosubstances, Biotechnol. Adv. 33 (2015) 1162–1176. https://doi.org/ 10.1016/j.biotechadv.2015.02.003

[83]   H. Cai, X. An, J. Cui, J. Li, S. Wen, K. Li, M. Shen, L. Zheng, G. Zhang, X. Shi, Facile hydrothermal synthesis and surface functionalization of polyethyleneimine-coated iron oxide nanoparticles for biomedical applications., ACS Appl. Mater. Interfaces 5 (2013) 1722–31. https://doi.org/10.1021/am302883m

[84]   J. Zeng, L. Jing, Y. Hou, M. Jiao, R. Qiao, Q. Jia, C. Liu, F. Fang, H. Lei, M. Gao, Anchoring Group Effects of Surface Ligands on Magnetic Properties of Fe3O4 Nanoparticles: Towards High Performance MRI Contrast Agents, Adv. Mater. 26 (2014) 2694–2698. https://doi.org/10.1002/adma.201304744

[85] R.M. Bezerra, D.M.A. Neto, W.S. Galvão, N.S. Rios, A.C.L. d. M. Carvalho, M.A. Correa, F. Bohn, R. Fernandez-Lafuente, P.B.A. Fechine, M.C. de Mattos, J.C.S. dos Santos, L.R.B. Gonçalves, Design of a lipase-nano particle biocatalysts and its use in the kinetic resolution of medicament precursors, Biochem. Eng. J. 125 (2017) 104-115. https://doi.org/10.1016/j.bej.2017.05.024

[86] C. Monteil, N. Bar, B. Moreau, R. Retoux, A. Bee, D. Talbot, D. Villemin, Phosphonated polyethylenimine-coated nanoparticles: Size- and zeta-potential-adjustable nanomaterials, Part. Part. Syst. Charact. 31 (2014) 219–227. https://doi.org/10.1002/ppsc.201300185

[87] S.A. McCarthy, G.-L. Davies, Y.K. Gun'ko, Preparation of multifunctional nanoparticles and their assemblies, Nat. Protoc. 7 (2012) 1677-1693. https://doi.org/10.1038/nprot.2012.082

[88] T. Gillich, C. Acikgöz, L. Isa, A.D. Schlüter, N.D. Spencer, M. Textor, PEG-stabilized core–shell nanoparticles: Impact of linear versus dendritic polymer shell architecture on colloidal properties and the reversibility of temperature-induced aggregation, ACS Nano 7 (2013) 316–329. https://doi.org/10.1021/nn304045q

[89] C. Grüttner, K. Müller, J. Teller, F. Westphal, A. Foreman, R. Ivkov, Synthesis and antibody conjugation of magnetic nanoparticles with improved specific power absorption rates for alternating magnetic field cancer therapy, J. Magn. Magn. Mater. 311 (2007) 181–186. https://doi.org/ 10.1016/j.jmmm.2006.10.1151

[90] C. Fang, O. Veiseh, F. Kievit, N. Bhattarai, F. Wang, Z. Stephen, C. Li, D. Lee, R.G. Ellenbogen, M. Zhang, Functionalization of iron oxide magnetic nanoparticles with targeting ligands: their physicochemical properties and in vivo behavior, Nanomedicine 5 (2010) 1357–1369. https://doi.org/10.2217/nnm.10.55

[91] D.L.J. Thorek, ew R. Elias, A. Tsourkas, Comparative analysis of nanoparticle-antibody conjugations: Carbodiimide versus click chemistry, Mol. Imaging. 8 (2009) 221-229. https://doi.org/10.2310/7290.2009.00021

[92] A.Z. Wang, V. Bagalkot, C.C. Vasilliou, F. Gu, F. Alexis, L. Zhang, M. Shaikh, K. Yuet, M.J. Cima, R. Langer, P.W. Kantoff, N.H. Bander, S. Jon, O.C. Farokhzad, Superparamagnetic iron oxide nanoparticle–aptamer bioconjugates for combined prostate cancer imaging and therapy, ChemMedChem 3 (2008) 1311–1315. https://doi.org/10.1002/cmdc.200800091

[93] N. Kohler, G.E. Fryxell, M. Zhang, A bifunctional poly(ethylene glycol) silane immobilized on metallic oxide-based nanoparticles for conjugation with cell targeting agents, J. Am. Chem. Soc. 126 (2004) 7206–7211. https://doi.org/10.1021/ja049195r

[94] K. Kluchova, R. Zboril, J. Tucek, M. Pecova, L. Zajoncova, I. Safarik, M. Mashlan, I. Markova, D. Jancik, M. Sebela, H. Bartonkova, V. Bellesi, P. Novak, D. Petridis, Superparamagnetic maghemite nanoparticles from solid-state synthesis – Their functionalization towards peroral MRI contrast agent and magnetic carrier for trypsin immobilization, Biomaterials 30 (2009) 2855–2863. https://doi.org/ 10.1016/j.biomaterials.2009.02.023

[95] M. Pereira, E.P.C. Lai, Capillary electrophoresis for the characterization of quantum dots after non-selective or selective bioconjugation with antibodies for immunoassay, J. Nanobiotechnol. 6 (2008) 10. https://doi.org/10.1186/1477-3155-6-10

[96] L. Johansson, K. Gunnarsson, S. Bijelovic, K. Eriksson, A. Surpi, E. Göthelid, P. Svedlindh, S. Oscarsson, A magnetic microchip for controlled transport of attomole levels of proteins, Lab Chip 10 (2010) 654–661. https://doi.org/10.1039/B919893H

[97] V. Biju, Chemical modifications and bioconjugate reactions of nanomaterials for sensing, imaging, drug delivery and therapy, Chem. Soc. Rev. 43 (2014) 744–764. https://doi.org/10.1039/C3CS60273G

[98] J. He, M. Huang, D. Wang, Z. Zhang, G. Li, Journal of Pharmaceutical and Biomedical Analysis Magnetic separation techniques in sample preparation for biological analysis : A review, J. Pharm. Biomed. Anal. 101 (2014) 84–101. https://doi.org/10.1016/j.jpba.2014.04.017

[99] N. Kishikawa, N. Kuroda, Analytical techniques for the determination of biologically active quinones in biological and environmental samples, J. Pharm. Biomed. Anal. 87 (2014) 261–270. https://doi.org/ 10.1016/j.jpba.2013.05.035

[100] F. Aflatouni, M. Soleimani, Preparation of a new polymerized ionic liquid-modified magnetic nano adsorbent for the extraction and preconcentration of nitrate and nitrite anions from environmental water samples, Chromatographia 81 (2018) 1475–1486. https://doi.org/10.1007/s10337-018-3590-5

[101] B. Zargar, A. Khazaeifar, Synthesis of an ion-imprinted sorbent by surface imprinting of magnetized carbon nanotubes for determination of trace amounts of cadmium ions, Microchim. Acta. 184 (2017) 4521–4529. https://doi.org/10.1007/s00604-017-2489-4

[102] L. Chen, W. Lu, J. You, J. Li, X. Zhang, Y. Sheng, One-pot synthesis of magnetic iron oxide nanoparticle-multiwalled carbon nanotube composites for enhanced removal of Cr(VI) from aqueous solution, J. Colloid Interface Sci. 505 (2017) 1134–1146. https://doi.org/10.1016/j.jcis.2017.07.013

[103] S.A. Mohamedsaid, M. Soylak, S. Ozdemir, E. Kilinc, A. Yıldırım, Application of magnetized fungal solid phase extractor with Fe_2O_3 nanoparticle for determination and

preconcentration of Co(II) and Hg(II) from natural water samples, Microchem. J. 143 (2018) 198–204. https://doi.org/10.1016/j.microc.2018.07.032

[104] Y. Zhang, R. Liu, Y. Hu, G. Li, Microwave heating in preparation of magnetic molecularly imprinted polymer beads for trace triazines analysis in complicated samples, Anal. Chem. 81 (2009) 967–976. https://doi.org/10.1021/ac8018262

[105] C. Zhou, Z. Du, G. Li, Y. Zhang, Z. Cai, Oligomers matrix-assisted dispersion of high content of carbon nanotubes into monolithic column for online separation and enrichment of proteins from complex biological samples, Analyst 138 (2013) 5783–5790. https://doi.org/10.1039/C3AN00951C

[106] J.R. Wiśniewski, D.F. Zielinska, M. Mann, Comparison of ultrafiltration units for proteomic and N-glycoproteomic analysis by the filter-aided sample preparation method, Anal. Biochem. 410 (2011) 307–309. https://doi.org/10.1016/j.ab.2010.12.004

[107] X. Xu, R.A. Sherry, S. Niu, D. Li, Y. Luo, Net primary productivity and rain-use efficiency as affected by warming, altered precipitation, and clipping in a mixed-grass prairie, Glob. Chang. Biol. 19 (2013) 2753–2764. https://doi.org/10.1111/gcb.12248

[108] Y. Liu, G. Yan, M. Gao, X. Zhang, Magnetic capture of polydopamine-encapsulated Hela cells for the analysis of cell surface proteins, J. Proteomics 172 (2018) 76–81. https://doi.org/ 10.1016/j.jprot.2017.10.009

[109] C. Rejeeth, X. Pang, R. Zhang, W. Xu, X. Sun, B. Liu, J. Lou, J. Wan, H. Gu, W. Yan, K. Qian, Extraction, detection, and profiling of serum biomarkers using designed Fe3O4@SiO2@HA core–shell particles, Nano Res. 11 (2018) 68–79. https://doi.org/10.1007/s12274-017-1591-6

[110] V. Natarov, D. Kotsikau, V. Survilo, A. Gilep, V. Pankov, Facile bulk preparation and structural characterization of agglomerated γ-Fe_2O_3/SiO_2 nanocomposite particles for nucleic acids isolation and analysis, Mater. Chem. Phys. 219 (2018) 109–119. https://doi.org/ 0.1016/j.matchemphys.2018.08.011

[111] S. Carinelli, C. Xufré, S. Alegret, M. Martí, M.I. Pividori, Talanta CD4 quanti fi cation based on magneto ELISA for AIDS diagnosis in low resource settings, Talanta 160 (2016) 36–45. https://doi.org/10.1016/j.talanta.2016.06.055

[112] G. Yao, D. Qi, C. Deng, X. Zhang, Functionalized magnetic carbonaceous microspheres for trypsin immobilization and the application to fast proteolysis, J. Chromatogr. A. 1215 (2008) 82–91. https://doi.org/ 10.1016/j.chroma.2008.10.114

[113] S. Padash Hooshyar, R.Z. Mehrabian, H. Ahmad Panahi, M. Habibi Jouybari, H. Jalilian, Synthesis and characterization of PEGylated dendrimers based on magnetic nanoparticles for letrozole extraction and determination in body fluids and

pharmaceutical samples, Microchem. J. 143 (2018) 190–197. https://doi.org/
10.1016/j.microc.2018.08.012

[114] P.-C. Lin, P.-H. Chou, S.-H. Chen, H.-K. Liao, K.-Y. Wang, Y.-J. Chen, C.-C.
Lin, Ethylene Glycol-Protected Magnetic Nanoparticles for a Multiplexed
Immunoassay in Human Plasma, Small. 2 (2006) 485–489.
https://doi.org/10.1002/smll.200500387

[115] J. Gao, K. Meyer, K. Borucki, P.M. Ueland, Multiplex Immuno-MALDI-TOF MS
for Targeted Quantification of Protein Biomarkers and Their Proteoforms Related to
Inflammation and Renal Dysfunction, Anal. Chem. 90 (2018) 3366–3373.
https://doi.org/10.1021/acs.analchem.7b04975

[116] P.-C. Lin, M.-C. Tseng, A.-K. Su, Y.-J. Chen, C.-C. Lin, Functionalized magnetic
nanoparticles for small-molecule isolation, identification, and quantification, Anal.
Chem. 79 (2007) 3401–3408. https://doi.org/10.1021/ac070195u

[117] S. Tang, G.H. Chia, Y. Chang, H.K. Lee, Automated dispersive solid-phase
extraction using dissolvable Fe_3O_4-layered double hydroxide core-shell microspheres
as sorbent, Anal. Che. 86 (2014)11070-11076. https://doi.org/10.1021/ac503323e

[118] W. Zhang, Y. Zhang, Q. Jiang, W. Zhao, A. Yu, H. Chang, X. Lu, F. Xie, B. Ye,
S. Zhang, Tetraazacalix[2]arence[2]triazine Coated Fe_3O_4/SiO_2 magnetic nanoparticles
for simultaneous dispersive solid phase extraction and determination of trace
multitarget analytes, Anal. Chem. 88 (2016) 10523–10532.
https://doi.org/10.1021/acs.analchem.6b02583

[119] C. Bendicho, C. Bendicho-Lavilla, I. Lavilla, Nanoparticle-assisted chemical
speciation of trace elements, TrAC Trends Anal. Chem. 77 (2016) 109–121.
https://doi.org/ 10.1016/j.trac.2015.12.015

[120] K. Aguilar-Arteaga, J.A. Rodriguez, E. Barrado, Magnetic solids in analytical
chemistry: A review, Anal. Chim. Acta. 674 (2010) 157–165. https://doi.org/
10.1016/j.aca.2010.06.043

[121] L. Xie, R. Jiang, F. Zhu, H. Liu, G. Ouyang, Application of functionalized
magnetic nanoparticles in sample preparation, Anal. Bioanal. Chem. 406 (2014) 377–
399. https://doi.org/10.1007/s00216-013-7302-6

[122] A.A. Hernández-hernández, G.A. Álvarez-romero, E. Contreras-lópez, K. Aguilar-
arteaga, A. Castañeda-ovando, Food analysis by microextraction methods based on the
use of magnetic nanoparticles as supports: Recent advances, Food Anal. Methods 10
(2017) 2974–2993. https://doi.org/10.1007/s12161-017-0863-9

[123] M. Hasanzadeh, N. Shadjou, M. de la Guardia, Iron and iron-oxide magnetic nanoparticles as signal-amplification elements in electrochemical biosensing, TrAC - Trends Anal. Chem. 72 (2015) 1–9. https://doi.org/10.1016/j.trac.2015.03.016

[124] V. Urbanova, M. Magro, A. Gedanken, D. Baratella, F. Vianello, R. Zboril, Nanocrystalline iron oxides, composites, and related materials as a platform for electrochemical, magnetic, and chemical biosensors, Chem. Mater. 26 (2014) 6653–6673. https://doi.org/10.1021/cm500364x

[125] F. Scholz, Electroanalytical methods, Springer Heidelberg Dordrecht London New York, 2010

[126] P. Yáñez-Sedeño, S. Campuzano, J.M. Pingarrón, Electrochemical sensors based on magnetic molecularly imprinted polymers: A review, Anal. Chim. Acta. 960 (2017) 1–17. https://doi.org/https://doi.org/10.1016/j.aca.2017.01.003

[127] S. Ansari, Application of magnetic molecularly imprinted polymer as a versatile and highly selective tool in food and environmental analysis: Recent developments and trends, TrAC Trends Anal. Chem. 90 (2017) 89–106. https://doi.org/https://doi.org/10.1016/j.trac.2017.03.001

[128] Q. Han, X. Shen, W. Zhu, C. Zhu, X. Zhou, H. Jiang, Magnetic sensing film based on Fe_3O_4@Au-GSH molecularly imprinted polymers for the electrochemical detection of estradiol, Biosens. Bioelectron. 79 (2016) 180–186. https://doi.org/https://doi.org/10.1016/j.bios.2015.12.017

[129] J.E. Lofgreen, G.A. Ozin, Controlling morphology and porosity to improve performance of molecularly imprinted sol–gel silica, Chem. Soc. Rev. 43 (2014) 911–933. https://doi.org/10.1039/C3CS60276A

[130] N. Wongkaew, M. Simsek, C. Griesche, A.J. Baeumner, Functional nanomaterials and nanostructures enhancing electrochemical biosensors and lab-on-a-chip performances : Recent progress , applications , and future perspective, Chem. Rev. 119 (2018) 120–194. https://doi.org/10.1021/acs.chemrev.8b00172

[131] Z. Sun, W. Wang, H. Wen, C. Gan, H. Lei, Y. Liu, Sensitive electrochemical immunoassay for chlorpyrifos by using flake-like Fe_3O_4 modified carbon nanotubes as the enhanced multienzyme label, Anal. Chim. Acta. 899 (2015) 91–99. https://doi.org/10.1016/j.aca.2015.09.057

[132] Z. Zhu, C. Yang, Recent progress in micro fl uidics-based biosensing, Anal. Chem. 9 (2019) 388-404. https://doi.org/10.1021/acs.analchem.8b05007

[133] Y. Song, B. Lin, T. Tian, X. Xu, W. Wang, Q. Ruan, J. Guo, Z. Zhu, C. Yang, Recent progress in microfluidics-based biosensing, Anal. Chem. 91 (2019) 388–404. https://doi.org/10.1021/acs.analchem.8b05007

[134] J.-R. Lee, D.J.B. Bechstein, C.C. Ooi, A. Patel, R.S. Gaster, E. Ng, L.C. Gonzalez, S.X. Wang, Magneto-nanosensor platform for probing low-affinity protein–protein interactions and identification of a low-affinity PD-L1/PD-L2 interaction, Nat. Commun. 7 (2016) 12220. https://doi.org/ 10.1038/ncomms12220

[135] I. Giouroudi, G. Kokkinis, Recent Advances in Magnetic Microfluidic Biosensors, Nanomaterials 7 (2017) 171. https://doi.org/10.3390/nano7070171

[136] M.A. Brown, R.C. Semelka, B.M. Dale, MRI: basic principles and applications, John Wiley & Sons, 2015

[137] E. Peng, F. Wang, J.M. Xue, Nanostructured magnetic nanocomposites as MRI contrast agents, J. Mater. Chem. B. 3 (2015) 2241–2276. https://doi.org/10.1039/C4TB02023E

[138] N. Lee, T. Hyeon, Designed synthesis of uniformly sized iron oxide nanoparticles for efficient magnetic resonance imaging contrast agents, Chem. Soc. Rev. 41 (2012) 2575–2589. https://doi.org/10.1039/c1cs15248c

[139] J. Wahsner, E.M. Gale, A. Rodríguez-Rodríguez, P. Caravan, Chemistry of MRI contrast agents: Current challenges and new frontiers, Chem. Rev. 119 (2019) 957–1057. https://doi.org/10.1021/acs.chemrev.8b00363

[140] J. Lu, J. Sun, F. Li, J. Wang, J. Liu, D. Kim, C. Fan, T. Hyeon, D. Ling, Highly Sensitive Diagnosis of Small Hepatocellular Carcinoma Using pH-Responsive Iron Oxide Nanocluster Assemblies, J. Am. Chem. Soc. 140 (2018) 10071–10074. https://doi.org/10.1021/jacs.8b04169

[141] T. Atanasijevic, A. Jasanoff, Preparation of iron oxide-based calcium sensors for MRI, Nat. Protoc. 2 (2007) 2582. https://doi.org/ 10.1038/nprot.2007.377

[142] T. Atanasijevic, M. Shusteff, P. Fam, A. Jasanoff, Calcium-sensitive MRI contrast agents based on superparamagnetic iron oxide nanoparticles and calmodulin, Proc. Natl. Acad. Sci. 103 (2006) 14707 LP-14712. https://doi.org/10.1073/pnas.0606749103

[143] J. Choi, S. Kim, D. Yoo, T.-H. Shin, H. Kim, M.D. Gomes, S.H. Kim, A. Pines, J. Cheon, Distance-dependent magnetic resonance tuning as a versatile MRI sensing platform for biological targets, Nat. Mater. 16 (2017) 537

[144] D. Alcantara, S. Lopez, M.L. García-Martin, D. Pozo, Iron oxide nanoparticles as magnetic relaxation switching (MRSw) sensors: Current applications in nanomedicine, nanomedicine nanotechnology, Biol. Med. 12 (2016) 1253–1262. https://doi.org/https://doi.org/10.1016/j.nano.2016.01.005

[145] Y. Zhang, H. Yang, Z. Zhou, K. Huang, S. Yang, G. Han, Recent Advances on Magnetic Relaxation Switching Assay-Based Nanosensors, Bioconjug. Chem. 28 (2017) 869–879. https://doi.org/10.1021/acs.bioconjchem.7b00059

[146] Y. Li, J. Liu, Y. Fu, Q. Xie, Y. Li, Magnetic-core@dual-functional-shell nanocomposites with peroxidase mimicking properties for use in colorimetric and electrochemical sensing of hydrogen peroxide, Microchim. Acta 186 (2018) 20. https://doi.org/10.1007/s00604-018-3116-8

[147] N. Lu, M. Zhang, L. Ding, J. Zheng, C. Zeng, Y. Wen, G. Liu, A. Aldalbahi, J. Shi, S. Song, X. Zuo, L. Wang, Yolk–shell nanostructured Fe_3O_4@C magnetic nanoparticles with enhanced peroxidase-like activity for label-free colorimetric detection of H_2O_2 and glucose, Nanoscale 9 (2017) 4508–4515. https://doi.org/10.1039/C7NR00819H

[148] J. Shen, Y. Yang, Y. Zhang, H. Yang, Z. Zhou, S. Yang, Functionalized Au-Fe_3O_4 nanocomposites as a magnetic and colorimetric bimodal sensor for melamine, Sens. Actuators B Chem. 226 (2016) 512–517. https://doi.org/10.1016/j.snb.2015.12.029

[149] H. Lai, F. Xu, L. Wang, A review of the preparation and application of magnetic nanoparticles for surface-enhanced Raman scattering, J. Mater. Sci. 53 (2018) 8677–8698. https://doi.org/10.1007/s10853-018-2095-9

[150] Z. Rong, C. Wang, J. Wang, D. Wang, R. Xiao, S. Wang, Magnetic immunoassay for cancer biomarker detection based on surface-enhanced resonance Raman scattering from coupled plasmonic nanostructures, Biosens. Bioelectron. 84 (2016) 15–21. https://doi.org/ 10.1016/j.bios.2016.04.006

Magnetochemistry - Materials and Applications
Materials Research Foundations **66** (2020) 217-235

Materials Research Forum LLC
https://doi.org/10.21741/9781644900611-6

Chapter 6

Magnetic Nanomaterials for Fuel Cells

Tuerxun Duolikun[1,2], Paul Thomas[2], Chin Wei Lai[2*], Bey Fen Leo[2,3*]

[1]Department of Mechanical Engineering, Faculty of Engineering, University of Malaya, 50603 Kuala Lumpur, Malaysia

[2]Nanotechnology & Catalysis Research Centre (NANOCAT), Institute for Advanced Studies, University of Malaya, 50603 Kuala Lumpur, Malaysia

[3]Faculty of Medicine, University of Malaya, 50603 Kuala Lumpur, Malaysia

*beyfenleo@um.edu.my, cwlai@um.edu.my

Abstract

In modern days, sustainable energy demand and its storage are the most challenging concern in the modern world. To meet the rising energy demands, there is a need to diversify energy sources which require extensively altered and sustainable materials for energy conversion, storage, generation, distribution and applications. There is significant progress in the field of energy generation, storage and conversion, in particular batteries, supercapacitors and fuel cells. The emergence of magnetic nanomaterials has resulted in considerable contributions towards the advancement in the energy industry. Hence, magnetic nanocomposites are introduced to high-performance fuel cells. This book chapter discusses the importance of magnetic nanomaterials for fuel cell applications. We mainly address the magnetic nanomaterials' synthesis and their applications to fuel cells. As our society upgraded to industrial 4.0, alternative greener and cleaner energy to fossil fuels are the goal. Starting from the first commercialization of fuel cells by NASA for space vehicle, R&D work continually discovers new potential applications for fuel cells and recently, pays much attention to materials being able to decrease the price, increase work efficiency and being eco-friendly.

Keywords

Magnetic Nanomaterial, Fuel Cell, Energy Production, Electrochemistry, Catalysts

Contents

1.   Introduction

In spite of the rising demand for energy alternatives, energy remains one of the essential factors that determine the sustainable development goals of human beings. It is one of the critical issues related to reducing carbon dioxide emissions and global warming, causing researchers to find a new mode of energy generation [1]. In 1839, scientist Sir William Robert Grove who anticipated the general theory of the conservation of energy, as well as first demonstrated the fuel cell technology. The 1950s, fuel cell technology was successfully developed by British engineer Francis Thomas Bacon's pioneering work for the NASA Apollo space program, and this modernisation is continually used for contemporary shuttle missions [2, 3]. Fuel cells are electrochemical cells using endearing power-generation technology that permutes chemical energy directly into electrical energy with imposing efficiency and are extensively thought of as green energy sources and increased energy security [3-5].

Magnetochemistry - Materials and Applications

Materials Research Forum LLC

Materials Research Foundations **66** (2020) 217-235

https://doi.org/10.21741/9781644900611-6

Fuel cells come in myriad varieties but operate in the same regular mode. They consist of an electrolyte, a cathode and an anode [6]. Nowadays, different sorts of fuel cells can be imminent from the kind of electrolyte: alkaline fuel cells (AFC), proton exchange membrane fuel cells (PEMFC), solid oxide fuel cells (SOFC), phosphoric acid fuel cells (PACF), molten carbonate fuel cells (MCFC), solid acid fuel cell (SAFC), high-temperature fuel cells and electric storage fuel cell, etc. [3]. The most common types of fuel cells information are compiled in Table 1.

Table 1. Comparison of the major fuel cell types

Fuel cell type	Mobile ion	working temperature	Efficiencies (cell)
AFC	OH^-	50-200°C	reach 70%
PEMFC	H^+	50 to 100 °C	50–60%
SOFC	O^{2-}	500 to 1000 °C	~50%
PAFC	H^+	150 to 200 °C	up to 70%
MCFC	CO_3^{2-}	600 °C and above	60%
DMFC	H^+	20-90 °C	20–30%

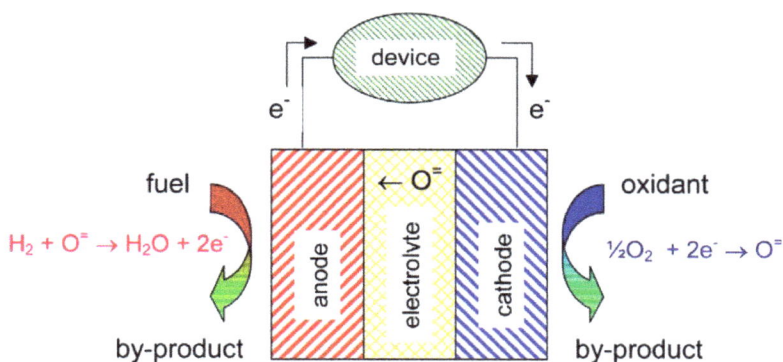

Figure 1. Illustrative of a fuel cells working mechanism, consisted of an electrolyte, a cathode and an anode. The complete chemical reaction is $H_2 + \frac{1}{2}O_2 \rightarrow H_2O$. Anode and cathode reactions given are convenient only for oxide ion directing electrolytes [5].

Mainly, the fuel cell can be split into two categories based on the manipulation temperature:

a)   Low Temperature Fuel Cells (LTFCs). The operating temperature is < 200 °C. The advantages are high efficiency and inexpensive.

b)   High-Temperature Fuel Cells (HTFCs). The operating temperature is generally above 500 °C. The shortages are low powers with high cost [7].

A fuel cell is favoured in energy research due to its excellent supremacy such as efficiency, simplicity, almost 'zero emission' and the quiet working environment — however, lower cost for some of the fuel cell types is needed. To date, research works continues on a fuel cell system as a true alternative to fossil fuels.

Fuel cells have attracted attention among researchers and are considered to become potential commercial products. First commercialisation of fuel cells has been introduced through the National Aeronautics and Space Administration (USA). Thereafter, fuel cells have been used in numerous other applications, including power, cogeneration, fuel cell electric vehicles and portable power system, etc.[6].

In January 2002, the Department of Energy (USA) and Council for Automotive Research drew a Blueprint of fuel cell vehicles (FCV) technology [8]. Some of the famous vehicle manufacturers introduce the use of FCV in the worldwide market, such as the Honda Clarity, Toyota Mirai, the Hyundai Nexo, Nissan X-Trail, and Elon Musk's Tesla Motors which are being developed as greener and eco-friendly vehicles [9, 10]. Besides, the fuel cell has been used in mobile power systems, portable computers, and communications parts [2].

Recently, magnetic NPs have been introduced in fuel cell applications ascribable to the promising properties, especially magnetic, electroconductive, catalytic functions, more convenient processing methods, size and shape, surface chemistry and most importantly are eco-friendly [11-13].

2.   Magnetic nanomaterials

Magnetic nanomaterials (MNMs) have tempted influential interest in the past due to the outstanding properties and wide-range potential applications [14]. The MNMs have received intensive focus recently, as their attractive properties which could be used in the direct methanol fuel cell application, resonance imaging, fluids, biomedicine, data storage and catalysis [14-16]. Synthesis method and promising applications of MNMs are summarised in Fig.2, as shown below:

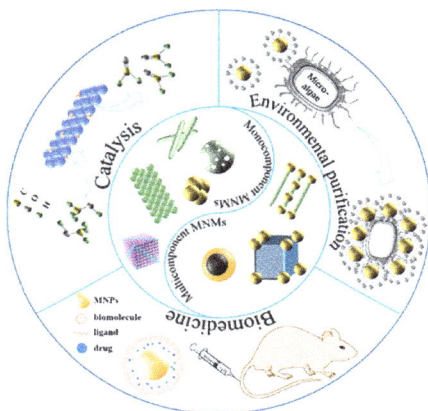

Figure 2. MNMs: Synthesis Methods and Promising Applications[14]

2.1   Monometallic magnetic nanomaterials

As the name implicates the monometallic nanoparticle embodies of a single metal. The subsists atom signifies the characteristics and properties of the NPs. The monometallic nanomaterial can be synthesised through various techniques relying on the sorts of metal atom presents like metallic, magnetic etc. However, the chemical synthesis approach is the most commonly utilised. Last few decades observed tremendous growth in research on the sector of magnetic metallic NPs due to the superior chemical and physical characteristics. Subsequently, they are utilised for numerous applications such as catalysis, electronics, fuel cell etc.

2.2   Bimetallic magnetic nanomaterials

The bimetallic nanoparticles (BNPs) constitute two of the different metals to enhance or to balance the requirements from both technological as well as a scientific point of view [17]. BNPs are synthesised via the amalgamation of dissimilar architectures of metallic nanomaterials. Only in recent decades bimetallic magnetic nanomaterials gained much popularity focused on the synthesis of bimetallic nanomaterials in disparate form namely contact aggregate, core/shell and alloys. As a consequence, the metallization leads to the strengthening of the catalytic characteristics compared to monometallic catalysts. In

Magnetochemistry - Materials and Applications   Materials Research Forum LLC
Materials Research Foundations **66** (2020) 217-235   https://doi.org/10.21741/9781644900611-6

bimetallic catalysts, electronic effects play a crucial role in charge transfer and electric features.

MNMs can be grouped as mono-component and multi-component magnetic nanostructures. The standard synthetic methods to produce MNMs, as shown in table 2 [14]. Each single reaction parameters in the synthesis process will significantly affect the final structure and performances of the MNMs [14]

MNMs can be used for highly efficient and lower pricing fuel cells development. However, it is difficult to obtain the monodisperse magnetic nanostructures due to the dipolar interactivities, surface and anti-oxidation perform, as well as the grains controlling techniques.

Table 2. The main chemical synthesis strategies of MNMs

Synthesis strategies	Type	Examples	Developed methods
Mono-component	Metal NPs	Fe, Co, Ni	High-temperature organic-phase method
	Metal Alloy	FePd, FePt, CoPt$_3$	Facile high-temperature organic-phase thermal
	Metallic Oxides	MnFe$_2$O$_4$, CoFe$_2$O$_4$, MgFe$_2$O$_4$	Organic-phase methods
	Metal Carbides	Fe$_2$C, Fe$_3$C, Fe$_5$C$_2$	Versatile solution chemistry route
Multi-component	Heterostructures	Fe$_3$O$_4$@Au@Ag, FePt−Au	The organic-phase method, Core-Shel method
	Exchange-Coupled	fct-FePt@Co, Nd$_2$Fe$_{14}$B/α-Fe	Seed-mediated synthesis

2.3   Chemical design and synthesis

The synthesis of NPs/MNPs has a vital role in all dimensions devoted to enhancing their suitability towards practical applicability. However, the commercial availability of nanomaterials eases the efforts of laboratories and research teams, which assures higher efficiency and reproducibility of the materials [20]. In spite of this fact, the synthesis of

NPs plays a vital role in research studies and is undoubtedly indispensable for the foundation in discovering new materials and its synthesis.

Materials focus on the synthesis of the magnetic segment can be further categorised while materials based upon compounds mainly nickel oxides, Fe, Co and other elements by amalgamating a few metals [21]. Generally, the utilised materials for the combination such as barium, zinc strontium and copper. Magnetic nanomaterials also consist of metallic nanoalloys and NPs. Various synthesis approaches have been widely discussed to synthesise magnetic NPs with submicron diameter without compromising its composition and characteristics. Recently Grasset and his colleagues reported the synthesis of $Y_3Fe_5O_{12}$ NPs using various synthesis approach: coprecipitation (microemulsion and hydroxide), metal alkoxides hydrolysis, glycothermal synthesis, citrate gel, glass crystallisation etc. [22-27]. Among the MNMs, iron oxide-based NPs, especially the ferrimagnetic categories maghemite and maghemite are the prominent ones due to their potential and suitability.

2.3.1 Thermal decomposition

As the name suggests thermal treatment technique such as pyrolysis of precursors in the solvent by overheating, the core limitation of this technique is the high-temperature conditions the detachment of unsteady nanocrystal phase from the unstable phase is challenging. This method is endothermal, while bond breaking requires high energy. Laser pyrolysis is another modern technique used to synthesise magnetic nanomaterials; the infrared laser has been employed as a heating source. Laser energy can be transferred to the precursors directly or indirectly through an inert photosensitizer. Laser-induced pyrolytic reactions have been reported in numerous articles regarding the production of magnetic nanomaterials with high purity magnetic nanomaterials with ultrafine size distribution [28, 29].

2.3.2 Micro-emulsion method

Micro-emulsions definite being a solution which is comprised of three constituents, to be specific, surfactant, polar and nonpolar component. The primary role of the surfactant is to configuration a layer within the nonpolar and polar component. Microemulsion is mainly classified into two types of oil in water or reverse depending on the continuous and dispersed phase. Generally, water in the oil-based microemulsion is used for synthesis; however, oil in water microemulsions has not been extensively utilised. Water in oil-based microemulsions, the NPs are synthesised using tiny reactors so-called reverse micelles; in reverse micelles, several water droplets scattered in the organic phase to aid of surfactant able to synthesise NPs with controllable morphology and size [30]. Hence morphology and size of the synthesised NPsNPs are proportional to the shape and size of the water droplets. Therefore NPs with dissimilar morphologies and sizes are able to be synthesised through optimising parameters (stirring time, stirring methods, rate of mixing, temperature and pH) of microemulsion method [31].

2.3.3 Sol-gel method

The sol-gel (SG) technique is a combination of gel and sol. The sol is a steady suspension of colloidal solid particles in the solvent, whereas the gel constitutes a 3D ceaseless network of the sol particles which encapsulate the liquid phase [32]. The gel composed of more solid concentration compares to liquid. It is a semi-rigid mass after the vaporisation particles start to develop a ceaseless network. The majority of the gel system exists the covalent interactivity. The fusion of these network effects known as SG synthesis, the synthesis process composed of both condensation and hydrolysis.

2.3.4 Hydrothermal method

In hydrothermal synthesis, the magnetic NPs are synthesised at high-temperature at 470 °C with pressure with less than 300 MPa. This hydrothermal approach enhances the dilution of the components which are insolubility under standard circumstances. However, the characteristics of the synthesised NPs depend upon the pressure, temperature and pH value of the medium. This technique is advantageous compared to other techniques as the monitoring of crystal growth can lead to the production of crystals of high quality and purity.

2.3.5 Solvothermal method

In hydrothermal synthesis, the solution is aqueous, while in solvothermal synthesis, the solution is non-aqueous. The solvothermal synthesis gains benefits from both hydrothermal and sol-gel method. The process carried out using solvent under moderate to high temperature (ranges between 100 to 1000°C) and pressure (ranges between 1 atm and 10,000 atm) that helps for better interaction of precursors during the synthesis.

Glycothermal synthesis is a solvothermal synthesis which utilises glycols as solvent [33]. The glycol molecules interact to the surface of the particles not only acts as a growing inhibitor but also acts as a dispersal stabiliser while glycol is being utilised as a solvent in solvothermal synthesis [34]. Numerous journals have reported the synthesis of ferromagnetic particles using glycothermal, and still new reports continue to come as solvothermal synthesis using glycol as a solvent allows better and precise control over structure and morphology of the synthesised NPs [35, 36]. The citrate gel technique is also utilised for the synthesis of magnetic nanometal oxides. The process involves pyrolysis at maximum temperature to synthesis quaternary, ternary and binary nanocomposites in the both amorphous and crystalline form [37]. Glycothermal synthesis technique is more advantageous as this technique produces more homogeneity. As the metal citrate gel gets heated, the organic component undergoes combustion at higher temperature ranges from 300 to 400°C, depending on the metal and additives.

2.4   Iron oxide magnetic nanoparticles

Iron oxide nanoparticles (IONPs) were widely used for numerous applications because of their magnetic behaviour, chemical stability, as well as biocompatibility makes it a suitable candidate [38]. There are multiple techniques used to synthesise IONPs, upon all existing various synthesis approaches, coprecipitation is a highly effective and most straightforward technique to synthesise IONPs. In this synthesis approach, an alkaline source (sodium hydroxide or ammonia) is added to a mixture of Fe^{3+} and Fe^{2+} salts. The synthesis approach is followed by washing the flocculate with a base or acid solutions. The particles shape, size and morphology can be coordinated by modifying various experimental factors, in particular ionic strength, pH, temperature, stoichiometry and nature of salts etc. [39, 40]. Stabilised magnetite colloids can be synthesised with carrying out the reaction in the existence of polymers like polyacrylic acid or dextran [41, 42]. The magnetic properties and physical characteristics of iron oxide-based NPs can be controlled using surfactants. Alkanolamines acts as a complexing agent, and the alkaline source was used to prepare small sized ferrites. This synthesis technique could control the size of nanoparticle synthesis and to improve the magnetic characteristics compare to the traditional approach [43]. Commonly, two main approaches and strategies were used to modify the shape and dimensions of the NPs. The seed meditated growth approach follows the adding of Fe precursor to pre-existent nuclei. This technique can be used to synthesise NPs of various sizes ranges from 4 nm to 16 nm. With another alternative approach the dimensions and morphologies of the NPs can be synthesised by optimising parameters, for instance, ligand/precursor ratio, reaction time, temperature and solvent with various boiling-points [44-46]. This technique of heating process produces NPs size ranges from 4 to 28 nm even if the coprecipitation could diversify the size and shape but is not able to control shape and size distribution.

The high-temperature decomposition technique can be utilised to yield uniform shaped and sized magnetic NPs with high crystallinity. However, this technique has certain limitations, such as it is expensive and toxic in nature [47]. The synthesised particles cannot be transferred directly to the aqueous media as a capping layer is surrounded by NPs. To maintain transfer to the aqueous media it needs the adding of amphiphilic ligands, e.g. lipids, alkylammonium salts and polymers or the exchange goes from hydrophobic to hydrophilic. The industrial-scale production needs to meet the challenges associated with cost-effectiveness, reproducibility and usage of innocuous reagents which still need to be overcome.

2.5 Other magnetic nanoparticles

Depending on the suitability of application, ultrafine magnetic NPs based on γ-Fe2O3 and Fe_3O_4 are commonly utilised [48]. Meanwhile, other oxides and alloys with appropriate characteristics, for example, $SrFe_{12}O_{19}$, $Y_3Fe_5O_{12}$ rarely are used for energy storage applications [49]. The synthesis of these materials using convention techniques is always challenging and could be synthesised using submicron particle preparation [22]. The synthesis of Ni-based MNPs with distinctive microstructure was reported and examined its suitability towards energy applications [50]. Conventional techniques for the synthesis of NiO MNPs have been reported, including sol-gel, spray pyrolysis, thermal decomposition, coprecipitation and polymer matrix assisted synthesis [51-53].

The liquid-phase chemical method is widely used to prepare MNMs. The common synthesis methods of MNMs are mentioned in the above section, except the methods including 1) Coprecipitation; 2) Flame spray synthesis, and 3) Laser pyrolysis techniques [16]. The synthesis methods of magnetic nanomaterials are clarified in Table 3.

3. Important considerations and limitation

The main issue in the process of the commercialisation is the weak transformation of fuel cell from chemical energy to electrical energy [54]. The significant challenges in the commercialisation of the fuel cell are the cost, durability, lower operating temperatures, materials designs and high power density issues. In the transportation application, the systems face further challenges of allotting and reserve hydrogen [7]. Since recent work observed that fuel starvation was the culprit of the declining voltages, a fuel starvation prevention has been included in the fuel cells system to resolve this issue [55].

The heat and liquid transporting in polymeric membrane fuel cells have evolved for appraisal of the structure in order to convert hydrogen into electricity directly. The dynamic simulation (*e.g.* mass transport limitation) allows simulation of the transitory state after variation of electrical load or flow rate of gas [56]. Besides, Takahisa *et al.* [57] proposed the model to investigate the oxygen transport limitation in the catalytical layer of polymeric electrolyte fuel cells. The fuel cell suffers from single cell output energy ascribed to the intrinsic cell size-limitation as microscale geometries are required to avert reactant crossing among the anode and the cathode. Li *et al.* [58] added a detached layer within the porous electrodes of the ordinary plate-frame microfluidic fuel cell to highlight the size-limitation issue over allowing significantly higher flow rates while retaining high fuel utilisation for the practical applications [58].

Table 3. Synthesis methods of magnetic nanoparticles

MNMs	Synthesis methods	References
Ferric oxide (Fe_3O_4)	$FeCl_2$, $FeCl_3$: 1) NaOH solutions were made via dissolving in DI H_2O with different ratios, and 2) iron solutions were added to NaOH solution.	[11]
Rod-shaped Fe_3O_4	$FeCl_3 \cdot 6H_2O$ and sodium hydroxide were mixture with ultra-pure water with 1-Butyl-3-methylimidazolium Chloride. The solution was heating to 380 – 400 ℃ until a dark-red rod-shaped Fe_3O_4 solution was formed.	[18]
Fe_3O_4@AuNPs	$Fe(NO_3)_3$ was diminished by ascorbic acid at ambient temperature under a nitrogen atmosphere. A dark solid showed that iron oxide core was successfully coated with an Au shell. Then, the sample was separated by a magnet and washed.	[18]
$L1_0$-CoPt/Pt NPs	Iron and Platinum in almost equal ratio, and Iron (3d) and Platinum (5d) atomic orbitals are coupled utterly along the crystallographic c direction, making the sample a hard magnet with Iron being effectually stabilised counter acid etching.	[19]

4.   Working principle of magnetic nanomaterials in fuel cells application

4.1   Electrochemical measurements of MNMs

Electroanalytical methods are part of the analytical chemistry techniques which measuring the potential (volts) and current (amperes). The cyclic voltammetry was used to study the electrochemical characterisation of the electrodes.

The polarisation curve assessed the effectiveness of the fuel cells system. This curve was gained by external resistance. Power (P) and current (I) were calculated derived from the following equations 1 and 2:

$$P = I \times E \tag{EQ. 1}$$

$$I = \frac{E}{R_{EXT}} \tag{EQ. 2}$$

where P is the power; E is the voltage of cell; R_{ext} is the external resistance, and I is the current.

The electrochemical behaviour of MNMs can be analysed at various scanning rates within the range of 10 to 100 mV s^{-1}. The selection of the benchmark redox system should be carefully carried out to analyse the surface sensitivity which is requires towards the electrochemical response. All the experiments were carried out at ambient temperature. In per electrochemical study involving the sensitivity, scanning rate and the manifestation of electrode need to be repeated at least three times [59]. The statistical studies of the electrochemical characterisation were done *via* GraphPad Prism 8 using one-way ANOVA for multiple comparisons.

5. Markets with Research and Development

The fuel cell technology attracts enormous R&D interest due to the descending possesses of the fossil oil and raised environmentally-pollution has requested of sustainable development goals to current automobile market which is petroleum based [9]. The FCV as an environmentally-friendly and being high efficient could change the automobile market. On the other hand, the government policy to supporting the eco-friendly sustainability strategies will empower the FCV for increasing market share. However, R&D continues to tackle the critical issues and the FCV by major automakers indicating growing interest. This interest coming out beside their broaden driving range and quick refuelling time comparative to FCV [7].

Table 4. Overall synthesis methods of MNPs

Methods	Appropriately	Comments	Main process
Coprecipitation	Iron oxides	A facile and convenient way	Aqueous ferric/ferrous salt solutions via adding the base under an inert atmosphere at ambient temperature.
Thermal Decomposition	Cr_2O_3, MnO, Co_3O_4, Magnetic alloys	Control the size and shape	The alkaline organo-metallic compounds in high-boiling organic solvents, including supporting surfactants.
Micro-emulsion	Metallic Co, Co/Pt alloys, Au coated Co/Pt NPs	The yield of NPs is low, inefficient and not easy to scale-up.	Reverse micelles of cetyltrimethylammonium bromide, used 1-butanol and octane.
Flame spray	Oxides, metal or carbon coated NPs	Thick coatings the sizable area with considerable deposition rate.	The melted materials are sprayed on the surface.
Hydrothermal	Fe_3O_4, $CoFe_2O_4$	can be formed the wide range of nanomaterials	The system consists of metal linoleate, an ethanol–linoleic acid liquid phase, and a H_2O_2-EtOH solution
Laser pyrolysis techniques	A large variety of oxide nanomaterials (TiO_2, SiO_2, Al_2O_3), nonoxide (Si_3N_4, MoS_2) and ternary composites (Si/C/N and Si/Ti/C)	High-quality, small-sized with the narrow size distribution	Using the heat flowing reactant gases, in molecular decay to initiate nucleation

Conclusion

After almost a century of slow research progress on the fuel cells, finally, the attention increased recently, due to its superior advantages. New engineered magnetic nanomaterials with excellent performances for wide range fuel cells application and other discovery work on process development/revolutions grow the interest. Nevertheless, the reduction of cost, system complexity, low conductive conditions and short life cycle remain as significant challenges in the development of fuel cells technology. Thus, fuel cell manufacturing R&D should focus on quality inspection for high volume (continuous) manufacturing processes to obtain higher yields, lower costs and increased reliability.

Acknowledgements

This study was supported by the University of Malaya Research Fund Assistance BK095-2016, the UMRG (RP045B-17AET & RP045D-17AET) and Global Collaborative Programme - SATU Joint Research Scheme (No. ST009-2018).

References

[1] E. Ogungbemi, O. Ijaodola, F. Khatib, T. Wilberforce, Z. El Hassan, J. Thompson, M. Ramadan, A. Olabi, Fuel cell membranes–Pros and cons, Energy 172 (2019) 155-172. https://doi.org/10.1016/j.energy.2019.01.034

[2] J. Larminie, A. Dicks, M.S. McDonald, Fuel cell systems explained: J. Wiley Chichester, UK, 2003. https://doi.org/10.1002/9781118878330

[3] G. Merle, M. Wessling, K. Nijmeijer, Anion exchange membranes for alkaline fuel cells: A review, J. Membrane Sci. 377 (2011) 1-35. https://doi.org/10.1016/j.memsci.2011.04.043

[4] E.P. Murray, T. Tsai, S.A. Barnett, A direct-methane fuel cell with a ceria-based anode, Nature 400 (1999) 649. https://doi.org/10.1038/23220

[5] S. M. Haile, Fuel cell materials and components, Acta Materialia 51 (2003) 5981-6000. https://doi.org/10.1016/j.actamat.2003.08.004

[6] Wikipedia. "Fuel cell," 06 May, 2019; https://en.wikipedia.org/wiki/Fuel_cell.

[7] J.-S. Lai, M. W. Ellis, Fuel cell power systems and applications, Proceedings of the IEEE, 105 (2017) 2166-2190. https://doi.org/10.1109/JPROC.2017.2723561

[8] V. Mehta, J. S. Cooper, Review and analysis of PEM fuel cell design and manufacturing, J. Power Sources 114 (2003) 32-53. https://doi.org/10.1016/S0378-7753(02)00542-6

[9]   H. Wang, A. Gaillard, D. Hissel, Online electrochemical impedance spectroscopy detection integrated with step-up converter for fuel cell electric vehicle, Int. J. Hydrogen Energy 44 (2019) 1110-1121. https://doi.org/10.1016/j.ijhydene.2018.10.242

[10]   L. NISSAN MOTOR CO. New X-TRAIL Fuel Cell Vehicle (FCV), 8 May, 2019; https://www.nissan-global.com/EN/TECHNOLOGY/OVERVIEW/fcv.html.

[11]   M. Rahimnejad, M. Ghasemi, G. Najafpour, M. Ismail, A. W. Mohammad, A. Ghoreyshi, S. H. Hassan, Synthesis, characterization and application studies of self-made Fe3O4/PES nanocomposite membranes in microbial fuel cell, Electrochim. Acta 85 (2012) 700-706. https://doi.org/10.1016/j.electacta.2011.08.036

[12]   X. Teng, X. Liang, S. Rahman, H. Yang, Porous nanoparticle membranes: Synthesis and application as fuel-cell catalysts, Adv. Mater. 17 (2005) 2237-2241. https://doi.org/10.1002/adma.200500614

[13]   M.F. Tai, C.W. Lai, S.B. Abdul Hamid, Facile synthesis polyethylene glycol coated magnetite NPs for high colloidal stability, J. Nanomater. 2016 (2016) 8612505. https://doi.org/10.1155/2016/8612505

[14]   K. Zhu, Y. Ju, J. Xu, Z. Yang, S. Gao, Y. Hou, Magnetic nanomaterials: Chemical design, synthesis, and potential applications, Acc. Chem. Res. 51 (2018) 404-413. https://doi.org/10.1021/acs.accounts.7b00407

[15]   N. Hasanabadi, S.R. Ghaffarian, M.M. Hasani-Sadrabadi, Magnetic field aligned nanocomposite proton exchange membranes based on sulfonated poly (ether sulfone) and Fe_2O_3 NPs for direct methanol fuel cell application, Int. J. Hydrogen Energy 36 (2011) 15323-15332. https://doi.org/10.1016/j.ijhydene.2011.08.068

[16]   A.H. Lu, E.L. Salabas, F. Schüth, Magnetic NPs: Synthesis, protection, functionalization, and application, Angew. Chem. Int. Ed. 46 (2007) 1222-1244. https://doi.org/10.1002/anie.200602866

[17]   G. Sharma, V. K. Gupta, S. Agarwal, A. Kumar, S. Thakur, and D. Pathania, Fabrication and characterization of Fe@MoPO NPs: Ion exchange behavior and photocatalytic activity against malachite green, J. Molecular Liquids 219 (2016) 1137-1143. https://doi.org/10.1016/j.molliq.2016.04.046

[18]   N. Atar, T. Eren, M. L. Yola, H. Karimi-Maleh, B. Demirdögen, Magnetic iron oxide and iron oxide@gold nanoparticle anchored nitrogen and sulfur-functionalized reduced graphene oxide electrocatalyst for methanol oxidation, RSC Adv. 5 (2015) 26402-26409. https://doi.org/10.1039/C5RA03735B

[19] J. Li, S. Sharma, X. Liu, Y.-T. Pan, J. S. Spendelow, M. Chi, Y. Jia, P. Zhang, D. A. Cullen, Z. Xi, Hard-magnet L10-CoPt NPs advance fuel cell catalysis, Joule 3 (2019) 124-135. https://doi.org/10.1016/j.joule.2018.09.016

[20] M. Mahmoudi, S. Sant, B. Wang, S. Laurent, T. Sen, Superparamagnetic iron oxide NPs (SPIONs): Development, surface modification and applications in chemotherapy, Adv. Drug Delivery Rev. 63 (2011) 24-46. https://doi.org/10.1016/j.addr.2010.05.006

[21] W. Wu, Q. He, C. Jiang, Magnetic iron oxide NPs: Synthesis and surface functionalization strategies, Nanoscale Res. Lett. 3 (2008) 397. https://doi.org/10.1007/s11671-008-9174-9

[22] F. Grasset, S. Mornet, A. Demourgues, J. Portier, J. Bonnet, A. Vekris, E. Duguet, Synthesis, magnetic properties, surface modification and cytotoxicity evaluation of $Y_3Fe_{5-x}Al_xO_{12}$ ($0 \leqslant x \leqslant 2$) garnet submicron particles for biomedical applications, J. Magn. Magn. Mater. 234 (2001) 409-418. https://doi.org/10.1016/S0304-8853(01)00386-9

[23] P. Vaqueiro, M.A. López-Quintela, Synthesis of yttrium aluminium garnet by the citrate gel process, J. Mater. Chem. 8 (1998) 161-163. https://doi.org/10.1039/a705635d

[24] M. Inoue, T. Nishikawa, T. Inui, Glycothermal synthesis of rare earth iron garnets, J. Mater. Res. 13 (1998) 856-860. https://doi.org/10.1557/JMR.1998.0114

[25] P. Vaqueiro, M. A. López-Quintela, J. Rivas, Synthesis of yttrium iron garnet NPsviacoprecipitation in microemulsion, J. Mater. Chem. 7 (1997) 501-504. https://doi.org/10.1039/a605403j

[26] S. Taketomi, Y. Ozaki, K. Kawasaki, S. Yuasa, H. Miyajima, Transparent magnetic fluid: Preparation of YIG ultrafine particles, J. Magn. Magn. Mater. 122 (1993) 6-9. https://doi.org/10.1016/0304-8853(93)91027-5

[27] P. Grosseau, A. Bachiorrini, B. Guilhot, Elaboration de poudres de yig par coprecipitation, J. Thermal Anal. Calorimetry 46 (1996) 1633-1644. https://doi.org/10.1007/BF01980769

[28] O. B. Miguel, M. Morales, C. Serna, S. Veintemillas-Verdaguer, Magnetic NPs prepared by laser pyrolysis, IEEE transactions on Magnetics 38 (2002) 2616-2618. https://doi.org/10.1109/TMAG.2002.801961

[29] F. Bensebaa, Nanoparticle technologies: From lab to market: Academic Press, 2012.

[30]   V. Uskoković, M. Drofenik, Synthesis of materials within reverse micelles, Surf. Rev. Lett. 12 (2005) 239-277. https://doi.org/10.1142/S0218625X05007001

[31]   H. Beygi, A. Babakhani, Microemulsion synthesis and magnetic properties of $Fe_xNi_{(1-x)}$ alloy NPs, J. Magn. Magn. Mater. 421 (2017) 77-183. https://doi.org/10.1016/j.jmmm.2016.07.071

[32]   G. Sharma, A.Kumar, S. Sharma, M. Naushad, R.P. Dwivedi, Z.A. ALOthman, G. T. Mola, Novel development of NPs to bimetallic NPs and their composites: A review, J. King Saud Uni. Sci. 31 (2017) 257-269. https://doi.org/10.1016/j.jksus.2017.06.012

[33]   M. Inoue, Glycothermal synthesis of metal oxides, J. Phys. Condensed Matter 16 (2004) S1291. https://doi.org/10.1088/0953-8984/16/14/042

[34]   H. Hara, S. Takeshita, T. Isobe, Y. Nanai, T. Okuno, T. Sawayama, S. Niikura, Glycothermal synthesis and photoluminescent properties of Ce^{3+}-doped YBO_3 mesocrystals, J. Alloys Compds 577 (2013) 320-326. https://doi.org/10.1016/j.jallcom.2013.05.203

[35]   W. Chen, J. Zhang, F. Chen, Glycothermal synthesis of fluorinated Fe_3O_4 microspheres with distinct peroxidase-like activity, Advanced Powder Technol. 30 (2019) 999-1005. https://doi.org/10.1016/j.apt.2019.02.014

[36]   D.S. Bae, K.S. Han, S.B. Cho, S.H. Choi, Synthesis of ultrafine Fe_3O_4 powder by glycothermal process, Mater. Lett. 37 (1998) 255-258. https://doi.org/10.1016/S0167-577X(98)00101-3

[37]   A.E. Danks, S.R. Hall, Z. Schnepp, The evolution of 'sol–gel'chemistry as a technique for materials synthesis, Mater. Horizons 3 (2016) 91-112. https://doi.org/10.1039/C5MH00260E

[38]   R. Jin, B. Lin, D. Li, H. Ai, Superparamagnetic iron oxide NPs for MR imaging and therapy: design considerations and clinical applications, Current opinion in pharmacology, 18 (2014) 18-27. https://doi.org/10.1016/j.coph.2014.08.002

[39]   C. Hui, C. Shen, T. Yang, L. Bao, J. Tian, H. Ding, C. Li, H.J. Gao, Large-scale Fe_3O_4 NPs soluble in water synthesized by a facile method, J. Phys. Chem. C 112 (2008) 11336-11339. https://doi.org/10.1021/jp801632p

[40]   J.P. Jolivet, É. Tronc, C. Chanéac, Synthesis of iron oxide-based magnetic nanomaterials and composites, Comptes Rendus Chimie, 5 (2002) 659-664. https://doi.org/10.1016/S1631-0748(02)01422-4

[41] Y.V. Kolen'ko, M. Bañobre-López, C. Rodríguez-Abreu, E. Carbó-Argibay, A. Sailsman, Y. Piñeiro-Redondo, M.F. Cerqueira, D.Y. Petrovykh, K. Kovnir, O.I.

Lebedev, Large-scale synthesis of colloidal Fe_3O_4 NPs exhibiting high heating efficiency in magnetic hyperthermia, J. Phys. Chem. C 118 (2014) 8691-8701. https://doi.org/10.1021/jp500816u

[42]   R. Borny, T. Lechleitner, T. Schmiedinger, M. Hermann, R. Tessadri, G. Redhammer, J. Neumüller, D. Kerjaschki, G. Berzaczy, G. Erman, Nucleophilic cross-linked, dextran coated iron oxide NPs as basis for molecular imaging: synthesis, characterization, visualization and comparison with previous product, Contrast Media & Molecular Imaging 10 (2015) 18-27. https://doi.org/10.1002/cmmi.1595

[43]   M. Filippousi, M. Angelakeris, M. Katsikini, E. Paloura, I. Efthimiopoulos, Y. Wang, D. Zamboulis, G. Van Tendeloo, Surfactant effects on the structural and magnetic properties of iron oxide NPs, J. Phys. Chem. C, 118 (2014) 16209-16217. https://doi.org/10.1021/jp5037266

[44]   N.N. Song, H.T. Yang, X. Ren, Z.A. Li, Y. Luo, J. Shen, W. Dai, X.Q. Zhang, Z.H. Cheng, Non-monotonic size change of monodisperse Fe_3O_4 NPs in the scale-up synthesis, Nanoscale 5 (2013) 2804-2810. https://doi.org/10.1039/c3nr33950e

[45]   A. Demortiere, P. Panissod, B. Pichon, G. Pourroy, D. Guillon, B. Donnio, S. Begin-Colin, Size-dependent properties of magnetic iron oxide nanocrystals, Nanoscale 3 (2011) 225-232. https://doi.org/10.1039/C0NR00521E

[46]   B. Qi, L. Ye, R. Stone, C. Dennis, T. M. Crawford, O.T. Mefford, Influence of ligand–precursor molar ratio on the size evolution of modifiable iron oxide NPs, J. Phys. Chem. C 117 (2013) 5429-5435. https://doi.org/10.1021/jp311509v

[47]   J. Park, K. An, Y. Hwang, J.-G. Park, H.-J. Noh, J.-Y. Kim, J.-H. Park, N.-M. Hwang, T. Hyeon, Ultra-large-scale syntheses of monodisperse nanocrystals, Nat. Materi. 3 (2004) 89. https://doi.org/10.1038/nmat1251

[48]   J. Kudr, Y. Haddad, L. Richtera, Z. Heger, M. Cernak, V. Adam, O. Zitka, Magnetic NPs: From design and synthesis to real world applications, Nanomaterials 7 (2017) 243. https://doi.org/10.3390/nano7090243

[49]   W. Matizamhuka, The impact of magnetic materials in renewable energy-related technologies in the 21st century industrial revolution: The case of South Africa, Adv. Mater. Sci. Eng. 2018 (2018) 3149412. https://doi.org/10.1155/2018/3149412

[50]   B. Moazzenchi, M. Montazer, Click electroless plating of nickel NPs on polyester fabric: Electrical conductivity, magnetic and EMI shielding properties, Colloids Surf. A: Physicochemical and Engineering Aspects, 571 (2019) 110-124. https://doi.org/10.1016/j.colsurfa.2019.03.065

[51]   K.C. Liu, M.A. Anderson, Porous nickel oxide/nickel films for electrochemical capacitors, J. Electrochem. Soc. 143 (1996) 124-130. https://doi.org/10.1149/1.1836396

[52]   S. Deki, H. Yanagimoto, S. Hiraoka, K. Akamatsu, and K. Gotoh, NH_2-terminated poly (ethylene oxide) containing nanosized NiO particles: Synthesis, characterization, and structural considerations, Chem. Mater. 15 (2003) 4916-4922. https://doi.org/10.1021/cm021754a

[53]   L. Xiang, X. Deng, Y. Jin, Experimental study on synthesis of NiO nano-particles, Scripta Materialia, 47 (2002) 219-224. https://doi.org/10.1016/S1359-6462(02)00108-2

[54] P. You, S.K. Kamarudin, Recent progress of carbonaceous materials in fuel cell applications: An overview, Chem. Eng. J. 309 (2017) 489-502. https://doi.org/10.1016/j.cej.2016.10.051

[55]   W. R. W. Daud, R. Rosli, E. Majlan, S. Hamid, R. Mohamed, T. Husaini, PEM fuel cell system control: A review, Renewable Energy 113 (2017) 620-638. https://doi.org/10.1016/j.renene.2017.06.027

[56]   M. Wöhr, K. Bolwin, W. Schnurnberger, M. Fischer, W. Neubrand, G. Eigenberger, Dynamic modelling and simulation of a polymer membrane fuel cell including mass transport limitation, Int. J. Hydrogen Energy 23 (1998) 213-218. https://doi.org/10.1016/S0360-3199(97)00043-8

[57]   T. Suzuki, K. Kudo, and Y. Morimoto, Model for investigation of oxygen transport limitation in a polymer electrolyte fuel cell, J. Power Sources 222 (2013) 379-389, 2013. https://doi.org/10.1016/j.jpowsour.2012.08.068

[58]   L. Li, S. Bei, R. Liu, Q. Xu, K. Zheng, Y. She, Y. He, Design of a radial vanadium redox microfluidic fuel cell: A new way to break the size limitation, Int. J. Energy Res. 43 (2019) 3028-3037. https://doi.org/10.1002/er.4473

[59]   Z. Abdul Rahim, N. Yusof, M. Mohammad Haniff, F. Mohammad, M. Syono, N. Daud, Electrochemical measurements of multiwalled carbon nanotubes under different plasma treatments, Materials, 11 (2018) 1902. https://doi.org/10.3390/ma11101902

Materials Research Forum LLC
https://doi.org/10.21741/9781644900611-7

Chapter 7

Magnetic Nanomaterials for Hydrogen Storage

Ertuğrul Kaya[1], Haydar Göksu[2]*, Husnu Gerengi[1], Kubilay Arikan[3], Mohd Imran Ahamed[4], Fatih Şen[3]*

[1] Corrosion Research Laboratory, Department of Mechanical Engineering, Faculty of Engineering, Duzce University, 81620, Duzce, Turkey

[2] Kaynasli Vocational College, Duzce University, 81900 Duzce, Turkey

[3] Sen Research Group, Department of Biochemistry, Dumlupınar University, 43100 Kütahya, Turkey

[4] Department of Chemistry, Faculty of Science, Aligarh Muslim University, Aligarh-202 002, India

*haydargoksu@duzce.edu.tr, fatih.sen@dpu.edu.tr

Abstract

Nowadays, technological developments have increased with the help of magnetic materials in many fields. These materials are mostly preferred in areas such as magnetic cooling and storage as well as environmental improvements. It has also become preferable in the field of magnetic detection and in the use of medical care properties of magnetic nanoparticles. It is also preferred in areas including magnetic resonance imaging (MRI) and drug delivery systems (DDS). In the use of magnetic nanomaterials, magnetic properties, size, surface properties, synthesis methods should be paid attention. The magnetic properties of nanomaterials on the nanoscale depend on the particle size of the material used. Materials having particle size below a critical value lead to forming monodisperse particles. Many studies have been done on the production of magnetic nanomaterials. Most of these applications are based on the efficiency of particle moment and field distortion and change in the relevant variables. Nanomaterials are preferred due to their properties such as mass absorption and surface adsorption, which have recently become important in hydrogen storage. They also act by regulating the diffusion rate in hydrogen decomposition and absorption. Magnetic nanomaterials have high adsorption capacity due to atomic groups on the ligand surface. This shows that it can be used in energy transformation and storage. In this section, the latest studies on the hydrogen storage capacity of magnetic nanomaterials are conveyed and provide an essential point of view in hydrogen storage.

Keywords

Hydrogen Storage, Magnetic Nanomaterials, Hydrogen Energy

Contents

1.   Introduction

Nanomaterials are used in a wide range area including data storage, catalysts, magnetic imaging resonance, magnetic cooling, magnetic fluids, and biomedical, so recently search of new magnetic nanomaterials have been enhanced [1–15]. Among these nanomaterials, magnetic nanoparticles have an important place in the field of medical care and magnetic detection, especially suitable for use in drug delivery systems (DDS) and magnetic resonance imaging (MRG) [4]. Another potential use of these materials is in energy storage and solar cells (Fig 1). The electrodes produced by nanomaterials have advantages such as high surface-volume ratio, electronic transport, and ion transmission, and high charge rate [16]. Magnetic nanoparticles are affected by various parameters such as particle size, magnetic properties, surface properties, and synthesis method of nanoparticles [17].

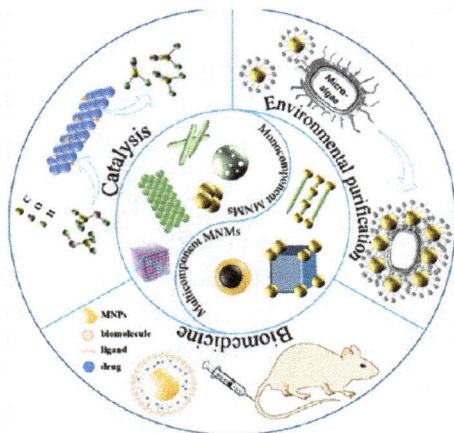

Figure 1. Applications of magnetic nanomaterials [7].

Nanoscale magnetic materials show different physical effects. These effects distinguish them from other materials and provide selectivity for use. This effect is defined as nanomagnetism, and magnetic nanomaterials exhibit this type of superparamagnetism. However, magnetic nanomaterials are difficult to obtain magnetic nanostructures due to nano-scale dipolar interactions and superficial effects [7]. Magnetic properties of nanomaterials on a nanoscale depend on the particle size of the material [18]. When the particle size falls below the critical value, a single-domain particle is formed. Control of the particle size of the magnetic nanomaterials is a key point for the production of high-performance nanomagnetic materials (Fig 2) [19].

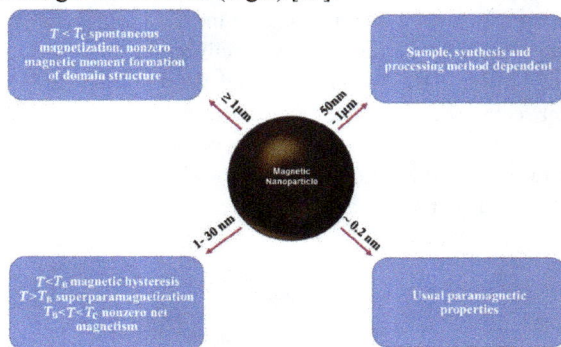

Figure 2. Effect of particle size on magnetic properties [19].

Magnetochemistry - Materials and Applications   Materials Research Forum LLC
Materials Research Foundations **66** (2020) 236-258   https://doi.org/10.21741/9781644900611-7

Many studies have been carried out on the production of magnetic nanomaterials in various applications based on particle effectiveness time, particles degradation sides, and the change in the related variables [20]. The crystallization states for iron during the synthesis process are shown in Figure 3. As seen in Figure 3, magnetic moments formed randomly are structured in paramagnetic crystal framework [21]. The magnetic moments interact with each other magnetically and when the applied field is removed, they should be zero. The atomic moments of the ferromagnetic materials show very strong interaction.

The interactions among materials are produced due to the changes in electronic force which lead to forming parallel and nonparallel atomic alignment moments. During forming the crystalline framework, some ionic oxides having magnetic features shousing ferrimagnetism are ionic. Schematic representation showing spins oxide compounds are given in Figure 3 (d)[22,23].

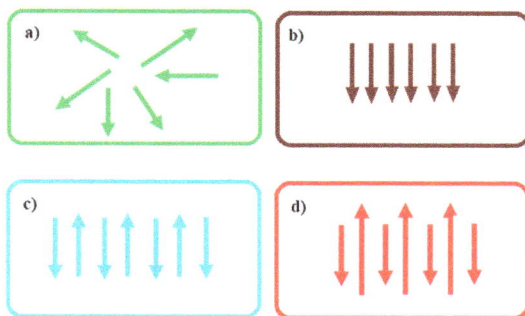

Figure 3. a) Paramagnetism, b) Ferromagnetism, c) Antiferromagnetism, d) Ferrimagnetism, Magnetic moment of atoms of iron.

Magnetic nanomaterials are synthesized as mono-component and multi-component by physical [24–26], chemical [27–29], and biological methods [30–32].

2.   Monocomponent magnetic nanomaterials

2.1   Metal magnetic nanomaterials

Commonly used magnetic metallic materials are transition metals such as iron, cobalt, and nickel. One reason for this is the stability of iron oxide in the air and strong magnetization. Metallic nanomaterials resist oxidizing agents [33]. In addition, there are advantages to magnetism such as saturation, chemical stability, and high magnetic

Magnetochemistry - Materials and Applications   Materials Research Forum LLC
Materials Research Foundations **66** (2020) 236-258   https://doi.org/10.21741/9781644900611-7

moment. However, they are reactive against oxidative agents in various degrees. This disadvantage causes unwanted side reactions and makes magnetic carrier problmes [34]. Usage areas for these materials are magnetic storage, magnetic cooling system, ferrofluids, biomedical, high-density data storage, catalysts and batteries etc. [35–41].

2.2   Metallic Oxides Magnetic Nanomaterials

The most common types of oxide magnetic nanoparticles are magnetite (Fe_3O_4) and maghemite (γ-Fe_2O_3). Magnetite, unlike other iron oxides, carries both two and three valent iron ions in its structure. All Fe^{2+} ions and half of the Fe^{3+} ions fill the octahedral areas while the remaining Fe^{3+} ions occupy the tetrahedral areas. Oxygen ions are routinely located within the cubic closed system along the axis [42]. Magnetite nanoparticles have received great attention recently from researchers and researches are being carried out in order to use them effectively in different areas such as biotechnology [43], drug loading [44], and magnetic resonance imaging [45]. Furthermore, magnetite nanoparticles can be used as an effective adsorbent in the removal of toxic and heavy metals from aqueous media, and can easily be separated from the aqueous medium by an external magnetic field due to magnetic susceptibility without the need for such processes as filtering and centrifugation (Fig. 4)[46].

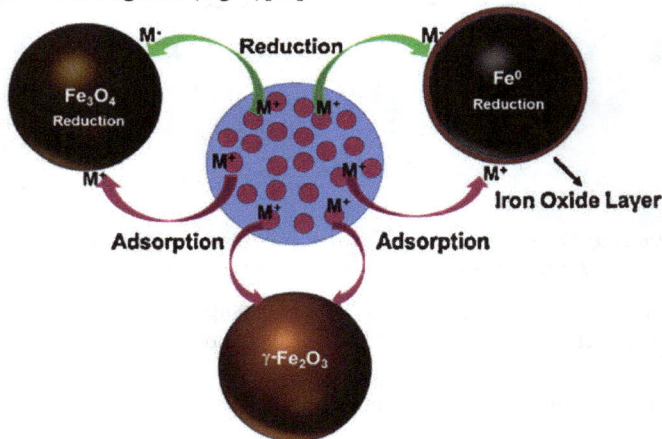

Figure 4. Adsorption and reduction potential of the magnetic nanomaterials pollutants[19].

By replacing the Fe^{2+} cations with other metal, cations can be made of different types of ferrite metals such as $MnFe_2O_4$ [47], $ZnFe_2O_4$ [48]. Although maghemite is similar to magnetite in structure, it contains empty positions due to lack of Fe^{2+} charged cations.

Metallic oxides magnetic nanomaterials have high magnetic susceptibility. In addition, there are advantages to low toxicity, high oxidative stability, colloidal stability, high electrical resistance, high magnetic permeability, superparamagnetism, high-quality monodispersity, large surface area to volume ratio [49,50]. Usage areas are labeling cells, magnetic data storage, and the gene therapy, gas sensors, batteries, catalysts, electromagnetic and optical devices [51–57].

2.3　Metal Alloy Magnetic Nanomaterials

Recently, due to the improved magnetic features and multifunctional processes, the focus is on developing magnetic alloy nanomaterials instead of metal oxide magnetic nanomaterials. Different magnetic alloy nanomaterials such as CoPt [58], NiPt [59], FePt [60,61] have been synthesized. FePt has been most interesting among these alloys. Due to the favorable magnetic properties of ferromagnets, FePt has great potential for permanent magnets, electrocatalysis, and magnetic data storage applications [62].

The magnetic properties of FePt can be adjusted not only by controlling the size and shape of FePt nanomaterials but also by changing the crystal structure. In general, FePt has two different crystalline structures such as chemically disordered face-centered cubic phase structure (FCC) and chemically regulated face ordered tetragonal phase structure (FCT). Typically, low-temperature synthesis of FePt nanomaterials produces FCC phase FePt nanoparticles, and high-temperature annealing is required to convert the FCC phase to the FCT phase [63,64].

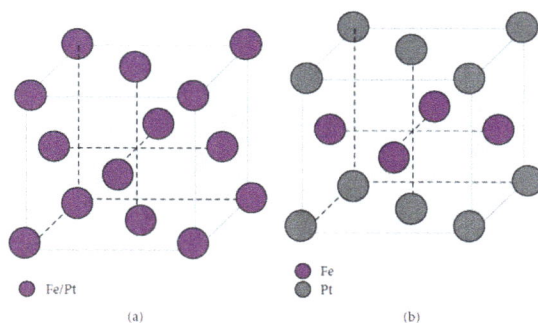

Figure 5. FePt crystal structure (a) FCC structure; (b) FCT structure [63].

2.4   Metal Carbides

Iron carbides such as Fe_5C_2 [65], Fe_3C [66], and Fe_2C [67] have been investigated for their stability and high magnetic saturation compared to iron oxides. Researchers are in particular on the control of phase, size, and morphology [68].

3.   Multicomponent Magnetic Nanomaterials

Magnetic heterostructures are composites that exhibit unique properties thanks to their various chemical components, which contain a functionally magnetic part. The combination of different components and the close contact of each other in the interface add new features to these substances. The magnetic nanomaterials in this structure are normally synthesized through the process of seed-mediated growth, where the predetermined nanomaterials are used as seeds for the nucleation and growth of other inorganic components [18]. The predetermined nanomaterials have been synthesized to design a new form and change their magnetic core-shell features using nanomaterials in an organic phase. New requirements such as high magnetic saturation, high resistance, high-frequency magnetic properties home been achieved through these syntheses [69,70]. The Fe / Fe_3O_4 core-shell structure was in-situ synthesized soft as magnetic composites oxidation method, by Zhang et al., and investigated the effect on magnetic properties (Fig 6). The shell thickness obtained by oxidizing the crude iron powders for 40 minutes showed optimum magnetic properties. Microstructure and magnetic properties have been reported by controlling the thickness of the insulation layer [71].

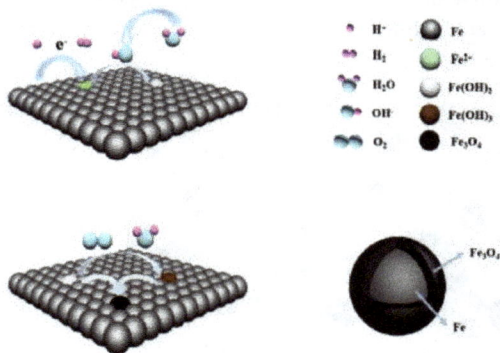

Figure 6. The formation process of the core-shell structured Fe/Fe₃O₄ composites [71].

Unlike the Core-shell structure, which is formed by the nanomaterials evenly, there are dumbbell nanomaterials formed by anisotropic nucleation and the growth of more shells [18]. The most studied dumbbell nanomaterials are magnetic oxide nanomaterials of noble metals. Magnetic dumbbell nanomaterials are formed by thermal decomposition and oxidation of Fe and its derivatives on the surface of nanomaterials of preformed noble metals [72]. The parameters affecting this formation are reported solvent polarity, seed-collector ratio, surfactants, temperature, and growth rate. Wei et al. reported dumbbell-like Au-Fe_3O_4 electrochemical immune sensors from nanomaterials. In order to increase the ability of nanomaterials to be reduced to H_2O_2, they have increased the sensitivity of the immunosensor with the signal amplification strategy of the Au-Fe_3O_4 nanomaterial (Fig. 7) [73].

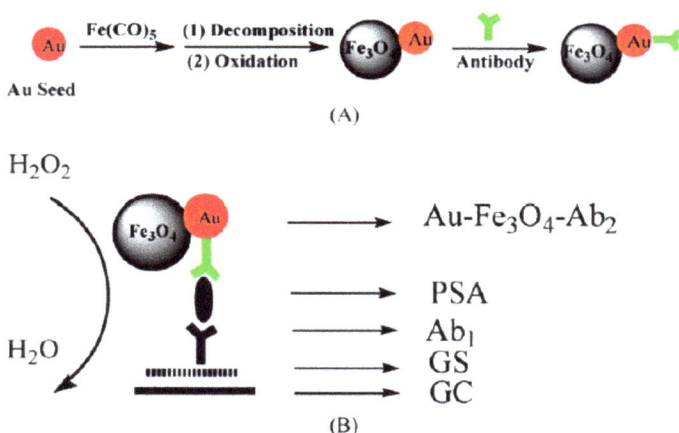

Figure 7. Preparation of Au-Fe_3O_4-Ab_2 nanostructure and (A) and immunosensor working mechanism [73].

4.   Use of Magnetic Nanomaterials in Hydrogen Storage

Presently, global warming-induced threats and the depletion of fossil fuels require the use of renewable and environmentally friendly energy sources [74]. Researches on renewable and clean energy sources are rapidly increasing. One of the most studied of these sources is hydrogen. Hydrogen is a renewable, environmentally friendly, safe, easily convertible, versatile, and extremely good energy carrier [75]. In addition, greenhouse gas emissions are significantly reduced, and internal combustion engines and proton exchange membranes increase the efficiency of energy conversion processes in fuel cells. In the fuel cells, the chemical energy in hydrogen is converted directly into electricity, water, and heat [76].

According to fossil fuel sources, hydrogen fuel cells are very strong. Although hydrogen energy has a high cost in terms of initial installation costs, subsequent maintenance costs are relatively inexpensive [77]. Chemical and polymer synthesis in the chemical industry, microelectronics, metallurgical processes are used [78]. The automotive sector also has quite cryptic properties. These are lightness, high volumetric and gravimetric densities, low decomposition, rapid kinetic activation, high level of reversibility, cost and availability, and suitable thermodynamic properties.

Hydrogen is an element with high compatibility with fuel cells. (9.5 kg of hydrogen is equivalent to 25 kg of gasoline's energy content) [79]. Hydrogen storage studies are being investigated in systems such as metal hydrides [80], complex and chemical hydrides [81], adsorbents [82] and nanomaterials (nanotubes, nanoparticles) [83], polymer nanocomposites [84] and, metal organic structures etc. [85]. Since hydrogen is not directly found in nature, it is obtained by using various methods [86,87]. These methods, include steam reforming [88], electrolysis [89], thermochemical cycle [90], photobiological water splitting [91], photocatalytic water splitting [92], fermentative hydrogen production [93] and enzymatic hydrogen generation [94].

Nanomaterials in hydrogen storage are recently preferred due to their properties such as surface adsorption, inter-intergrain boundaries, and bulk absorption [95]. Nanomaterials regulate the adsorption and decomposition of hydrogen by affecting the diffusion rate. Among the nanomaterials, core-shell magnetic nanomaterials are characterized by their superior properties. Due to the atomic groups on the surface of the ligand, the adsorption potential of the material is high. Because of the changes in load transfer, interconnections and band structures between core-shell are effective. These physical and chemical properties cause widespread use of core-shell magnetic materials in some applications such as conversion and energy storage materials [96,97]. New techniques are being developed to increase the storage capacity of hydrogen. The studies in energy fields revealed that nanoparticles composed of carbon atoms have a core-shell structure that have a significant effect in enhancing hydrogen storage capacity. Chang et al. reported that the CNF® Co and CNT® Co magnetic nanoparticles produced by the ball milling method showed superior properties in hydrogen storage (Fig 8). CNF/Co and CNT/Co, obtained as a result of their research, have a storage capacity of 739.4 and 717.3 mAh g^{-1} hydrogen under mild conditions [98].

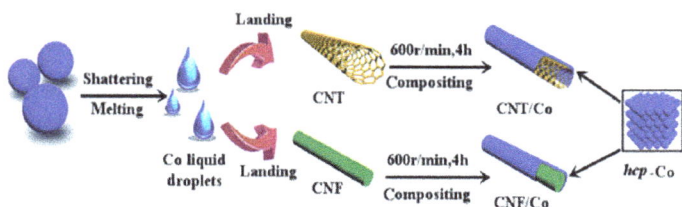

Figure 8. The formation process of magnetic core/shell structured CNT/Co and CNF/Co with ball-milling method [98].

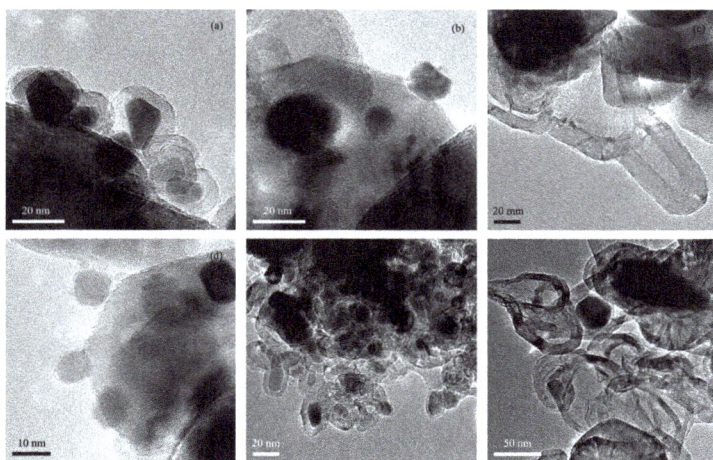

Figure 9. TEM patterns for Ni@CNOs, $Fe_{0.64}Ni_{0.36}$@CNOs and Fe_3C@CNOs prepared at 850 °C including M/MgO catalysts with 10% (a,b,c) and 20% metal loading (d,e,f) [102].

Bao et al. prepared the C60/Co magnetic nanocomposite by the ball-milling process. The hydrogen storage capacity of the C60/Co nanocomposite at room temperature and ambient pressure was very high at 907 mAh /g (3.32% hydrogen), close to the theoretical hydrogen storage capacities of individual metal Co (3.33 wt % hydrogen) [99]. Nanomaterials having core-shell structure have some drawbacks that they have high hydrogen storage capacities, due to their high reactivity in practical applications and the melting of the relevant complex hydride [96]. As an alternative to inorganic metal oxides and sulfides, magnetic nanoparticles such as MnO_2, Fe_3O_4, NiO, which are similar in environmental behavior with electrochemical behavior, have been used in energy storage

[100]. In particular, iron oxide has attracted attention in magnetic nanoparticles due to its low cost and eco-frendly to natūre [101].

Zhang et al. used three types of magnetic carbon nano anion: Ni@CNOs, Fe$_3$C@CNOs, and Fe$_{0.64}$Ni$_{0.36}$@CNOs. (Figure 3 a and d) Ni@CNO using 10% and 20% metal loading M/MgO catalysts in TEM images in Figure 9 a-b Ni@CNO, Fe$_{0.64}$Ni$_{0.36}$@CNOs (Figure 3 b and e) and (Figure 9 c-f) Fe$_3$C@CNOs appear to be synthesized as the main product.

In literature, it has been reported that the growth of CNOs depends on the different catalysts. The properties of the catalyst related to the chemical composition of the nanoparticles used, and also magnetic CNOs in potential fuel cells. Thus, by controlling the catalysts, the growth of magnetic CNOs can be controlled [102]. As a summary, the researchers have systematically studied the features of nanoparticles, their effectiveness of storage, synthesis, crystallinity, and catalyst composition [103–118].

Conclusions and Outlook

The use of magnetic nanomaterials has increased in many areas with advances in technology. It has also been used for storage, magnetic cooling and environmental improvement. Magnetic nanomaterials can also be synthesized by different synthesis methods.

Today, new sources of energy have been tested due to the rapid decline in fossil fuels and the damages caused by these fuels. The synthesized magnetic nanomaterials can also be used in hydrogen energy. Hydrogen is an important source of energy for research that avoids this cost due to high storage costs. The use of hydrogen and its incorporation with magnetic nanomaterials to eliminate high storage costs is an important study of the future use of this energy source. The use of magnetic nanomaterials in such areas is very important and is used not only for energy purposes, but also for many environmental and biomedical purposes.

References

[1]   T.F. Massoud, S.S. Gambhir, Molecular imaging in living subjects: Seeing fundamental biological processes in a new light., Genes Dev. 17 (2003) 545–580. https://doi.org/10.1101/gad.1047403.

[2]   H. Bin Na, I.C. Song, T. Hyeon, Inorganic Nanoparticles for MRI Contrast Agents, Adv. Mater. 21 (2009) 2133–2148. https://doi.org/10.1002/adma.200802366.

[3]   Z. Yang, T. Zhao, X. Huang, X. Chu, T. Tang, Y. Ju, Q. Wang, Y. Hou, S. Gao, Modulating the phases of iron carbide nanoparticles: from a perspective of interfering

with the carbon penetration of $Fe@Fe_3O_4$ by selectively adsorbed halide ions, Chem. Sci. 8 (2017) 473–481. https://doi.org/10.1039/C6SC01819J.

[4] S. Chandra, M.D. Patel, H. Lang, D. Bahadur, Dendrimer-functionalized magnetic nanoparticles: A new electrode material for electrochemical energy storage devices, J. Power Sources 280 (2015) 217–226. https://doi.org/10.1016/J.JPOWSOUR.2015.01.075.

[5] N. Tran, T.J. Webster, Magnetic nanoparticles: biomedical applications and challenges, J. Mater. Chem. 20 (2010) 8760. https://doi.org/10.1039/c0jm00994f.

[6] S. He, L. Zhong, J. Duan, Y. Feng, B. Yang, L. Yang, Bioremediation of Wastewater by Iron Oxide-Biochar Nanocomposites Loaded with Photosynthetic Bacteria, Front. Microbiol. 8 (2017) 823. https://doi.org/10.3389/fmicb.2017.00823.

[7] K. Zhu, Y. Ju, J. Xu, Z. Yang, S. Gao, Y. Hou, Magnetic nanomaterials: Chemical design, synthesis, and potential applications, Acc. Chem. Res. 51 (2018) 404–413. https://doi.org/10.1021/acs.accounts.7b00407.

[8] K. Saikia, K. Bhattacharya, D. Sen, S.D. Kaushik, J. Biswas, S. Lodha, B. Gogoi, A.K. Buragohain, W. Kockenberger, P. Deb, Solvent evaporation driven entrapment of magnetic nanoparticles in mesoporous frame for designing a highly efficient MRI contrast probe, Appl. Surf. Sci. 464 (2019) 567–576. https://doi.org/10.1016/J.APSUSC.2018.09.117.

[9] Y. Zhu, H. Da, X. Yang, Y. Hu, Preparation and characterization of core-shell monodispersed magnetic silica microspheres, Colloids Surfaces A Physicochem. Eng. Asp. 231 (2003) 123–129. https://doi.org/10.1016/J.COLSURFA.2003.08.020.

[10] S.V. Gopal, R. Mini, V.B. Jothy, I.H. Joe, Synthesis and Characterization of Iron Oxide Nanoparticles using DMSO as a Stabilizer, Mater. Today Proc. 2 (2015) 1051–1055. https://doi.org/10.1016/J.MATPR.2015.06.036.

[11] M. Takafuji, S. Ide, H. Ihara, Z. Xu, Preparation of Poly(1-vinylimidazole)-Grafted Magnetic Nanoparticles and Their Application for Removal of Metal Ions, Chem. Mater. 16 (2004) 1977–1983. https://doi.org/10.1021/cm030334y.

[12] Y. Hwang, J.-S. Park, Y.J. Choi, Y.J. Suh, H.-S. Lee, D.S. Kang, J.K. Lee, Prevention of Coalescence During Annealing of FePt Nanoparticles Assembled by Convective Coating, J. Nanosci. Nanotechnol. 10 (2010) 3516–3520. https://doi.org/10.1166/jnn.2010.2285.

[13]   R. Kumar, Magnetocaloric Effect in Cryocooler Regenerator Materials and its Applications, Mater. Today Proc. 4 (2017) 5544–5551. https://doi.org/10.1016/J.MATPR.2017.06.011.

[14]   Y. Ao, J. Xu, X. Shen, D. Fu, C. Yuan, Magnetically separable composite photocatalyst with enhanced photocatalytic activity, J. Hazard. Mater. 160 (2008) 295–300. https://doi.org/10.1016/J.JHAZMAT.2008.02.114.

[15]   J. Ge, T. Huynh, Y. Hu, Y. Yin, Hierarchical magnetite/silica nanoassemblies as magnetically recoverable catalyst–supports, Nano Lett. 8 (2008) 931–934. https://doi.org/10.1021/nl080020f.

[16]   Y. Li, X. Zhao, Q. Xu, Q. Zhang, D. Chen, Facile preparation and enhanced capacitance of the polyaniline/sodium alginate nanofiber network for supercapacitors, Langmuir. 27 (2011) 6458–6463. https://doi.org/10.1021/la2003063.

[17]   M. Mikhaylova, D.K. Kim, N. Bobrysheva, M. Osmolowsky, V. Semenov, T. Tsakalakos, M. Muhammed, Superparamagnetism of magnetite nanoparticles: dependence on surface modification, Langmuir. 20 (2004) 2472–2477. https://doi.org/10.1021/la035648e.

[18]   L. Wu, A. Mendoza-Garcia, Q. Li, S. Sun, Organic phase syntheses of magnetic nanoparticles and their applications, Chem. Rev. 116 (2016) 10473–10512. https://doi.org/10.1021/acs.chemrev.5b00687.

[19]   L. Mohammed, H.G. Gomaa, D. Ragab, J. Zhu, Magnetic nanoparticles for environmental and biomedical applications: A review, Particuology. 30 (2017) 1–14. https://doi.org/10.1016/j.partic.2016.06.001.

[20]   A.K. Andriola Silva, R. Di Corato, F. Gazeau, T. Pellegrino, C. Wilhelm, Magnetophoresis at the nanoscale: tracking the magnetic targeting efficiency of nanovectors, Nanomedicine. 7 (2012) 1713–1727. https://doi.org/10.2217/nnm.12.40.

[21]   A.S. Teja, P.Y. Koh, Synthesis, properties, and applications of magnetic iron oxide nanoparticles, Prog. Cryst. Growth Charact. Mater. 55 (2009) 22–45. https://doi.org/10.1016/J.PCRYSGROW.2008.08.003.

[22]   M. Getzlaff, Solid State Magnetism, in: Fundam. Magn., Springer Berlin Heidelberg, Berlin, Heidelberg, 2008: pp. 25–39. https://doi.org/10.1007/978-3-540-31152-2_3.

[23]   M.D. Simon, A.K. Geim, Diamagnetic levitation: Flying frogs and floating magnets (invited), J. Appl. Phys. 87 (2000) 6200–6204. https://doi.org/10.1063/1.372654.

[24]   J.-H. Lee, J. Jang, J. Choi, S.H. Moon, S. Noh, J. Kim, J.-G. Kim, I.-S. Kim, K.I. Park, J. Cheon, Exchange-coupled magnetic nanoparticles for efficient heat induction, Nat. Nanotechnol. 6 (2011) 418–422. https://doi.org/10.1038/nnano.2011.95.

[25]   S. Sun, H. Zeng, Size-Controlled synthesis of magnetite nanoparticles, J. Am. Chem. Soc. 124 (2002) 8204–8205. https://doi.org/10.1021/ja026501x.

[26]   D. Ho, X. Sun, S. Sun, Monodisperse magnetic nanoparticles for theranostic applications, Acc. Chem. Res. 44 (2011) 875–882. https://doi.org/10.1021/ar200090c.

[27]   K. Ulbrich, K. Holá, V. Šubr, A. Bakandritsos, J. Tuček, R. Zbořil, Targeted drug delivery with polymers and magnetic Nanoparticles: Covalent and noncovalent approaches, release control, and clinical studies, Chem. Rev. 116 (2016) 5338–5431. https://doi.org/10.1021/acs.chemrev.5b00589.

[28]   R. Tietze, S. Lyer, S. Dürr, T. Struffert, T. Engelhorn, M. Schwarz, E. Eckert, T. Göen, S. Vasylyev, W. Peukert, F. Wiekhorst, L. Trahms, A. Dörfler, C. Alexiou, Efficient drug-delivery using magnetic nanoparticles — biodistribution and therapeutic effects in tumour bearing rabbits, Nanomedicine Nanotechnology, Biol. Med. 9 (2013) 961–971. https://doi.org/10.1016/J.NANO.2013.05.001.

[29]   S. Majidi, F. Zeinali Sehrig, S.M. Farkhani, M. Soleymani Goloujeh, A. Akbarzadeh, Current methods for synthesis of magnetic nanoparticles, Artif. Cells, Nanomedicine, Biotechnol. 44 (2016) 722–734. https://doi.org/10.3109/21691401.2014.982802.

[30]   M. Timko, A. Dzarova, J. Kovac, A. Skumiel, A. Józefczak, T. Hornowski, H. Gojżewski, V. Zavisova, M. Koneracka, A. Sprincova, O. Strbak, P. Kopcansky, N. Tomasovicova, Magnetic properties and heating effect in bacterial magnetic nanoparticles, J. Magn. Magn. Mater. 321 (2009) 1521–1524. https://doi.org/10.1016/J.JMMM.2009.02.077.

[31]   J. Baumgartner, L. Bertinetti, M. Widdrat, A.M. Hirt, D. Faivre, Formation of Magnetite nanoparticles at low temperature: From superparamagnetic to stable single domain particles, PLoS One. 8 (2013) e57070. https://doi.org/10.1371/journal.pone.0057070.

[32]   D. Faivre, D. Schüler, Magnetotactic bacteria and magnetosomes, Chem. Rev. 108 (2008) 4875–4898. https://doi.org/10.1021/cr078258w.

[33]   A. Aseri, G.S. Kumar, N. Anjali, S.K. Trivedi, A. Jawed, Magnetic nanoparticles: Magnetic nano-technology using biomedical applications and future prospects, Int. J. Pharm. Sci. Rev. Res. 31 (2015) 119–131.

[34]   Z. Hedayatnasab, F. Abnisa, W.M.A.W. Daud, Review on magnetic nanoparticles for magnetic nanofluid hyperthermia application, Mater. Design 123 (2017) 174-196. https://doi.org/10.1016/j.matdes.2017.03.036.

[35]   G. Leteba, C. Lang, G.M. Leteba, C.I. Lang, Synthesis of bimetallic platinum nanoparticles for biosensors, Sensors. 13 (2013) 10358–10369. https://doi.org/10.3390/s130810358.

[36]   I. Astefanoaei, A. Stancu, H. Chiriac, Magnetic hyperthermia with Fe - Cr- Nb - B magnetic particles, in: AIP Conf. Proc., AIP Publishing LLC , 2017: p. 040006. https://doi.org/10.1063/1.4972384.

[37]   J. Jakobi, S. Petersen, A. Menéndez-Manjón, P. Wagener, S. Barcikowski, Magnetic alloy nanoparticles from laser ablation in cyclopentanone and their embedding into a photoresist, Langmuir. 26 (2010) 6892–6897. https://doi.org/10.1021/la101014g.

[38]   I. Robinson, S. Zacchini, L.D. Tung, S. Maenosono, N.T.K. Thanh, Synthesis and characterization of magnetic nanoalloys from bimetallic carbonyl clusters, Chem. Mater. 21 (2009) 3021–3026. https://doi.org/10.1021/cm9008442.

[39]   Q. Zhou, Y. Wang, J. Xiao, H. Fan, C. Chen, Preparation and characterization of magnetic nanomaterial and its application for removal of polycyclic aromatic hydrocarbons, J. Hazard. Mater. 371 (2019) 323–331. https://doi.org/10.1016/J.JHAZMAT.2019.03.027.

[40]   J. Wang, S. Zheng, Y. Shao, J. Liu, Z. Xu, D. Zhu, Amino-functionalized Fe_3O_4@SiO_2 core–shell magnetic nanomaterial as a novel adsorbent for aqueous heavy metals removal, J. Colloid Interface Sci. 349 (2010) 293–299. https://doi.org/10.1016/J.JCIS.2010.05.010.

[41]   S. Su, B. Chen, M. He, B. Hu, Z. Xiao, Determination of trace/ultratrace rare earth elements in environmental samples by ICP-MS after magnetic solid phase extraction with Fe_3O_4@SiO_2@polyaniline–graphene oxide composite, Talanta 119 (2014) 458–466. https://doi.org/10.1016/J.TALANTA.2013.11.027.

[42]   R.M. Cornell, U. Schwertmann, The iron oxides : structure, properties, reactions, occurrences, and uses, Wiley-VCH, 2003.

M. Tamaddon, S. Javadpour, I.J. Bruce, PEG conjugated citrate-capped magnetite nanoparticles for biomedical applications, J. Magn. Magn. Mater. 328 (2013) 91–95. https://doi.org/10.1016/J.JMMM.2012.09.042.

[44]   S. Sundar, R. Mariappan, S. Piraman, Synthesis and characterization of amine modified magnetite nanoparticles as carriers of curcumin-anticancer drug, Powder Technol. 266 (2014) 321–328. https://doi.org/10.1016/J.POWTEC.2014.06.033.

[45]   S. Zhang, X. Liu, L. Zhou, W. Peng, Magnetite nanostructures: One-pot synthesis, superparamagnetic property and application in magnetic resonance imaging, Mater. Lett. 68 (2012) 243–246. https://doi.org/10.1016/J.MATLET.2011.10.070.

[46]   M.R. Shishehbore, A. Afkhami, H. Bagheri, Salicylic acid functionalized silica-coated magnetite nanoparticles for solid phase extraction and preconcentration of some heavy metal ions from various real samples, Chem. Cent. J. 5 (2011) 41. https://doi.org/10.1186/1752-153X-5-41.

[47]   E. Kang, J. Park, Y. Hwang, M. Kang, J.G. Park, T. Hyeon, Direct synthesis of highly crystalline and monodisperse manganese ferrite nanocrystals, J. Phys. Chem. B. 108 (2004) 13932–13935. https://doi.org/10.1021/jp049041y.

[48]   J. Jang, H. Nah, J.-H. Lee, S.H. Moon, M.G. Kim, J. Cheon, Critical enhancements of mri contrast and hyperthermic effects by dopant-controlled magnetic nanoparticles, Angew. Chem. Int. Ed. 121 (2009) 1260–1264. https://doi.org/10.1002/ange.200805149.

[49]   R. Ali, A. Mahmood, M.A. Khan, A.H. Chughtai, M. Shahid, I. Shakir, M.F. Warsi, Impacts of Ni–Co substitution on the structural, magnetic and dielectric properties of magnesium nano-ferrites fabricated by micro-emulsion method, J. Alloys Compd. 584 (2014) 363–368. https://doi.org/10.1016/J.JALLCOM.2013.08.114.

[50]   N. Kang, D. Xu, Y. Han, X. Lv, Z. Chen, T. Zhou, L. Ren, X. Zhou, Magnetic targeting core/shell Fe_3O_4/Au nanoparticles for magnetic resonance/photoacoustic dual-modal imaging, Mater. Sci. Eng. C. 98 (2019) 545–549. https://doi.org/10.1016/J.MSEC.2019.01.013.

[51]   N.T. Lan, N.P. Duong, T.D. Hien, Influences of cobalt substitution and size effects on magnetic properties of coprecipitated Co–Fe ferrite nanoparticles, J. Alloys Compd. 509 (2011) 5919–5925. https://doi.org/10.1016/J.JALLCOM.2011.03.050.

[52]   R. S., P. M., Multi-functional core-shell Fe_3O_4@Au nanoparticles for cancer diagnosis and therapy, Colloids Surf. B Biointerf. 174 (2019) 252–259. https://doi.org/10.1016/J.COLSURFB.2018.11.004.

[53]   D. Alcantara, S. Lopez, M.L. García-Martin, D. Pozo, Iron oxide nanoparticles as magnetic relaxation switching (MRSw) sensors: Current applications in nanomedicine,

Nanomedicine Nanotechnology, Biol. Med. 12 (2016) 1253–1262.
https://doi.org/10.1016/J.NANO.2016.01.005.

[54]   F. Yan, R. Sun, Facile synthesis of bifunctional Fe_3O_4/Au nanocomposite and their
application in catalytic reduction of 4-nitrophenol, Mater. Res. Bull. 57 (2014) 293–
299. https://doi.org/10.1016/J.MATERRESBULL.2014.06.012.

[55]   Z. Beji, M. Sun, L.S. Smiri, F. Herbst, C. Mangeney, S. Ammar, Polyol synthesis
of non-stoichiometric Mn–Zn ferrite nanocrystals: structural /microstructural
characterization and catalytic application, RSC Adv. 5 (2015) 65010–65022.
https://doi.org/10.1039/C5RA07562A.

[56]   L. Zhang, X. Zhu, H. Sun, G. Chi, J. Xu, Y. Sun, Control synthesis of magnetic
Fe3O4–chitosan nanoparticles under UV irradiation in aqueous system, Curr. Appl.
Phys. 10 (2010) 828–833. https://doi.org/10.1016/J.CAP.2009.10.002.

[57]   L. Qian, J. Peng, Z. Xiang, Y. Pan, W. Lu, Effect of annealing on magnetic
properties of Fe/Fe3O4 soft magnetic composites prepared by in-situ oxidation and
hydrogen reduction methods, J. Alloys Compd. 778 (2019) 712–720.
https://doi.org/10.1016/J.JALLCOM.2018.11.184.

[58]   S. Wodarz, T. Hasegawa, S. Ishio, T. Homma, Structural control of ultra-fine CoPt
nanodot arrays via electrodeposition process, J. Magn. Magn. Mater. 430 (2017) 52–
58. https://doi.org/10.1016/J.JMMM.2017.01.061.

[59]   V.V. Pham, V.-T. Ta, C. Sunglae, Synthesis of NiPt alloy nanoparticles by
galvanic replacement method for direct ethanol fuel cell, Int. J. Hydrogen Energy. 42
(2017) 13192–13197. https://doi.org/10.1016/J.IJHYDENE.2017.01.236.

[60]   Z. Meng, Z. Wei, K. Fu, L. Lv, Z.-Q. Yu, W.-Y. Wong, Amphiphilic bimetallic
polymer as single-source precursors for the one-pot synthesis of L10-phase FePt
nanoparticles, J. Organomet. Chem. 892 (2019) 83–88.
https://doi.org/10.1016/J.JORGANCHEM.2019.04.015.

[61]   M.M. Goswami, A. Das, D. De, Wetchemical synthesis of FePt nanoparticles:
Tuning of magnetic properties and biofunctionalization for hyperthermia therapy, J.
Magn. Magn. Mater. 475 (2019) 93–97. https://doi.org/10.1016/J.JMMM.2018.11.024.

[62]   S. Srivastava, N.S. Gajbhiye, Exchange coupled L10-FePt/fcc-FePt nanomagnets:
Synthesis, characterization and magnetic properties, J. Magn. Magn. Mater. 401
(2016) 969–976. https://doi.org/10.1016/J.JMMM.2015.10.064.

[63]   Y. Shi, M. Lin, X. Jiang, S. Liang, Recent advances in FePt Nanoparticles for
biomedicine, J. Nanomater. 2015 (2015) 1–13. https://doi.org/10.1155/2015/467873.

[64]   S. Sun, C.B. Murray, D. Weller, L. Folks, A. Moser, Monodisperse FePt nanoparticles and ferromagnetic FePt nanocrystal superlattices, Science 287 (2000) 1989–92. https://doi.org/10.1126/science.287.5460.1989.

[65]   W. Ge, W. Gao, J. Zhu, Y. Li, In situ synthesis of Hägg iron carbide (Fe_5C_2) nanoparticles with a high coercivity and saturation magnetization, J. Alloys Compd. 781 (2019) 1069–1073. https://doi.org/10.1016/J.JALLCOM.2018.12.154.

[66]   A. Schneider, G. Inden, Carbon diffusion in cementite (Fe_3C) and Hägg carbide (Fe_5C_2), Calphad 31 (2007) 141–147. https://doi.org/10.1016/J.CALPHAD.2006.07.008.

[67]   C.M. Fang, M.A. van Huis, H.W. Zandbergen, Structure and stability of Fe_2C phases from density-functional theory calculations, Scr. Mater. 63 (2010) 418–421. https://doi.org/10.1016/J.SCRIPTAMAT.2010.04.042.

[68]   S. Li, P. Ren, C. Yang, X. Liu, Z. Yin, W. Li, H. Yang, J. Li, X. Wang, Y. Wang, R. Cao, L. Lin, S. Yao, X. Wen, D. Ma, Fe_5C_2 nanoparticles as low-cost HER electrocatalyst: the importance of Co substitution, Sci. Bull. 63 (2018) 1358–1363. https://doi.org/10.1016/J.SCIB.2018.09.016.

[69]   J.M. Silveyra, E. Ferrara, D.L. Huber, T.C. Monson, Soft magnetic materials for a sustainable and electrified world., Science 362 (2018) eaao0195. https://doi.org/10.1126/science.aao0195.

[70]   A.M. Leary, P.R. Ohodnicki, M.E. McHenry, Soft magnetic materials in high-frequency, high-power conversion applications, JOM: J. Minerals Metals Mater. Soc. 64 (2012) 772–781. https://doi.org/10.1007/s11837-012-0350-0.

[71]   Q. Zhang, W. Zhang, K. Peng, In-situ synthesis and magnetic properties of core-shell structured Fe/Fe3O4 composites, J. Magn. Magn. Mater. 484 (2019) 418–423. https://doi.org/10.1016/J.JMMM.2019.04.053.

[72]   Y. Lee, M.A. Garcia, N.A. Frey Huls, S. Sun, Synthetic tuning of the catalytic properties of Au-Fe_3O_4 nanoparticles, Angew. Chem. Int. Ed. 49 (2010) 1271–1274. https://doi.org/10.1002/anie.200906130.

[73]   Q. Wei, Z. Xiang, J. He, G. Wang, H. Li, Z. Qian, M. Yang, Dumbbell-like Au-Fe_3O_4 nanoparticles as label for the preparation of electrochemical immunosensors, Biosens. Bioelectron. 26 (2010) 627–631. https://doi.org/10.1016/J.BIOS.2010.07.012.

[74]   M. Kaur, K. Pal, Review on hydrogen storage materials and methods from an electrochemical viewpoint, J. Energy Storage 23 (2019) 234–249. https://doi.org/10.1016/J.EST.2019.03.020.

[75]   T. Zhao, X. Li, H. Yan, Metal catalyzed preparation of carbon nanomaterials by hydrogen–oxygen detonation method, Combust. Flame 196 (2018) 108–115. https://doi.org/10.1016/J.COMBUSTFLAME.2018.06.011.

[76]   B. Sakintuna, F. Lamari-Darkrim, M. Hirscher, Metal hydride materials for solid hydrogen storage: A review, Int. J. Hydrogen Energy 32 (2007) 1121–1140. https://doi.org/10.1016/j.ijhydene.2006.11.022.

[77]   T.A. Hamad, A.A. Agll, Y.M. Hamad, S. Bapat, M. Thomas, K.B. Martin, J.W. Sheffield, Hydrogen recovery, cleaning, compression, storage, dispensing, distribution system and End-Uses on the university campus from combined heat, hydrogen and power system, Int. J. Hydrogen Energy 39 (2014) 647–653. https://doi.org/10.1016/J.IJHYDENE.2013.10.111.

[78]   Y. Hu, J. Lei, Z. Wang, S. Yang, X. Luo, G. Zhang, W. Chen, H. Gu, Rapid response hydrogen sensor based on nanoporous Pd thin films, Int. J. Hydrogen Energy 41 (2016) 10986–10990. https://doi.org/10.1016/J.IJHYDENE.2016.04.101.

[79]   L.M. Das, On-board hydrogen storage systems for automotive application, Int. J. Hydrogen Energy 21 (1996) 789–800. https://doi.org/10.1016/0360-3199(96)00006-7.

[80]   K. Manickam, P. Mistry, G. Walker, D. Grant, C.E. Buckley, T.D. Humphries, M. Paskevicius, T. Jensen, R. Albert, K. Peinecke, M. Felderhoff, Future perspectives of thermal energy storage with metal hydrides, Int. J. Hydrogen Energy 44 (2019) 7738–7745. https://doi.org/10.1016/J.IJHYDENE.2018.12.011.

[81]   B.D. Salih, M.A. Alheety, A.R. Mahmood, A. Karadag, D.J. Hashim, Hydrogen storage capacities of some new Hg(II) complexes containing 2-acetylethiophene, Inorg. Chem. Commun. 103 (2019) 100–106. https://doi.org/10.1016/J.INOCHE.2019.03.019.

[82]   B. Hardy, D. Tamburello, C. Corgnale, Hydrogen storage adsorbent systems acceptability envelope, Int. J. Hydrogen Energy 43 (2018) 19528–19539. https://doi.org/10.1016/J.IJHYDENE.2018.08.140.

[83]   S.S. Mao, S. Shen, L. Guo, Nanomaterials for renewable hydrogen production, storage and utilization, Prog. Nat. Sci. Mater. Int. 22 (2012) 522–534. https://doi.org/10.1016/J.PNSC.2012.12.003.

[84]   S. Fu, Z. Sun, P. Huang, Y. Li, N. Hu, Some basic aspects of polymer nanocomposites: A critical review, Nano Mater. Sci. 1 (2019) 2–30. https://doi.org/10.1016/J.NANOMS.2019.02.006.

[85]   R. Colorado-Peralta, R. Peña-Rodríguez, M.A. Leyva-Ramírez, A. Flores-Parra, M. Sánchez, I. Hernández-Ahuactzi, L.E. Chiñas, D.J. Ramírez, J.M. Rivera, Metal-organic structures with formate and sulfate anions: Synthesis, crystallographic studies and hydrogen storage by PM7 and ONIOM, J. Mol. Struct. 1189 (2019) 210–218. https://doi.org/10.1016/J.MOLSTRUC.2019.03.102.

[86]   C. Acar, I. Dincer, Comparative assessment of hydrogen production methods from renewable and non-renewable sources, Int. J. Hydrogen Energy 39 (2014) 1–12. https://doi.org/10.1016/J.IJHYDENE.2013.10.060.

[87]   J. Andersson, S. Grönkvist, Large-scale storage of hydrogen, Int. J. Hydrogen Energy 44 (2019) 11901–11919. https://doi.org/10.1016/J.IJHYDENE.2019.03.063.

[88]   E.R. López, F. Dorado, A. de Lucas-Consuegra, Electrochemical promotion for hydrogen production via ethanol steam reforming reaction, Appl. Catal. B Environ. 243 (2019) 355–364. https://doi.org/10.1016/J.APCATB.2018.10.062.

[89]   Q. Lei, B. Wang, P. Wang, S. Liu, Hydrogen generation with acid/alkaline amphoteric water electrolysis, J. Energy Chem. 38 (2019) 162–169. https://doi.org/10.1016/J.JECHEM.2018.12.022.

[90]   C. Herradón, R. Molina, J. Marugán, J.Á. Botas, Experimental assessment of the cyclability of the Mn2O3/MnO thermochemical cycle for solar hydrogen production, Int. J. Hydrogen Energy 44 (2019) 91–100. https://doi.org/10.1016/J.IJHYDENE.2018.06.158.

[91]   Y. Guan, M. Deng, X. Yu, W. Zhang, Two-stage photo-biological production of hydrogen by marine green alga Platymonas subcordiformis, Biochem. Eng. J. 19 (2004) 69–73. https://doi.org/10.1016/J.BEJ.2003.10.006.

[92]   X. Gan, D. Lei, K.-Y. Wong, Two-dimensional layered nanomaterials for visible-light-driven photocatalytic water splitting, Mater. Today Energy 10 (2018) 352–367. https://doi.org/10.1016/J.MTENER.2018.10.015.

[93]   J. Wang, Y. Yin, Fermentative hydrogen production using various biomass-based materials as feedstock, Renew. Sustain. Energy Rev. 92 (2018) 284–306. https://doi.org/10.1016/J.RSER.2018.04.033.

[94]   S.E. Hosseini, M.A. Wahid, Hydrogen production from renewable and sustainable energy resources: Promising green energy carrier for clean development, Renew. Sustain. Energy Rev. 57 (2016) 850–866. https://doi.org/10.1016/J.RSER.2015.12.112.

[95] L. Schlapbach, A. Züttel, Hydrogen-storage materials for mobile applications, Nature 414 (2001) 353–358. https://doi.org/10.1038/35104634.

[96] H. Feng, L. Tang, G. Zeng, Y. Zhou, Y. Deng, X. Ren, B. Song, C. Liang, M. Wei, J. Yu, Core-shell nanomaterials: Applications in energy storage and conversion, Adv. Colloid Interface Sci. 267 (2019) 26–46. https://doi.org/10.1016/J.CIS.2019.03.001.

[97] X. Guo, P. Brault, G. Zhi, A. Caillard, G. Jin, X. Guo, Structural evolution of plasma-sputtered core–shell nanoparticles for catalytic combustion of methane, J. Phys. Chem. C. 115 (2011) 24164–24171. https://doi.org/10.1021/jp206606r.

[98] C. Chang, P. Gao, D. Bao, L. Wang, Y. Wang, Y. Chen, X. Zhou, S. Sun, G. Li, P. Yang, Ball-milling preparation of one-dimensional Co–carbon nanotube and Co–carbon nanofiber core/shell nanocomposites with high electrochemical hydrogen storage ability, J. Power Sources 255 (2014) 318–324. https://doi.org/10.1016/J.JPOWSOUR.2014.01.034.

[99] D. Bao, P. Gao, X. Shen, C. Chang, L. Wang, Y. Wang, Y. Chen, X. Zhou, S. Sun, G. Li, P. Yang, Mechanical ball-milling preparation of fullerene/cobalt core/shell nanocomposites with high electrochemical hydrogen storage ability, ACS Appl. Mater. Interfaces 6 (2014) 2902–2909. https://doi.org/10.1021/am405458u.

[100] G. Wang, L. Zhang, J. Zhang, A review of electrode materials for electrochemical supercapacitors, Chem. Soc. Rev. 41 (2012) 797–828. https://doi.org/10.1039/C1CS15060J.

[101] Q. Wang, L. Jiao, H. Du, Y. Wang, H. Yuan, Fe_3O_4 nanoparticles grown on graphene as advanced electrode materials for supercapacitors, J. Power Sources 245 (2014) 101–106. https://doi.org/10.1016/J.JPOWSOUR.2013.06.035.

[102] C. Zhang, J. Li, C. Shi, C. He, E. Liu, N. Zhao, Effect of Ni, Fe and Fe-Ni alloy catalysts on the synthesis of metal contained carbon nano-onions and studies of their electrochemical hydrogen storage properties, J. Energy Chem. 23 (2014) 324–330. https://doi.org/10.1016/S2095-4956(14)60154-6.

[103] S. Taçyıldız, B. Demirkan, Y. Karataş, M. Gulcan, F. Sen, Monodisperse Ru Rh bimetallic nanocatalyst as highly efficient catalysts for hydrogen generation from hydrolytic dehydrogenation of methylamine-borane, J. Mol. Liq. 285 (2019) 1–8. https://doi.org/10.1016/j.molliq.2019.04.019.

[104] B. Şen, A. Aygün, A. Şavk, F. Gülbağça, S.K. Gülbay, M.H. Çalımlı, F. Şen, Binary Palladium–Nickel/Vulcan carbon-based nanoparticles as highly efficient

catalyst for hydrogen evolution reaction at room temperature, J. Taiwan Inst. Chem. Eng. 101 (2019) 92–98. https://doi.org/10.1016/J.JTICE.2019.04.040.

[105] S. Eris, Z. Daşdelen, F. Sen, Enhanced electrocatalytic activity and stability of monodisperse Pt nanocomposites for direct methanol fuel cells, J. Colloid Interface Sci. (2018). https://doi.org/10.1016/j.jcis.2017.11.085.

[106] S. Eris, Z. Daşdelen, F. Sen, Investigation of electrocatalytic activity and stability of Pt@f-VC catalyst prepared by in-situ synthesis for Methanol electrooxidation, Int. J. Hydrogen Energy. 43 (2018) 385–390. https://doi.org/10.1016/J.IJHYDENE.2017.11.063.

[107] B. Şen, B. Demirkan, M. Levent, A. Şavk, F. Şen, Silica-based monodisperse PdCo nanohybrids as highly efficient and stable nanocatalyst for hydrogen evolution reaction, Int. J. Hydrogen Energy. 43 (2018) 20234–20242. https://doi.org/10.1016/j.ijhydene.2018.07.080.

[108] B. Şen, A. Aygün, A. Şavk, S. Akocak, F. Şen, Bimetallic palladium–iridium alloy nanoparticles as highly efficient and stable catalyst for the hydrogen evolution reaction, Int. J. Hydrogen Energy. 43 (2018) 20183–20191. https://doi.org/10.1016/J.IJHYDENE.2018.07.081.

[109] S. Günbatar, A. Aygun, Y. Karataş, M. Gülcan, F. Şen, Carbon-nanotube-based rhodium nanoparticles as highly-active catalyst for hydrolytic dehydrogenation of dimethylamineborane at room temperature, J. Colloid Interface Sci. (2018). https://doi.org/10.1016/j.jcis.2018.06.100.

[110] B. Sen, A. Şavk, F. Sen, Highly efficient monodisperse Pt nanoparticles confined in the carbon black hybrid material for hydrogen liberation, J. Colloid Interface Sci. 520 (2018) 112–118. https://doi.org/10.1016/j.jcis.2018.03.004.

[111] B. Şen, A. Aygün, A. Şavk, M.H. Çalımlı, S.K. Gülbay, F. Şen, Bimetallic palladium-cobalt nanomaterials as highly efficient catalysts for dehydrocoupling of dimethylamine borane, Int. J. Hydrogen Energy (2019). https://doi.org/10.1016/J.IJHYDENE.2019.01.215.

[112] Y. Karataş, A. Aygun, M. Gülcan, F. Şen, A new highly active polymer supported ruthenium nanocatalyst for the hydrolytic dehydrogenation of dimethylamine-borane, J. Taiwan Inst. Chem. Eng. 99 (2019) 60–65. https://doi.org/10.1016/J.JTICE.2019.02.032.

[113] B. Şen, A. Aygün, A. Şavk, S. Duman, M.H. Calimli, E. Bulut, F. Şen, Polymer-graphene hybrid stabilized ruthenium nanocatalysts for the dimethylamine-borane

dehydrogenation at ambient conditions, J. Mol. Liq. 279 (2019) 578–583. https://doi.org/10.1016/J.MOLLIQ.2019.02.003.

[114] Y. Karatas, M. Gülcan, F. Sen, Catalytic methanolysis and hydrolysis of hydrazine-borane with monodisperse Ru NPs@nano-CeO_2 catalyst for hydrogen generation at room temperature, Int. J. Hydrogen Energy 44 (2019) 13432–13442. https://doi.org/10.1016/j.ijhydene.2019.04.012.

[115] B. Sen, A. Aygün, M. Ferdi Fellah, M. Harbi Calimli, F. Sen, Highly monodispersed palladium-ruthenium alloy nanoparticles assembled on poly(N-vinyl-pyrrolidone) for dehydrocoupling of dimethylamine–borane: An experimental and density functional theory study, J. Colloid Interface Sci. 546 (2019) 83–91. https://doi.org/10.1016/j.jcis.2019.03.057.

[116] B. Şen, A. Aygün, A. Şavk, C. Yenikaya, S. Cevik, F. Şen, Metal-organic frameworks based on monodisperse palladiumcobalt nanohybrids as highly active and reusable nanocatalysts for hydrogen generation, Int. J. Hydrogen Energy 44 (2019) 2988–2996. https://doi.org/10.1016/J.IJHYDENE.2018.12.051.

[117] B. Şen, B. Demirkan, A. Şavk, S. Karahan Gülbay, F. Şen, Trimetallic PdRuNi nanocomposites decorated on graphene Oxide: A superior catalyst for the hydrogen evolution reaction, Int. J. Hydrogen Energy 43 (2018) 17984–17992. https://doi.org/10.1016/j.ijhydene.2018.07.122.

[118] S. Eris, Z. Daşdelen, Y. Yıldız, F. Sen, Nanostructured polyaniline-rGO decorated platinum catalyst with enhanced activity and durability for Methanol oxidation, Int. J. Hydrogen Energy 43 (2018) 1337–1343. https://doi.org/10.1016/J.IJHYDENE.2017.11.051.

Magnetochemistry - Materials and Applications   Materials Research Forum LLC
Materials Research Foundations **66** (2020) 259-275   https://doi.org/10.21741/9781644900611-8

Chapter 8

Magnetic Nanomaterials for Supercapacitors

K. Srinivas[1*], K. Chandra Babu Naidu[1], G. Balakrishna[2], B. Venkata Shiva Reddy[1],
N. Suresh Kumar[3], S. Ramesh[1], Prasun Banerjee[1], D. Baba Basha[4]

[1]Dept. of Physics, GITAM Deemed to be University, Bangalore-562163, Karnataka, India

[2]Dept. of Electrical & Electronics Engineering, Srinivasa Ramanujan Institute of Technology, Anathapuram-515701, A.P., India

[3]Department of Physics, JNTUA, Anantapuramu-515002, A.P, India

[4]Department of Physics, College of Computer and Information Sciences, Majmaah University Al'Majmaah, K.S.A-11952

*srinkura@gmail.com

Abstract

Attractive ferrite nanoparticles have gained enthusiasm in recent years owing to their uncommon synthetic and physical properties with promising applications in ferro fluids, concoction sensors, impetuses etc. Notwithstanding these applications, there is similarly an expanding enthusiasm for vitality stockpiling research dependent on the quickly developing business sector for electronic gadgets which are being intended to be littler, lighter, and generally less expensive. Hence, an across the board gadget requires productive vitality stockpiling parts which will fit into such structure plan with improved vitality execution. In this part we have completely researched different attractive material put together supercapacitor and their performances with respect to different doping elements. The basic examination of the incorporated iron oxide (Fe_3O_4) nanocrystals uncovers the magnetite period of Fe_3O_4. Moreover, these Fe_3O_4 crystals indicated bi-practical superparamagnetic and ferromagnetic conduct beneath or more the impending temperature, individually. The utilization of the above said nanocrystals as a negative electrode (anode) for supercapacitor was examined by investigating the cyclic voltametry (CV) and galvanostatic charge– release tests of Fe_3O_4, the uniform nano size of Fe_3O_4 causes the high explicit capacitance. This chapter concentrates on an extreme simple strategy to incorporate nanostructured Fe_3O_4 for the application in cutting edge vitality stockpiling materials. Electrochemical properties of the undoped and aluminum-doped nickel copper ferrite supercapacitor anodes examined by CV and galvanostatic charge/release estimations in 1M KOH. A particular capacitance of 412.5 Fg^{-1} was seen with $Al_{0.2}Ni_{0.4}Cu_{0.4}Fe_2O_4$ at a present thickness of 1 Ag^{-1} with vitality thickness of 57.3

WhKg^{-1}. The vitality densities of the nanocomposite (magnetite/polypyrrole) capacitors having diverse substance nano magnetite particles accomplish an expansion of in excess of multiple times at a present thickness of 10.0 A/g, contrasted with the partners without an attractive field.

Keywords

Nanomaterials, Supercapacitors, Capacitance, Ferrite Composites

Contents

1.   Introduction

The expanding interest for large thickness has invigorated extraordinary research enthusiasm for the electrochemical devices like supercapacitors [1-3]. These supercapacitors are prominent because of the capacity thickness, magnificent cyclability and reversibility [4, 5]. These supercapacitors can be utilized in various fields for example burst control age, electric vehicles, memory back-up gadgets etc. [6,7]. Also, they can be utilized in the life cycle improvement and power upgradation of significant sources of power such as fuel cells and batteries [8, 9]. Depending on the charge stockpiling instrument supercapacitors are mainly classified as two types which are EDLC (electrical double layer capacitors) in which, the capacitance emerges from the electrolyte interface or the detachment of the charge at the terminal and pseudocapacitors in which the capacitance (pseudocapacitance) emerges due to the faradic responses happening at the terminal interface. Fig. 1 represents the mechanism of different kinds of supercapacitors. Different kinds of materials such as transition metal oxides (TMO), polymers and carbon are broadly utilized for such applications [2].

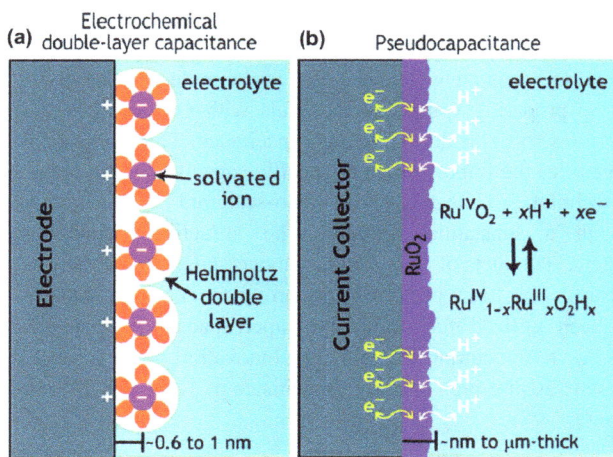

Figure 1. Mechanism of EDLC and Pseudocapacitors

As of late, different change metal oxides, for example, RuO_2, Co_3O_4, NiO, Fe_2O_3, Fe_3O_4, MnO_2, and so on, be examined for electrochemical applications [10-27]. Among these TMOs, RuO_2 is the utmost encouraging material as it can give large charge stockpiling capacity [10, 11]. Nevertheless, a few constraints of RuO_2, for example, surprising expense, poisonous quality and shortage. The conceivable answer for this issue is to improve the charge stockpiling limit of the ease, normally plenteous and earth friendly metal oxides. Fe_3O_4 (Iron oxide) can be a prominent contender for those appliances because of its simple redox response, ease and small ecological effect. The charge stockpiling limit of the Fe_3O_4 can be radically enhanced by expanding the dynamic surface territory. Hydrothermally developed Fe_3O_4 film was utilized as anode for supercapacitor appliances [24]. The Fe_3O_4 film demonstrated at 6 mA current with a specific capacitance of 118.2 F g^{-1}. The particular capacitance of 117.2 F g^{-1} was watched for the compound, which is multiple times greater than unadulterated Fe_3O_4. Improved or advanced electrochemical exhibitions might be ascribed to the great conductivity of carbon nano tubes (CNTs) just like the tied down Fe_3O_4 particles on the CNTs. Nitrogen-doped graphene–Fe_2O_3 compounds (NGFeCs) have been combined and used for supercapacitor applications. It was noticed that the arranged NGFeCs demonstrate a superior electrochemical act when compare to the graphene– Fe_2O_3 composites (GFeCs). Wang et al. [22] used a compound of Fe_3O_4 nanoparticles developed on decreased GO (graphene oxide) for applications in supercapacitor. The particular capacitances of 110.5, 220.1 and 65.4 F g^{-1} were watched for Fe_3O_4/decreased graphene oxide, Fe_3O_4/carbon nanotubes and Fe_3O_4, separately. The decrease in the molecule size of Fe_3O_4 from micro to nano level by preparing composite materials with GO and CNTs was one of the primary motivations to build the particular capacitance. Therefore, the reduction in the molecular size causes the improvement in specific capacitance of a redox dynamic metal oxide. A easy technique to integrate about mono-scattered Fe_3O_4 nanocrystals, which can be scaled-up in a substantial amount. The attractive estimations uncover the superparamagnetic conduct of the Fe_3O_4 nanocrystals. The itemized auxiliary and attractive portrayal has been performed to prove the nanostructure of Fe_3O_4. These crystals have been utilized for storage or supercapacitor applications to assess the plausibility of Fe_3O_4 for cutting edge vitality innovation. We report that at 1 mA the Fe_3O_4 nanocrystals exhibit high specific capacitance of 185 F g^{-1}. This is due to uniform nano-size of Fe_3O_4.

Electrochemical properties of the undoped and aluminum-doped nickel copper ferrite supercapacitor terminals have been studied by cyclicvoltammetry (CV) and galvanostatic charge/release estimations in 1M KOH. A particular capacitance of 412.5 Fg^{-1} was gotten with $Al_{0.2}Ni_{0.4}Cu_{0.4}Fe_2O_4$ at a present thickness of 1Ag^{-1} with vitality thickness of 57.3 WhKg^{-1} [28]. The vitality densities of the magnetite or polypyrrole nanocomposite capacitors [29] comprising diverse substance nanoparticles (magnetite) accomplish an expansion of in excess of multiple times at a large present thickness of 10 A/g, contrasted with the partners without an attractive field. The conceivable instrument is that the attractive field incites electrolyte particle development improvement and charge exchange obstruction decrease, which astoundingly cause the expansion of capacitance and vitality thickness. This work gives a creative methodology to altogether improve the rate abilities of current ECs by a basic physical procedure instead of synthetic procedure.

Cobalt ferrite attractive nanoparticles ($CoFe_2O_4$-MNPs) [30] were orchestrated by aqueous and coprecipitation techniques utilizing various forerunners, for example, nitrates, chlorides, and acetic acid derivations, at various focuses with or without emulsifier under various development conditions. The basic and morphological examinations uncover the development of a solitary stage $CoFe_2O_4$ in nanoplatelet-formed NPs with normal molecule estimate somewhere in the range of 11 and 26 nm relying upon blend disorder. The particular surface territory of these nano particles acquired through aqueous strategy was ~ 34 m^2g^{-1}. Electrochemical exhibitions of the got NPs in a 3-terminal arrangement with a electrolyte (6 M KOH) uncovered a particular capacitance (C_s) of 429.0 F/g at 0.5 A/g, with fantastic capacitance maintenance of 98.79% at 10 A/g after 6000 cycles for electroactive nanoparticles orchestrated by aqueous strategy at 200 °C/18 h.

2.   Magnetic NPs

Attractive NPs are extremely intriguing because of their different properties, for example, measure impacts, surface-to-volume proportion, association, attractive detachment, explicitness and surface science. Attractive NPs, specifically maghemite, magnetite and nano zero-valent iron have started the applications in prescription, atomic science, and reformation of contaminated H_2O (water) [31] Much of the time maghemite or magnetite is utilized to shape the center of attractive iron oxide (Fe_3O_4) nanoparticles (MION).

Fig 1. Charge hysteresis circles of superparamagnetic NPs. (M_r – zero remanence, M_s – high immersion polarization, H – coercivity) MION's might be extensively partitioned into three primary categories which are para, ferro and superparamagnetic [32]. In paramagnetic material the behavior indicates that the attractive dipoles are situated in arbitrary ways at ordinary temperatures because of unpaired electrons, this is due to a low positive vulnerability (feeble association) in an attractive field. In ferromagnetic materials rely upon their space structure to stay charged even without a connected attractive field yet measure diminishes to not exactly the area estimate when they experience a noteworthy change. Superparamagnetism will in general have bigger attractive vulnerability than paramagnets since the attractive snapshot of the whole nanoparticle adjusts toward the attractive field (Fig. 1).

2.1   Synthesis of attractive NPs

As of late an assortment of substance combination techniques have been created for planning monodisperse superparamagnetic nanoparticles with custom fitted structures, measure, surface science and condition of total. Especially amid the previous couple of years, numerous distributions have depicted productive union techniques including microemulsion synthesis, synthetic co-precipitation, aqueous synthesis, thermal decomposition, etc. [33].

2.1.1 Co-precipitation

Co-precipitation is a simple and helpful approach to blend Fe_3O_4 attractive NPs (MION) from fluid Fe^{2+} and Fe^{3+} salt arrangements by the expansion of a base at raised temperature or at room temperature. The magnetic NPs shape and organization rely on the sort of salt (for example sulfate, chloride etc.), the Fe^{2+} and Fe^{3+} proportion, the temperature responses, pH esteem and ionic quality of the media. By the by, if union circumstances are static, the nature of the attractive NPs is completely re-producible [34, 35]. The combination of attractive nano particles could be achieved by Eq. 1 through overall precipitation of Fe_3O_4 within the pH range 7.5-14 (for example pH in the scope of 7.5-11 are gotten when working with microemulsion frameworks) under idle condition. This technique includes readiness of huge amounts of NPs in a solitary group, yet the size circulation is extensive [36].

$$2FeCl_3 + FeCl_2 + 4H_2O + 8NH_3 \rightarrow Fe_3O_4 + 8NH_4Cl \tag{1}$$

2.1.2 Micro-emulsion

Micro-emulsion (ME) is a 3-part framework comprising of H_2O, surfactant and oil shaping a thermo dynamically steady isotropic arrangement. Contingent upon the centralizations of various segments, microemulsion (ME) can shape w/o (water-in-oil) ME [37] or an o/w (oil-in-water) ME [38]. The surfactants utilized in this strategy could be non-ionic – Berol 050, CTAB (cationic – certyltrimethylammonium bromide), polyethylenoxide and anionic – bis 2-ethylhexyl sulfosuccinate. The principle favorable position of this method is that it permits the arrangement of nanoparticles in a even size dispersion. The span of the particles could be custom fitted with water-surfactant molar ratio [39]. Through this technique, monodiperesed nanoparticles with different morphology and surface change might be set up for broad appliances [33].

2.1.3 Thermal decay

Thermal decay is a procedure that is utilized for the most part in combination of top notch semiconductor oxides and nanocrystals in non-fluid medium. Monodisperse attractive nanocrystals were accomplished via warm deterioration of organo-metallic mixes in large-bubbling natural solvent comprising balancing out surfactants. In this procedure, unsaturated fats, for example, oleic corrosive and hexadecylamine are frequently utilized as surfactants [40, 41]. Amid the union procedure, response time, maturing time and temperature response are basic factors for controlled morphology and size of attractive nano particles. The primary disadvantages in this strategy are the absence of nano particles suspension in fluid media because of large crystallinity. Also, it is a confused procedure which necessitates a dormant air and high temperature amid blend that goes on for a few hours.

2.1.4 Hydrothermal strategy

Hydrothermal strategy is considered as a standout amongst the most adaptable and neighborly technique [42]. The system depends on the general stage exchange and partition component happening at the fluid interface, strong and arrangement stage present amid the combination. The designing of nanoparticle (NP) surfaces can't be cultivated and thusly post-preparing steps are entailed. One of the primary downside in this technique is that at some random temperature, the amount of energy is smaller when contrasted with the different combination strategies [33]. The favorable circumstances and disservices of the four referenced amalgamation strategies are quickly outlined in Table 1 [33]. In addition to the techniques portrayed, microemulsion and co-precipitation are broadly used in the blend of superparamagnetic nano particles (NPs) for natural and biomedical applications. The active window is very vast for these two techniques and advances surface science change of the nanoparticle amid combination or after blend. Another preferred standpoint incorporates temperature response and planning times that are much lesser than aqueous and warm deterioration strategies.

2.4.4 Magnetic separation

Detachments utilizing attractive area are generally utilized in the zones of medication conveyance, sub-atomic science, diagnostics, immuno examines, catalysis and environmental remediation [43, 44]. Attractive NPs are made out of attractive center, specifically super-paramagnetic nano particles (NPs) are effectively isolated by the assistance of an outside magnet. To be sure attractive NPs are widely utilized in the recycling of waste, for instance the reformative of substantial metals from waste water [45]. This method is considered as one of the best strategy in which decontamination doesn't produce option waste and also the materials can be reused. The immersion components of attractive nano paticles (NPs) with impurities are ionic trade and further frail powers. This methodology is generally alluded to as attractively helped concoction division [46]. Lasting or electro-magnets (for example attractive separator with a pivoting plate and an attractive drum) are accessible financially for huge scale setups relying upon the application and quality of attractive NPs. Attractive NPs can be custom fitted utilizing regular or engineered polymers for particular adsorptions under research center conditions.

Table. 1 Different magnetic nanoparticle (NPs) synthesis techniques.

Method of synthesis	Co-precipitation	Microemulsion	Thermal decomposition	Hydrothermal synthesis
Technique	very simple, ambient condtions	simple	simple, inert atmosphere	simple, high pressure
Component	water	organic compound, water	organic compound	water-ethanol
Reaction period	hours	hours	hours-days	hours
Reaction temp. [°C]	20-90	20-50	100-320	220
Size distribution	relatively narrow	relatively narrow	very narrow	very narrow
Shape control	not good	good	very good	very good
Yield	high/scalable	low	high/scalable	medium

2.2   Functionalization and utilization of attractive NPs

Despite the fact that nano particles exhibit high surface area -to-volume proportion, soundness is an essential prerequisite once they are presented to air on account of unadulterated material particles such as cobalt, nickel, iron etc. [33]. Covering of nanomaterials incorporates polymer or surfactant, valuable metal, carbon and silica for different appliances. Subsequently, covering helps in securing the center shell as well as utilized moreover of utilitarian gatherings for explicit toxin expulsion in water treatment [47]. By and large, attractive nano materials are surface changed with hydroxyl, carboxyl and amino gatherings for their particular associations. For instance, attractive nano materials could be topped with either positively or negatively charged substance over surface science so as to build their dependability. Ongoing examinations uncovered that microemulsion arranged attractive iron oxide nanoparticles (ME-MION) with protein restricting brought about decrease of suspended particles and organisms [48, 49]. Singh et al. [50] described successful expulsion of >95% overwhelming materials, for example, Cu, Cd, Ni, Zn and Pb from fluid arrangement utilizing amino, thiol and carboxyl functionalized attractive nano materials. In this manner, functionally attractive nanomaterials contain a high level of cooperation and capacity to evacuate explicit impurities in WWTP.

The utilization of bi-metallic nanomaterials have been of extraordinary enthusiasm because of their upgraded security, restraint of oxidation and expanded response. Regularly when two different metals are in connection with various electrical possibilities, for example on account of Fe (iron) and Ni (nickel), iron goes about as a diminishing operator and nickel as an impetus with hydrogen produced from H_2O. In any case, Ni/Fe likewise behaves a decent consumption preservative and lesser the expense for on location waste H_2O treatment [51]. Iron fills in as electron benefactor to respond with the impurities and the respectable metal is ensured in the expulsion of chlorinated aliphatic. In general, TEM (transmission electron microscopy) is used to determine the size, XRD (X-ray diffraction) determines the structure also used to estimate the structural parameters; FT-IR (Fourier transform infrared spectroscopy) is used to describe the presence of M-O (metal oxide bonds) on the molecule surface; isoelectric point and surface change can be determined with the zeta potential. Polarization calculations and hydrodynamic breadth of the particles in suspension can be enumerated by using DLS (dynamic light scattering).

3. Some remarkable properties of Iron oxide nano particles

Iron oxide (Fe_3O_4) NPs happen normally in the world's covering by different ecological resources, for example, fires, volcanoes etc. It may exist in numerous structures in nature. Magnetite, γ-Fe_2O_3 (Maghemite) and α-Fe_2O_3 (Hematite) are the utmost well-known structures which are steady as a function of their synthetic composition at room temperature. Magnetite has a cubic converse spinal-structure with oxygen structure, face focused cubic pressing whereas Fe-cation possesses the interstitially tetrahedral & octahedral locales. Fe (III) ions possess the tetrahedral destinations, whereas remaining half involve the octahedral locales going with Fe (II) ions. AB_2O_4 is the magnetite general expression, Here A and B epitomizes Fe (II) and Fe (III). Metal ferrites, for example, $NiFe_2O_4$, $MnFe_2O_4$ etc. might be framed by replacing Fe (II) by the relating metal ions. Magnetite bit by bit oxidizes to maghemite in surrounding air.

Magnetochemistry - Materials and Applications   Materials Research Forum LLC
Materials Research Foundations **66** (2020) 259-275   https://doi.org/10.21741/9781644900611-8

3.1   Iron oxide nanoparticles

Coming of nanotechnology [52] upsets the entire logical world which gives chances to incorporate the materials in to nano scale going to 100 nm from 1nm. In this small-scale, the materials display exceptional qualities because of unbounded little size and large surface territory to volume proportion. Fe_3O_4 NPs have one increasingly exceptional property as super-paramagnetism particularly with the size running from 2 to 20 nm in width alongside little size and large surface region. It is a type of attraction particularly appeared by adequately little size NPs in which polarization could arbitrarily alter course affected by temperature and time required to flip the heading starting with one phase then onto the next is named as Neel's Relaxation time. Therefore, their charge esteem can't be estimated without outside attractive field. Within the sight of attractive field, they demonstrate a decent attractive vulnerability when contrast with paramagnets. The attractive conduct is a significant parameter [53] that should be considered genuinely while structuring and blending very iron oxide NPs so as to awfully encourage their proficiency in partition region as high polarization esteem is essential in this strategy. As of now Fe_3O_4 NPs with unique properties and capacities have been generally investigated in the area of partition innovations because of its savvy union, simple covering or surface change and capacity to regulate/controlling materials on a small-scale measurement give astonishing adaptability in division procedures [54]. Aside from these, biocompatibility, low poisonous quality and synthetic dormancy of Fe_3O_4 NPs demonstrate the promising applications in biomedicine. Excessively paramagnetic Fe_3O_4 NPs has vast surface zone, biocompatibility, low harmfulness and compound dormancy this is profoundly apt in adsorption innovation for ecological remediation.

3.2   Al doped ferrite nanomaterials

To abridge, we have blended aluminum-doped nickel copper ferrite nanomaterials as potential supercapacitor anodes [28]. Supposedly, it is the first occasion when that aluminum doping has been acquainted in ferrite nanomaterials with improvement in their electrochemical properties. A high explicit capacitance of 412.5 Fg^{-1} was acquired with $Al_{0.2}Ni_{0.4}Cu_{0.4}Fe_2O_4$ at a present thickness of 1 Ag^{-1}. A vitality thickness of 57.3 $WhKg^{-1}$ was recorded with $Al_{0.2}Ni_{0.4}Cu_{0.4}Fe_2O_4$ though with undoped NiCuF, a vitality thickness of 34.3 $WhKg^{-1}$ was gotten. The upsides of aluminum doping in this nanomaterial are: (1) simple and quick union; (2) great execution (high explicit capacitance). Every one of these attributes affirms that the aluminum doped nickel copper ferrite nanomaterial might be a promising possibility for supercapacitors.

3.3   Magnetite/polypyrrole nanocomposite capacitors

As shown in outline the attractive field can essentially improve the vitality densities of nanocomposite cathodes (magnetite/polypyrrole) at large rates [29]. The joined power under the magnetic and electric fields expanded the particular capacitance. In addition, along these lines ascribed to developed vitality densities over the control of the dynamic movement of the electrolyte particles and the decrease of the charge exchange opposition. The attractive field expanded the vitality thickness by a factor of 10.49, 11.11, and 19.31, at a high present thickness of 10.0 A/g for the nanocomposite (magnetite/polypyrrole) cathodes at a magnetite molecule stacking of 10.0, 20.0, and 40.0 wt%, separately. This basic methodology gives a creative option to essentially upgrading the vitality thickness of ECs at large rates with no trading off their capacity densities and is imagined to impact other capacity units, for example, LIBs and energy components also.

3.4   Cobalt based ferrite materials

$CoFe_2O_4$ spinel ferrite attractive MNPs [30] have been progressively blended by two distinct strategies co-precipitation and aqueous procedure utilizing diverse cobalt and ferrite antecedents with various fixations just as development conditions (temperature and time). The upgraded condition for the coprecipitation technique is utilizing nitrate-based forerunners below 353 K response temperature, though there was no perceptible contrast in outcomes for the hydrothermally arranged examples. The structural investigation affirms the development of mono-stage cubic spinel structure and the investigation on HRTEM & FESEM illustrate that the arrangement of the particles in nano range with a normal molecule measure somewhere in the range of 11.0 nm & 26.1 nm. The pore and surface zone estimate conveyance estimations uncover the nearness of a mesoporous structured in these Co-ferrite nanoparticles with a surface territory about 34.1 m^2g^{-1}. The NPs prepared by aqueous blend have an immersion charge of 63.0emu/g and a coercivity of 750 Oe at room-temperature. Electrochemical properties of these ferrite materials arranged through the aqueous strategy uncovers a prevalent faradaic type conduct, with a large explicit capacitance of roughly 428.9F/g at a current thickness of 0.5 A/g and amazing cycle existence with a coulombic efficiency of 98.79 % after 6000 cycles at a present thickness of 10 A/g. The outcomes acquired show the selection of the ferrite nanoparticles are prominent materials for the applications in supercapacitors.

Conclusions

We have completely researched different attractive materials put together supercapacitors and their performances with respect to different doping elements. The basic examination of the incorporated iron oxide (Fe_3O_4) nanocrystals uncovers the magnetite period of

Fe_3O_4. Moreover, these Fe_3O_4 crystals indicated bi-practical superparamagnetic and ferromagnetic conduct beneath or more the impending temperature, individually. This facile and green synthesis strategy offers an effective way to produce a high performance supercapacitor and shows a promising large-scale application in energy storage.

References

[1] R. Koetz and M. Carlen, Principles and applications of electrochemical capacitors, Electrochim. Acta, 45 (2000) 2483–2498. https://doi.org/10.1016/S0013-4686(00)00354-6

[2] Y. Zhang, H. Feng, X. Wu, L. Wang, A. Zhang, T. Xia, H. Dong, X. Li, L. Zhang, Progress of electrochemical capacitor electrode materials: A review, Int. J. Hydrogen Energy, 34 (2009) 4889–4899. https://doi.org/10.1016/j.ijhydene.2009.04.005

[3] A. Burke, R&D considerations for the performance and application of electrochemical capacitors, Electrochim. Acta 53 (2007) 1083–1091. https://doi.org/10.1016/j.electacta.2007.01.011

[4] H.R. Ghenaatian, M.F. Mousavi, M.S. Rahmanifar, High performance battery-supercapacitor hybrid energy storage system based on self-doped polyaniline nanofibers, Synth. Met. 161 (2011) 2017–2023. https://doi.org/10.1016/j.synthmet.2011.07.018

[5] B.E. Conway, V. Birss, J. Wojtowicz, The role and utilization of pseudocapacitance for energy storage by supercapacitors, J. Power Sources 66 (1997) 1–14. https://doi.org/10.1016/S0378-7753(96)02474-3

[6] A. Burke, Ultracapacitors: why, how, and where is the technology, J. Power Sources 91 (2000) 37–50. https://doi.org/10.1016/S0378-7753(00)00485-7

[7] C. Ashtiani, R. Wright, G. Hunt, Ultracapacitors for automotive applications, J. Power Sources 154 (2006) 561–566. https://doi.org/10.1016/j.jpowsour.2005.10.082

[8] A. Chu, P. Braatz, Comparison of commercial supercapacitors and high-power lithium-ion batteries for power assist applications in hybrid electric vehicles: I. Initial characterization, J. Power Sources 112 (2002) 236–246. https://doi.org/10.1016/S0378-7753(02)00364-6

[9] A. Burke, M. Miller, The power capability of ultracapacitors and lithium batteries for electric and hybrid vehicle applications, J. Power Sources 196 (2011) 514–522. https://doi.org/10.1016/j.jpowsour.2010.06.092

[10] V. Subramanian, S.C. Hall, P. H. Smith, B. Rambabu, Mesoporous anhydrous RuO_2 as a supercapacitor electrode material, Solid State Ionics 175 (2004) 511–515. https://doi.org/10.1016/j.ssi.2004.01.070

[11] U.M. Patil, S.B. Kulkarni, V.S. Jamadade, C.D. Lokhande, Chemically synthesized hydrous RuO_2 thin films for supercapacitor application, J. Alloys Compd. 509 (2011) 1677–1682. https://doi.org/10.1016/j.jallcom.2010.09.133

[12] T. Liu, W.G. Pell, B.E. Conway, Self-discharge and potential recovery phenomena at thermally and electrochemically prepared RuO_2 supercapacitor electrodes, Electrochim. Acta 42 (1997) 3541–3552. https://doi.org/10.1016/S0013-4686(97)81190-5

[13] Y.F. Yuan, X.H. Xia, J.B. Wu, X.H. Huang, Y.B. Pei, J.L. Yang, S.Y. Guo, Hierarchically porous Co_3O_4 film with mesoporous walls prepared via liquid crystalline template for supercapacitor application, Electrochem. Commun. 13 (2011) 1123–1126. https://doi.org/10.1016/j.elecom.2011.07.012

[14] C. Xiang, M. Li, M. Zhi, A. Manivannan, N. Wu, A reduced graphene oxide/Co_3O_4 composite for supercapacitor electrode, J. Power Sources 226 (2013) 65–70. https://doi.org/10.1016/j.jpowsour.2012.10.064

[15] J.H. Kwak, Y.W. Lee, J. H. Bang, Supercapacitor electrode with an ultrahigh Co_3O_4 loading for a high areal capacitance, Mater. Lett., 110 (2013) 237–240. https://doi.org/10.1016/j.matlet.2013.08.032

[16] G. He, J. Li, H. Chen, J. Shi, X. Sun, S. Chen, X. Wang, Hydrothermal preparation of Co_3O_4@graphene nanocomposite for supercapacitor with enhanced capacitive performance, Mater. Lett., 82 (2012) 61–63. https://doi.org/10.1016/j.matlet.2012.05.048

[17] J. Zhu, J. Jiang, J. Liu, R. Ding, H. Ding, Y. Feng, G. Wei and X. Huang, Direct synthesis of porous NiO nanowall arrays on conductive substrates for supercapacitor application, J. Solid State Chem. 184 (2011) 578–583. https://doi.org/10.1016/j.jssc.2011.01.019

[18] X. Yan, X. Tong, J. Wang, C. Gong, M. Zhang, L. Liang, Synthesis of mesoporous NiO nanoflake array and its enhanced electrochemical performance for supercapacitor application, J. Alloys Compd. 593 (2014) 184–189. https://doi.org/10.1016/j.jallcom.2014.01.036

[19] H. Gao, H. Zhang, G.-P. Cao, M.F. Han, Y.S. Yang, Spherical porous VN and NiO_x as electrode materials for asymmetric supercapacitor, Electrochim. Acta 87 (2013) 375–380. https://doi.org/10.1016/j.electacta.2012.09.075

[20] B.J. Lokhande, R.C. Ambare, S.R. Bharadwaj, Thermal optimization and supercapacitive application of electrodeposited Fe_2O_3 thin films, Measurement 47 (2014) 427–432. https://doi.org/10.1016/j.measurement.2013.09.005

[21] P.M. Kulal, D.P. Dubal, C.D. Lokhande, V.J. Fulari, Chemical synthesis of Fe_2O_3 thin films for supercapacitor application, J. Alloys Compd. 509 (2011) 2567–2571. https://doi.org/10.1016/j.jallcom.2010.11.091

[22] Q. Wang, L. Jiao, H. Du, Y. Wang, H. Yuan, Fe_3O_4 nanoparticles grown on graphene as advanced electrode materials for supercapacitors, J. Power Sources 245 (2014) 101–106. https://doi.org/10.1016/j.jpowsour.2013.06.035

[23] Y. H. Kim, S.J. Park, Roles of nanosized Fe_3O_4 on supercapacitive properties of carbon nanotubes, Curr. Appl. Phys. 11 (2011) 462–466. https://doi.org/10.1016/j.cap.2010.08.018

[24] J. Chen, K. Huang, S. Liu, Hydrothermal preparation of octadecahedron Fe_3O_4 thin film for use in an electrochemical supercapacitor, Electrochim. Acta 55 (2009) 1–5. https://doi.org/10.1016/j.electacta.2009.04.017

[25] X. Zhang, P. Yu, H. Zhang, D. Zhang, X. Sun, Y. Ma, Rapid hydrothermal synthesis of hierarchical nanostructures assembled from ultrathin birnessite-type MnO_2 nanosheets for supercapacitor applications, Electrochim. Acta 89 (2013) 523–529. https://doi.org/10.1016/j.electacta.2012.11.089

[26] H. Jin, G.T. Cao, J.Y. Sun, Hybrid supercapacitor based on MnO_2 and columned FeOOH using Li_2SO_4 electrolyte solution, J. Power Sources 175 (2008) 686–691. https://doi.org/10.1016/j.jpowsour.2007.08.115

[27] W.J. Ge, H.B. Yao, W. Hu, X.F. Yu, Y.X. Yan, L.B. Mao, H.H. Li, S.S. Li, S.H. Yu, Facile dip coating processed graphene/MnO_2 nanostructured sponges as high performance supercapacitor electrodes, Nano Energy, 2 (2013) 505–513. https://doi.org/10.1016/j.nanoen.2012.12.002

[28] B. Bhujun, A.S. Shanmugam, M.T.T. Tan, Aluminium-doped nickel copper ferrites for high-performance supercapacitors, Int. J. Res. Chem. Metallurgical Civil Engg. (IJRCMCE) 3 (2016) 37-41. https://doi.org/10.15242/IJRCMCE.E0316002

[29] H. Wei, H. Gu, J. Guo, D. Cui, X. Yan, J. Liu, D. Cao, X. Wang, S. Wei, Z. Guo, Significantly enhanced energy density of magnetite/polypyrrole nanocomposite capacitors at high rates by low magnetic fields, Adv. Compos. Hybrid Mater. 1 (2018) 127-134. https://doi.org/10.1007/s42114-017-0003-4

[30] H. Kennaz, A. Harat, O. Guellati, D.Y. Momodu, F. Barzegar, J.K. Dangbegnon, N. Manyala, M. Guerioune, Synthesis and electrochemical investigation of spinel cobalt ferrite magnetic nanoparticles for supercapacitor application, J. Solid State Electrochem. 22 (2018) 835-847. https://doi.org/10.1007/s10008-017-3813-y

[31] S.C.N. Tang, I.M.C. Lo, Magnetic nanoparticles: Essential factors for sustainable environmental applications. Water Research 47 (2013) 2613-2632. https://doi.org/10.1016/j.watres.2013.02.039

[32] Y.K. Gun'ko, D.F. Brougham, Magnetic nanomaterials as MRI contrast agents: In Magnetic Nanomaterials, Ed. Kumar, S.R. John Wiley & Sons, Inc. (2009).

[33] A.H. Lu, E.L. Salabas, F. Schüth, Magnetic nanoparticles: Synthesis, protection, functionalization, and application. Angew. Chem. Int. Ed. 46 (2007) 1222–1244. https://doi.org/10.1002/anie.200602866

[34] Y. Hu, Y. Du, J. Yang, J.F. Kennedy, X. Wang, L. Wang, Synthesis, characterization and antibacterial activity of guanidinylated chitosan, Carbohydrate Polym. 67 (2007) 66-72. https://doi.org/10.1016/j.carbpol.2006.04.015

[35] Z.Z. Guo, H.H. Liu, X.X. Chen, X.X. Ji, P.P. Li, Hydroxyl radicals scavenging activity of N-substituted chitosan and quaternized chitosan, Bioorg. Med. Chem. Lett. 16 (2006) 6348-6350. https://doi.org/10.1016/j.bmcl.2006.09.009

[36] G.D.L. Tran, X.P. Nguyen, D.H. Vu, N.T. Nguyen, V.H. Tran, T.T.T. Mai, N.T. Nguyen, Q.D. Le, T.N. Nguyen, T.C. Ba, Some biomedical applications of chitosan-based hybrid nanomaterials. Adv. Natural Sci. Nanosci. Nanotechnol. 2 (2011) 1-6. https://doi.org/10.1088/2043-6262/2/4/045004

[37] M. Boutonnet, S. Lögdberg, E.E. Svensson, Recent developments in the application of nanoparticles prepared from w/o microemulsions in heterogeneous catalysis, Curr. Opinion Colloid Interf. Sci. 13 (2008) 270-286. https://doi.org/10.1016/j.cocis.2007.10.001

[38] M.S. Dominguez, M. Boutonnet, C. Solans, A novel approach to metal and metal oxide nanoparticle synthesis: the oil-in-water microemulsion reaction method, J. Nanoparticle Res. 11 (2009) 1823–1829. https://doi.org/10.1007/s11051-009-9660-8

[39] M. Faraji, Y. Yamini, M. Rezaee, Magnetic nanoparticles: synthesis, stabilization, functionalization, characterization, and applications, J. Iranian Chem. Soc. 7 (2010) 1-37. https://doi.org/10.1007/BF03245856

[40] N.R. Jana, Y. Chen, X. Peng, Size-and shape-controlled magnetic (Cr, Mn, Fe, Co, Ni) oxide nanocrystals via a simple and general approach. Chem. Mater. 16 (2004) 3931-3935. https://doi.org/10.1021/cm049221k

[41] S.A. Samia, K.K. Hyzer, J.A.J. Schlueter, C.J.C. Qin, J.S.J. Jiang, S.D.S. Bader, X.M.X. Lin, Ligand effect on the growth and the digestion of Co

nanocrystals. J. Am. Chem. Soc.127 (2005) 4126–4127.
https://doi.org/10.1021/ja044419r

[42] X. Wang, J. Zhuang, Q. Peng, Y. Li, A general stratergies for nanocrystal synthesis, Nature 437 (2005) 121-124. https://doi.org/10.1038/nature03968

[43] Y. Hu, Y. Li, R. Liu, W. Tan, G. Li, Magnetic molecularly imprinted polymer beads prepared by microwave heating for selective enrichment of ß-agonists in pork and pigliver samples, Talanta 84 (2011) 462–470.
https://doi.org/10.1016/j.talanta.2011.01.045

[44] L. Zhang, H.Y. Niu, S.X. Zhang, Y.Q. Cai, Preparation of a chitosan-coated C18-functionalized magnetite nanoparticle sorbent for extraction of phthalate ester compounds from environmental water samples, Anal. Bioanal. Chem. 397 (2010) 791-798. https://doi.org/10.1007/s00216-010-3592-0

[45] A.F. Ngomsik, A. Bee, M. Draye, G. Cote, V. Cabuil, Magnetic nano- and microparticles for metal removal and environmental applications: A review. Comptes Rendus Chimie 8 (2005) 963–970.
https://doi.org/10.1016/j.crci.2005.01.001

[46] L. Nuñez, B.A. Buchholz, M. Kaminski, S.B. Aase, N.R. Brown, G.F. Vandegrift, Actinide Separation of High-level waste using solvent extractants on magnetic microparticles, Sep. Sci. Technol. 31 (1996) 1393–1407.
https://doi.org/10.1080/01496399608001403

[47] M. Faraji, Y. Yamini, M. Rezaee, Magnetic nanoparticles: Synthesis, stabilization, functionalization, characterization, and applications, J. Iranian Chem. Soc. 7 (2010) 1-37. https://doi.org/10.1007/BF03245856

[48] R. Lakshmanan, C. Okoli, M. Boutonnet, S. Järås, G.K. Rajarao, Effect of magnetic iron oxide nanoparticles in surface water treatment: Trace minerals and microbes, Bioresource Technol. 129 (2013) 612-615.
https://doi.org/10.1016/j.biortech.2012.12.138

[49] C. Okoli, M. Boutonnet, S. Järås, G.K. Rajarao, Protein-functionalized magnetic iron oxide nanoparticles: time efficient potential-water treatment. J. Nanoparticle Res. 14 (2012) 1194-1199. https://doi.org/10.1007/s11051-012-1194-9

[50] S. Singh, K . C . Barick, D. Bahadur, Surface engineered magnetic nanoparticles for removal of toxic metal ions and bacterial pathogens, J. Hazardous Mater. 192 (2011) 1539-1547.
https://doi.org/10.1016/j.jhazmat.2011.06.074

[51] B. Schrick, J.L. Blough, D.A. Jones, T.E. Mallouk, Hydrodechlorination of trichloroethylene to hydrocarbons using bimetallic nickel-iron nanoparticles, Chem. Mater. 14 (2002) 5140-5147. https://doi.org/10.1021/cm020737i

[52] A.B. Cundy, L. Hopkinson, R.L.D. Whit, Use of iron-based technologies in contaminated land and groundwater remediation, A review, Sci. Total Environ. 400 (2008) 42–51. https://doi.org/10.1016/j.scitotenv.2008.07.002

[53] R.D. Ambashta, M. Sillanpaa, Water purification using magnetic assistance, a review, J. Hazardous Mater. 180 (2010) 38–49. https://doi.org/10.1016/j.jhazmat.2010.04.105

[54] M. Baalousha, Aggregation and disaggregation of iron oxide nanoparticles, influence of particle concentration, pH and natural organic matter, Sci. Total Environ. 407 (2009) 2093–2101. https://doi.org/10.1016/j.scitotenv.2008.11.022

[55] M. Aghazadeha, I. Karimzadehb, M.G. Maragheha, M. Rez, Ganjali, Enhancing the supercapacitive and superparamagnetic performances of iron oxide nanoparticles through yttrium cations electrochemical doping, Mater. Res. 21 (2018) 1-10. https://doi.org/10.1590/1980-5373-mr-2018-0094

[56] X. Wu, L. Meng, Q. Wang, W. Zhang, Y. Wang, A novel inorganic-conductive polymer core-sheath nanowire arrays as bendable electrode for advanced electrochemical energy storage, Chem. Eng. J. 358 (2019) 1464–1470. https://doi.org/10.1016/j.cej.2018.10.162

[57] H. Pang, C. Wei, X. Li, G. Li, Y. Ma, S. Li, J. Chen, J. Zhang, Microwave-assisted synthesis of NiS_2 nanostructures for supercapacitors and cocatalytic enhancing photocatalytic H_2 production, Sci. Rep. 4 (2014) 3577. https://doi.org/10.1038/srep03577

Magnetochemistry - Materials and Applications
Materials Research Foundations **66** (2020) 276-322

Materials Research Forum LLC
https://doi.org/10.21741/9781644900611-9

Chapter 9

Iron Oxide based Magnetic Nanomaterials for Biomedical Applications

Rushikesh Fopase, Lalit M. Pandey*

Bio-interface and Environmental Engineering Lab, Department of Biosciences and Bioengineering, Indian Institute of Technology Guwahati, Guwahati, India- 781039

*lalitpandey@iitg.ac.in

Abstract

Over the vast area of applications, magnetic nanomaterials have great potential in the field of biomedicine due to their unique magnetic characteristics, which make them a suitable candidate for diagnostics and therapeutic applications. Superparamagnetic iron oxide nanoparticles (SPIONs) offers controllable and tunable magnetic properties via surface functionalization and the addition of impurities to their crystal structures. Size-dependent behaviors of SPIONs characteristics ease of surface modifications to make them a suitable candidate for the diverse applications. In this chapter, we have discussed the applications of SPIONs in the biomedical field. With the briefing of about the properties of iron oxide nanomaterials, synthesis methods and surface modifications for SPIONs are also discussed. The chapter focuses on the applications of SPIONs in hyperthermia, medical resonance imaging (MRI), drug delivery, and tissue engineering in order to get attention to the recent advancements in the field of biomedicine.

Keywords

Superparamagnetic Iron Oxide Nanoparticles, Hyperthermia, Magnetic Resonance Imaging, Drug Delivery, Tissue Engineering

Contents

1. Introduction

Magnetic nanomaterials are widely popular for applications in biomedicine, drug delivery, bioimaging, and the environment [1-4]. Understanding of the magnetic, electric, and optical properties of magnetic materials has significantly widened their purpose. Advancements in research have resulted in the application of magnetic nanoparticles (MNPs) in ferrofluids [5], permanent magnet manufacturing [6], room temperature magnetic refrigerants [7], and many more. Biomedical applications like bioimaging, drug delivery, and hyperthermia use different MNPs of pure metals Fe, Ni, Co, and Mn; oxides of Fe; and ferrites of Ni, Co, and Mn [8-11].

The scope of application depends on the type of magnetism delivered by the magnetic materials. Magnetic moments within the domain of the molecules define the type magnetism of material. Domains are the small locales having the same magnetic directions. There are five types of magnetic materials, depending on the orientation of magnetic moments [12]. Diamagnetic materials have an equal spin and angular momentum of electrons leading to zero magnetization. In ferromagnetic materials, all magnetic moments are aligned in the same direction giving higher magnetic properties and hence are used in the preparation of permanent magnets. On the other hand, antiferromagnetic materials magnetic moments cancel out each other, giving rise to zero magnetization value. Ferrimagnetic materials give weak magnetic properties because of unequal and opposite magnetic moments. Paramagnetic materials show the random orientation of the magnetic moments in zero magnetic field but rearrange its moments parallel to the applied field when applied externally. Apart from the major types of magnetic materials, there is one more type referred to as Superparamagnetic materials. These materials show spontaneous magnetization when exposed to an external magnetic field and complete demagnetization when the applied field is removed.

Further, the intrinsic and extrinsic properties define the magnetic characteristics of the MNPs. The intrinsic properties include saturation magnetization value (M_S), Curie temperature (T_C), Néel temperature, magnetic moment per unit volume, and crystalline anisotropy. The extrinsic properties involve coercivity (H_C), remanence (M_R), and magnetic susceptibility [13]. M_S decides the release of energy and thus defines the effectivity of the material. The H_C of magnetic materials determines the thermal stability and depends on the anisotropy of the crystal. For superparamagnetic characteristics, the H_C must be zero. A low degree of anisotropy and higher M_S are required to obtain superparamagnetism [14]. Curie temperature determines the range of energy released in the form of heat by the magnetic materials. Above the Curie temperature, MNPs lose their magnetic properties as thermal energy is high enough to overcome magnetic moments. Materials with T_C in the therapeutic range (42 to 45°C) are better for biological applications including hyperthermia [15-17]. In this chapter, a brief about iron oxide NPs, synthesis routes, and surface modifications are discussed. Role of these NPs in various biomedical applications such as hyperthermia, tissue engineering, medical imaging and drug delivery are elaborated with appropriate examples.

1.1   Role of iron oxide nanoparticles

Among the several types of nanomaterials investigated till date, iron oxide nanoparticles have shown sovereignty in the field of medicine with their fundamental and remarkable magnetic properties, biocompatibility, and stability [18, 19]. Based on the state of oxidation, iron oxides can be classified into eight classes, among which Magnetite (Fe_3O_4), Maghemite ($\gamma\text{-}Fe_2O_3$), and Hematite ($\alpha\text{-}Fe_2O_3$) are popularly used as they provide temperature-dependent phase transitions [20, 21]. The fundamental difference between the properties of these iron oxides is due to their magnetic behavior arising from valance states and vacancies in the sub-lattice structure [22].

In iron, unpaired electrons in the '$3d$' shell give strong magnetic characteristics. Fe^{2+} has four unpaired electrons, while Fe^{3+} contains five unpaired electrons in the '$3d$' shell. The distribution of unpaired electrons gives rise to crystal formation of para, dia, or ferromagnetic state [23, 24]. Fe_3O_4 contains all Fe^{2+} ions at half of the octahedral sites, and Fe^{3+} occupies the remaining half at octahedral and tetrahedral sites. The stoichiometric ratio of Fe^{2+}: Fe^{3+} is 1:2, and divalent ions are replaceable with other divalent cations. Maghemite is a cubic structure with 32 O^{2-} in the closed packed array, 21 *1/3* of Fe^{3+} in distributed over tetrahedral sites and with 2 *1/3* vacancies [25]. X-Ray

Diffraction pattern shows similar crystal structures for Magnetite and Maghemite with a difference of a slight shift in magnetite towards a higher diffraction angle. Fig. 1 represents the crystalline and atomic structure of oxides of iron.

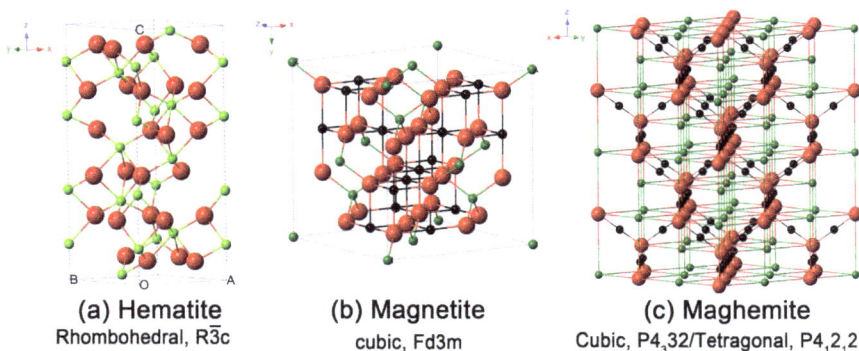

(a) Hematite	(b) Magnetite	(c) Maghemite
Rhombohedral, R$\bar{3}$c	cubic, Fd3m	Cubic, P4$_3$32/Tetragonal, P4$_1$2$_1$2

Figure 1. Crystal structure of iron oxides (Black: Fe^{2+}, Green: Fe^{3+} and Red: O^{2-})
(adapted from Sci. Technol. Adv. Mater. 16 (2015) 023501 [26])

Superparamagnetism behavior is size-dependent and observable for sizes up to 20 nm, where only one magnetic domain is present in the particle [27, 28]. With the application of a magnetic field, the domains arrange themselves with the direction of the field and get disoriented with the removal of the magnetic field. The number of domains reduces with the size, and single-domain particles show the highest possible magnetic moment for a particular size [29]. By the conventional definition, superparamagnetic particles do not possess any remanence or coercivity; therefore show no net magnetic field after removal of the external applied field [30]. Superparamagnetic nanoparticles show no or less agglomeration due to the presence of zero magnetic field between particles, and these materials are considered to be best suitable for biomedical applications. Superparamagnetic iron oxide nanoparticles (SPIONs) of size >20 nm with narrow size distribution have reported the highest efficiency in the biomedical field [31, 32]. Due to its superparamagnetism property, higher magnetic susceptibility of Magnetite makes it a preferred candidate in medical applications over maghemite [33]. Information of atomic structures of iron nanoparticles is of utmost importance for considering their applications in the Biomedical field [34-36].

1.2 Synthesis methods

The composition and morphology of nanoparticles determine the magnetic characteristics of the iron oxide nanoparticles. Thus, the selection of the synthesis method needs to be performed carefully to obtain the desired shape, size, and crystallinity of these particles. Approaches for the synthesis of SPIONs include physical, chemical, and biological methods. Physical methods include ball milling, gas deposition, electron beam lithography, and sputtering. However, the physical approach does not allow control over the particle size of the nanoparticle. Mechanisms of the biological method include a breakdown of precursor salts into their respective nanoparticles. Enzymes or phytochemicals of biological origin such as microbes and plants are responsible for the breakdown by redox reactions. This approach is eco-friendly and gives higher biocompatibility. However, low yield and broad size distribution make the approach less efficient. The chemical approach for SPIONs synthesis comprises of the co-precipitation, thermal decomposition, microemulsion, sol-gel, and hydrothermal method. Some of the highly used methods are explained in brief as follows:

1.2.1 Co-precipitation

Co-precipitation is a widely used method for the synthesis of SPIONs and involves simultaneous precipitation of iron salt solutions in the alkaline conditions [37]. This method offers better control over the size of the nanoparticle. Characteristics of SPIONs depend on the type of precursors, ratio of precursors, pH, temperature, and the type of the alkaline agent. The addition of certain modifying agents like Poly Vinyl Alcohol (PVA), starch, and talc to the reaction offers a reduction in aggregation and ease of separation [38-40].

1.2.2 Microemulsion

This technique synthesizes SPIONs by confining the precursors in micelles and initiating nucleation for the growth of particles [41, 42]. There are two types of emulsion formations- oil in water (regular micelles) [43] and water in oil (reverse micelles) [44]. The mixture of two immiscible liquids forms thermodynamically stable and isotropically dispersed micellar structures [45]. Amphiphilic surfactants like Sodium Dodecyl Sulfate (SDS), Dioctyl Sodium Dodecyl Sulfate (DSS), and Cetyl Trimethyl Ammonium Bromide (CTAB) are used in the microemulsion techniques. The dispersed phase offers a confined environment for the formation of nanoparticles. The ratio of oil-water-surfactant controls the size of micelles, and the homogeneity of micellar size reduces the polydispersity index of particles [46, 47]. The disadvantage of the microemulsion technique is its requirement of a low and small temperature range resulting in reduced crystallinity and yield. Retention of the micellar structure at high temperature and annealing of iron oxide can show an increase in the crystallinity. The effect of surfactant used on quality SPIONs and difficulties in scale-up is another disadvantage [48].

1.2.3 Sol-gel

This method for SPIONs synthesis includes hydrolysis and condensation reactions of the precursors [49-51]. Precursors are hydrolyzed with either water or ethanol and mixed thoroughly. The silica matrix is also used for the uniform distribution of NPs in the sol-gel [52]. The solvent is evaporated by heat resulting in the formation of gel-like a 3D network of precursors. The obtained 3D network gives segregated precursor particles. Further drying is required for the removal of solvents with additional crushing to separate the particles [53]. Otherwise, the addition of chelating agents like citric acid, malonic acid, or surfactants produces nanoparticles without network formation. Chelating agents decrease the system free energy and thus reduce the aggregation of particles [54, 55]. Citric acid, used in the synthesis of SPIONs, chelates Fe ions and gives separated nano-sized particles. Parameters like temperature, pH, precursor concentrations, and solvent used plays a significant role in the synthesis of nanoparticles and their structural properties [56-58]. The sol-gel method gives relatively large-sized NPs and monodispersed particles with good yield under ambient conditions. However, the method requires an additional heat treatment for better crystallinity. Also, by-products of the reaction are a significant problem with sol-gel reactions which subsequently requires further post-purification treatment.

1.3 Surface modifications

The stability of SPIONs is a major issue in biomedical applications as they tend to agglomerate and precipitate over time. Metallic nanoparticles of iron, cobalt, and Ni and their alloys undergo oxidation in contact with air. Additionally, SPIONs administered in the body for certain biomedical applications tend to agglomerate via hydrophobic interactions due to high surface energy and magnetic characteristics. The agglomerations lead to adsorption of blood plasma proteins and are rapidly released out of the body through phagocytes and macrophages [59]. Therefore, certain modifications are required to avoid aggregation of SPIONs and to improve their function as well.

Surface modifications can improve the chemical stability of MNPs against the oxidation and aggregations by introducing specific characteristics such as hydrophilicity over the surface to avoid agglomeration. The surface modification of MNPs serves purposes of i) stabilizing NPs in *in vivo* conditions at physiological conditions, ii) providing surface

Magnetochemistry - Materials and Applications   Materials Research Forum LLC
Materials Research Foundations **66** (2020) 276-322   https://doi.org/10.21741/9781644900611-9

functionalization for further applications, and iii) delaying the exclusion of administered SPIONs by the reticuloendothelial system [60].

Modifications of SPIONs create a coating over particle surface, resulting in the formation of a core-shell structure [61]. Coating isolates the magnetic core particle from the external environment and thus preserves its original characteristics. The coating material may consist of inorganic materials such as metals ions, oxides, silica, carbon, and organic compounds like polymers and surfactants [62-65]. Polymeric coating materials are added to the initial reaction mixture during the synthesis processes, such as co-precipitation and sol-gel [66]. In ferrofluids comprised of colloidal SPIONs, surface properties determine the colloidal stability. Attachment of polymers or surfactants on the SPIONs surface helps to improve colloidal stability by preventing agglomeration of particles. Surfactants and polymers anchored on the surface of SPIONs act against the van der Waal forces by steric repulsions. Nevertheless, surface modifications offer the functionalization of SPIONs by functional groups includes carboxyl, amine, phosphate and sulfate groups.

Surface modification of MNPs not only offers stability and biocompatibility but also widens the scope of application. Particular coated MNPs can be used for multiple applications, which make them efficient and economical. Surface functionalization improves the quality of treatment in targeted drug delivery as well as in hyperthermia applications [67]. However, the Stability of coating or modifications of SPIONs within the body fluid is also essential. Polymer coated SPIONs with small layers tend to lose stability over time due to the surrounding environmental conditions. The coatings are possibly leached out very quickly in an acidic environment and thus expose core SPIONs particles to surroundings to form aggregates and result in subsequent loss of characteristics [68]. Leached out conjugated materials mix into the body fluid and may affect body functions causing several health issues and disorders.

2.   Applications of MNPs

2.1   Hyperthermia

Hyperthermia is a phenomenon in which body temperature is elevated above its normal range (36.5-37.5°C). In clinical aspects, the elevation of temperature in order to kill cancer cells by induced apoptosis or by making them more susceptible to radiations or anticancer drugs is referred as hyperthermia therapy [69, 70]. There are three possible strategies for the application of hyperthermia therapy: 1. Whole-body hyperthermia, 2) Regional hyperthermia, and 3) Local hyperthermia [71]. Whole-body hyperthermia is used in the cases of metastatic cancers, while the other two are applied for localized malignant tumors [72]. Local hyperthermia is often implemented for the treatment of surface tumors [73, 74] and regional hyperthermia is applied for tumors that lie deep in tissues or over large areas [75-77].

Nanoparticles within size <20 nm show a single magnetic domain structure and are the most promising candidates for application in hyperthermia due to their superparamagnetic properties [78]. SPIONs have excellent superparamagnetic characteristics and can generate heat when exposed to Alternating Magnetic Field (AMF). Neel relaxation and Brownian relaxations are primarily responsible for the generation of heat by SPIONs [79, 80]. Typically, multi-domain particles release heat due to hysteresis loss, and single-domain particles release heat due to the relaxation of the particle [81]. However, single-domain nanoparticles show higher magnetization values than multi-domain particles [82]. Fig. 2a shows the schematic representation of Neel and Brownian relaxation.

Neel relaxation time τ_N is the internal rotation of magnetic moments given by Eq. 1 and Brownian relaxation time τ_B is the physical rotation of the MNPs given by Eq. 2 as follows [83]-

$$\tau_N = \tau_0 e^{\left(\frac{KV_M}{k_B T}\right)} \tag{1}$$

$$\tau_B = \frac{3\eta V_H}{k_B T} \tag{2}$$

where K is magnetic anisotropy constant, V is volumes of the magnetic particle cores, K_B is the Boltzmann constant, T is absolute temperature, η is the dynamic viscosity, V_H is the hydrodynamic volume and τ_0 is time constant ($\approx 10^{-9}$)

The final effective relaxation of MNP is expressed by Eq. 3-

$$\frac{1}{\tau} = \frac{1}{\tau_N} + \frac{1}{\tau_B} \tag{3}$$

When exposed to an external magnetic field, magnetic domains of the SPIONs align with the direction of the applied magnetic field, and saturation magnetization is achieved with the alignment of all the magnetic domains. Removal of the applied field misaligns the magnetic domains but do not lose their magnetization completely. The remained magnetism is referred to as remanent magnetization or remanence (M_R) and is responsible for the aggregation of particles [84]. A magnetic field applied to remove remanence is called as a coercive field or coercivity (H_C). Fig. 2b represents the schematic representation of the M-H curve (Hysteresis loop) [85]. The application of an AMF causes the cyclic alignment of magnetic domains with the direction of the applied magnetic field. However, the aligning of moments lags behind the frequency of the applied external field because of M_R, which creates a hysteresis. Lag in the synchronous alignment causes hysteresis loss in the form of heat and thus leads to a rise in the temperature [29, 86]. The integration of the area under the M-H curve gives the energy loss in the form of heat. The heating ability of MNPs is referred to as the Specific Absorption Rate (SAR) and is explained further by Rosenweig.

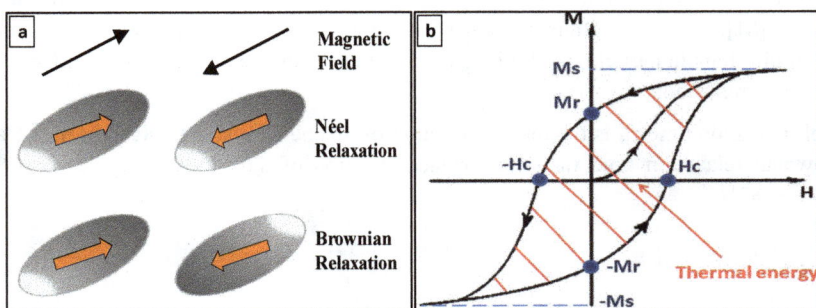

Figure 2. a) Neel and Brownian relaxation, b) Hysteresis loop of ferromagnetic material (M= magnetization, H= magnetic field intensity, M_S= Saturation magnetization, M_r= Remanence, H_C= Coercivity)
(Fig. 2 b is adapted from Nanoscale, 2014, 6, 11553–11573[85])

Rosenweig proposed the mechanism of heat generation by MNPs in 2002 explained in Linear Response Theory (LRT) [87, 88]. According to LRT, the application of an AMF over MNPs results in heat generation. Rotational relaxations of single domain NPs are responsible for the heat generated and linearly varies with AMF. The power dissipation (P) expression of LRT is given by Eq. 4-

$$P = \pi\mu_0\chi_0 H_0^2 f \frac{2\pi f\tau}{1+(2\pi f\tau)^2} \tag{4}$$

where μ_o is the free space permeability, χ_0 is equilibrium susceptibility, H_0 is amplitude, and f is the frequency of applied AMF. The expression shows the relation of heat generation and frequency, relaxation time, and parameters of AMF. The heating efficiency is the ratio of power dissipated per unit mass of MNPs and referred to as SAR or Specific Loss power (SLP) (Eq. 5) [89, 90].

$$SLP = \frac{P}{m_{MNP}} \tag{5}$$

where P is the power dissipated and m_{MNP} is a unit mass of magnetic nanoparticles. A small amount of heat is also generated due to the Eddy currents. When AMF is applied to a material, the induction of voltage results in the generation of circular currents. These currents affect back on the source magnetic field with the formation of the oppositely directed magnetic field. Subsequently, magnetic energy is lost in the form of heating referred as Eddy current loss [91].

The heat generated by the application of AMF is applied to elevate the temperature of the body or region and is referred as magnetic hyperthermia. Temperature raised within the range of 41- 44°C is very promising in cancer therapy [71]. Cancer cells are more susceptible to elevated temperature than healthy cells [92]. Cancer cells are unable to dissipate the additional heat due to the lack of proper vascular network. While healthy cells having an excellent blood flow network can dissipate the excess heat via convection and conduction within the surroundings, a rise in temperature causes irreversible damage to cancer cells, which is reversible for healthy cells [93]. The damage includes protein denaturation, apoptosis, and necrosis. Additionally, increased temperature triggers the immune system and increases the number of Heat Shock Protein 70 (HSP70) on the surface. Cancer cells having relatively higher porous membrane release HSP70, which acts as a signal for natural killer cells. This results in the activation of MHC I by dendritic cells, lymphocytes migration to the site of HSP70 protein presenting sites to eliminate the cells [94-97].

The amount of heat generated depends on the movement of magnetic domains and the particles themselves. The frequency of the external applied magnetic field determines the rate of particle movement occurring within the system. [98]. Excess increase in temperature may result in overheating and thus cause some damages to healthy cells. Therefore, maintaining the temperature in the secure range is of utmost importance. T_C can help to control the change in the temperature in a biologically secure range. T_C of material controls the release of energy from oscillating magnetic domains. At 'T_C' materials lose their internal magnetization and acquire induced magnetization. [99]. Material with Tc between 41°C to 47°C can self-regulate the heating rate and range and are suitable for hyperthermia treatments. Combining MNPs with nonmagnetic materials can reduce the Tc of materials. A study done on γ-Fe-Ni NPs showed the tunability of Tc to control the heating when applied in the external field. The synthesis was performed using chemical reduction of Fe and Ni with varying stoichiometric ratio and further reduced by the addition of Mn precursor [100]. Tc of the material was found to decrease from 82°C to 78°C with the increased percentage of Mn. The transition of magnetic properties was observed from ferromagnetic to paramagnetic, and no heating was observed after temperature exceeded Tc.

2.1.1 Delivery of SPIONs to the target site

The application of hyperthermia relies on the delivery of effective dosage to the tumors selectively with a desirable response without harming the nearby healthy cells and tissues. Among the strategies applied for performing hyperthermia, intracellular hyperthermia is proven popular and widely applied. The technique is ideal for killing the cancer cells even in the metastatic state. Delivery of SPIONs to the target site is critical for the success of applications. Magnetic Fluid Hyperthermia (MFH) is the mode of *in vivo* application of magnetic hyperthermia [101-103]. In MFH, a biocompatible colloidal solution containing SPIONs is injected within the target tissue and exposed to AMF. The heat generated due to Neel relaxations and Brownian relaxations damage local cancer cells. The amount of heat generated depends on the type of tissue, and the surrounding environment as the viscosity of the medium alters the movement of NPs [104].

Biocompatibility and stability of materials are vital in the application of materials in *in vivo* conditions. SPIONs are biocompatible up to a specific concentration range. The concentration range of 45-60 $\mu g/mL$ of blood is used for the animal model in drug delivery, while for humans, the SPIONs based contrast agent concentration range used is 33 $\mu g/mL$ of blood [105]. Accountable toxicity is observed around 100 $\mu g/mL$ of blood [106]. The surface chemistry of the material plays an essential role in the interaction with biomolecules within the body. SPIONs are unstable in the colloidal solution and also offer difficulties in synthesis due to hydrophobicity and large surface-area-to-volume ratio. Self-agglomeration behavior and sensitivity towards air and moisture make the application of SPIONs difficult [107]. However, tuning the characteristics by coating the surface of SPIONs can solve the problem by improving interacting forces between particles and thus improving biocompatibility. Surface modification of SPIONs is described briefly in section 1.3.

2.1.2 Nanomaterials used for hyperthermia

Local and regional hyperthermia procedure involves the administration of MNPs within the body in affected locations and exposing them to AMF. Many magnetic nanomaterials of pure metals such as Fe, Ni, and Co, and their oxides have been examined for their potential applications in hyperthermia. Table 1 lists the various iron-based magnetic nanomaterials used in hyperthermia applications.

A study on Fe_3O_4 NPs synthesized by the co-precipitation method at various temperature ranges showed the ability of Fe_3O_4 for hyperthermia [108]. SPIONs were prepared at different reaction temperatures 40°C, 60°C and 80°C and 5 mg/ml of each NPs samples were exposed to AMF of 400kHz for 30 minutes for hyperthermia applications. M_S values of the samples 40°C, 60°C and 80°C were observed ranging from 28.5 emu/g, 42 emu/g, and 54 emu/g, respectively with increasing particle diameter. Sample of 60°C showed the most prominent SAR value of 181 W/g with a change in temperature of 36°C. The study confirmed that with a decrease in size, the transition of SPIONs in the single-domain particle from a multidomain particle resulted in the achievement of higher superparamagnetic fraction values.

High oxidation tendency of pure metal MNPs induce cytotoxicity and instability and thus limit their biomedical applications [109-111]. However, alloyed nanomaterials have proven better characteristics and stabilities over metallic nanomaterials [112, 113]. Numerous studies confirmed the superiority of iron-based nanomaterials based on their excellent and tunable characteristics such as high SAR values and magnetic properties with better chemical stability [114]. Ferrites comprised of Co, Ni, Mn, Zn are popularly used for hyperthermia applications and showed improvement in the magnetic properties compared to iron oxides nanoparticles [115-119]. Studies have shown that ferrites possess higher M_S values compared to similar size magnetite NPs with negligible hysteresis loop making resultant MNPs prominent in hyperthermia and bioimaging [120].

In a study on ferrites, dextran $MgFeO_4$ nanoparticles (size 20 nm) showed their potentiality for hyperthermia application [121]. NPs showed a rise in temperature by 50°C and 73°C for the concentration of 5 mg/ml and 10 mg/ml, respectively, after exposure to AMF. The reported the SAR value was 85.57 W/g at 26.7 kA/m (265kHz). Dextran coating over the particles resulted in the improved cell viability of 86% 1.8 mg/ml compared to uncoated ferrites. Therefore, surface modifications of magnetic particles can improve potentiality for hyperthermia applications.

For a study by Kerroum *et al.*, $ZnFe_2O_4$ NPs were prepared by co-precipitation with the size ranging from 19 to 33 nm with a change in the pH of the reaction [122]. Small-sized ferrite NPs showed higher H_C values compared to large size well dispersed NPs. The observed superparamagnetic behavior illustrated maximum SAR values up to 125 W/g under an AMF of 5 to 65 kA/m at 355 kHz and capable of reaching the hyperthermia temperature range (42-45°C) in a short time. The study confirmed the better heat efficiency of well-dispersed $ZnFe_2O_4$ NPs over aggregated NPs emphasizing the advantage of reduced aggregation.

Another study on Fe_3O_4 NPs and Ni-Zn ferrite NPs confirmed the capability of these particles to provide reliable heat energy for rising temperatures for hyperthermia application [81]. NPs were investigated for SAR values by static and dynamic hysteresis loop areas and hyperthermia in water to establish the protocol and model in non-adiabatic conditions under the field of 60 mT at 69 kHz. Dynamic hysteresis loops showed accurate determination while static loops showed lesser values of SAR in actual conditions. The results showed the lower SAR values for small size (10-20 nm) Fe_3O_4 NPs due to dipolar interaction on Neel relaxation lowering the energy absorption of NPs.

A study for magnetite nanoclusters of Fe_3O_4 using the solvothermal method showed heat generation by exposing to AMF for particles synthesized by using 1% of PAA as a stabilizer [28]. Individual nanocrystals of Fe_3O_4 showed Neel relaxation of a 10 nm length scale, and ~250 nm length scale of Brownian relaxation was observed for nanoclusters itself. The highest SAR value obtained was 80.5 W/g under the field of 31.6 kA/m and resulted from the effective Brownian relaxation. Thus the study confirmed the effect of relaxation values of the SPIONs on the magnetic efficiency to generate heat. Mg-doped magnetite NPs were examined for desired heating efficiency synthesized by the co-precipitation method [123]. Doped Mg concentration up to 40% showed improvement in the magnetization values. The SAR value obtained under non-adiabatic conditions for the only Fe_3O_4 was 111 W/g, and for Mg dopant concentrations of 40%, the SAR value obtained was 83.1 W/g. The drop in the SAR value is due to increased anisotropy resulting in a reduction of Neel relaxations. The study showed the tuning of the magnetic properties of SPIONs via the addition of impurities in the SPIONs crystals to obtain the desired hyperthermia temperature.

Along with ferrites, rare earth materials are also used as a dopant in MNPs as they offer further improvement in the magnetic characteristics of materials. Rare earth elements have a common oxidation state of +3 and +4 and f- shell associated unique magnetic and optical properties [124]. Doping of the rare earth materials has reported increasing the electrical, structural, and magnetic properties in ferrites [125-128]. Ferrites doped with rare earth elements show variation in the interaction in sublattice, leading to a change in the structural and magnetic properties. Many studies have investigated the effect of doping of lanthanides such as Sm, Eu. Gd, Yt, Ce, La on properties of ferrites. Among the rare earth elements, Gd shows the highest magnetic moment due to its seven unpaired electrons, which makes it a preferable dopant for the improvement of magnetization of the materials and is popularly used in hyperthermia and bioimaging as a contrast agent [129-131].

Researchers have examined the effects of doping of Tb and Gd on SPIONs to manipulate its magnetic properties for its applications in drug delivery and hyperthermia [132]. SPIONs synthesized by the co-precipitation method were doped with 2.5 to 15% of Gd and 2.6 to 12.8% of Tb. The doping of Gd resulted in a decrease of M_S value with increasing dopant concentration from 60 to 27 emu/g, while for Tb, M_S value reduced from 59 to 22 emu/g. M_S value for pure Fe_3O_4 was observed as 78 emu/g. Lowered coercivity and controlled M_S value make the ferrites of rare earth metals of interest in drug delivery and hyperthermia.

Hyperthermia is regarded to be a promising treatment for cancer. MNPs with superparamagnetic properties are administered within the body and exposed to AMF. Neel relaxation and Brownian relaxation of the materials results in the generation of heat. The generated heat rises temperature within the range of 42-46°C, causing irreversible damages selectively to cancer cells. The superparamagnetic properties of the magnetic materials used are an important criterion for the generation of heat under AMF. Tuning of T_C of materials via combining magnetic materials with nonmagnetic materials helps to achieve a secure temperature range to avoid damage to healthy cells. SPIONs are the best suitable material for hyperthermia applications due to their extensive superparamagnetic properties and biocompatibility. However, certain surface modifications can further improve the efficiency of SPIONs performance. Other than SPIONs, different metal NPs are also used for the application. Ferrites offer better SAR values and biocompatibility and can be used for multiple purposes.

Table 1: Iron based magnetic materials used in the hyperthermia applications

Material	Method	Coating	Variant	Size (nm)	Magnetic Field (kA/m)	M_S (emu/g)	f (kHz)	Current field (kA/m)	Cnoc. (mg/ml)	SAR (W/g)	Temp raised (°C)	Time taken (s)	Ref
$Mn_{1-x}Co_xFe_2O_4$	Thermal decomposition	SiO₂ coating & Au@Fe₃O₄ NPs	Mn=1	70		43.2				269 ± 7	~30		
			Mn=0.75	160		31.1				219 ± 6	~31		
			Mn=0.5	90	0.4	30.9	454	4.38735	-	227 ± 6	~29	900	[133]
			Mn=0.25	170		30				302 ± 8	~34		
			Mn=0	210		81.4				291 ± 7	~34		
$Zn_xFe_{3-x}O_4$	Precipitation and hydrothermal treatment	NA	Zn =1	4.1		3.09				1.2			
			Zn = 0.8	8.0		43.36				2..0			
			Zn =0.6	8.3		49.2				11.1			
			Zn = 0.4	12.8	8	90.23	577	15.9	3	261.2	NA	120	[134]
			Zn = 0.2	15.2		88.23				135.1			
			Zn = 0.1	17.8		91.38				88.4			
			Zn = 0	19.8		77.29				83.8			
FePt	One-pot	rGO	NA	NA	128	11.93	700	-	-	-	43	300	[135]
$MnFe_2O_4$	Microwave refluxing	γ-Fe₂O	-	29±9	20	67	113	19.9425	0.75	~ 150	46	1800	[136]
							260	18.3470		~ 200			

Material	Method	Coating	Condition	Size										Ref
Fe_3O_4	Seeded growth method	-	-	13.9	4000	81.26	111	3		5.84	4.6	44	-	[137]
										6.90	2.7	-		
										7.08	7.3	-		
Fe_3O_4	Co-precipitation	uncoated	pH- 11.37	14		46.5				5	133	32		[138]
										10	146	34		
										20	167	37		
	Sono-chemical	PEG 400	pH 11	8	1595	47	290	0.75		5	188	42	900	
										10	217	45		
										20	234	47		
			pH 12	5		49				5	209	41		
										10	221	46		
										20	238	48		
Fe_3O_4	Co-precipitation	Uncoated	NA	3.4	10000	54	13560	4.5	0.5		169	~83	1800	[139]
		PAA		3.6		60					110	~78		
		Polystyrene (PS)		4.6		~59					72	~33		
		Si-PS		5.6		~57					70	~42		
Fe_3O_4	Co-precipitation	Uncoated		21.8	20000	51.68	265	26.7229	2		79.32	64	600	[140]
		Chitosan		15.1		49.96					118.85	67		
FeC	Hydrothermal	NA	NA	22	1595	88	261	18.3470		2.5	85	46	2700	[141]
										5	50	48	2700	
										10	25	50	2700	
										20	20	53	840	
							117	19.9425		2.5	9	40		
										5	35	50	1800	
										10	15	50		
										20	11	49		
Fe_3O_4	Co-precipitation	PEG- PBA-PEG	Fe_3O_4	15	798	37.14	13.56	0.4	0.5		781		600	[142]
			Fe_3O_4-FA	17		37.5					725	-		
			TMZ-Fe_3O_4	30		30.5					580			
			TMZ-Fe_3O_4-FA	35		28.5					530			

2.2 Tissue engineering

Magnetic nanomaterials are popularly used in the field of Tissue Engineering. Clinical applications, such as dental, orthopedic, and craniofacial treatments, use magnetic nanomaterials, especially SPIONs. The use of magnetic scaffolds and the integration of magnetic nanoparticles within the system are the two strategies applied for tissue engineering [143]. The success of tissue engineering relies on the extent to mimic the physiological condition in *in vitro* to achieve similar characteristics of the construct. Tissues like cartilages and the central nervous system and organs like bones, heart, skin are composed of different kinds of cellular phenotypes. Variation in cell type occurs

Materials Research Foundations **66** (2020) 276-322

https://doi.org/10.21741/9781644900611-9

because of the anisotropic distribution of stimulating growth factors (SGFs) within the surrounding environment in the embryonic and preadolescent stage. The gradient results in a spatial variance in cellular signaling, and thus the emergence of different phenotypes occurs. Achieving such a gradient of SGFs during the *in vitro* development of tissue helps in the replication of native complexity and functionality of the construct. The applied magnetic field offers the distribution of growth factor conjugated SPIONs throughout the system and thus create a gradient of growth factors [144]. Fig. 3 shows a schematic representation of magnetically driven SPIONs to create a gradient of growth factors for the development of the osteochondral tissue.

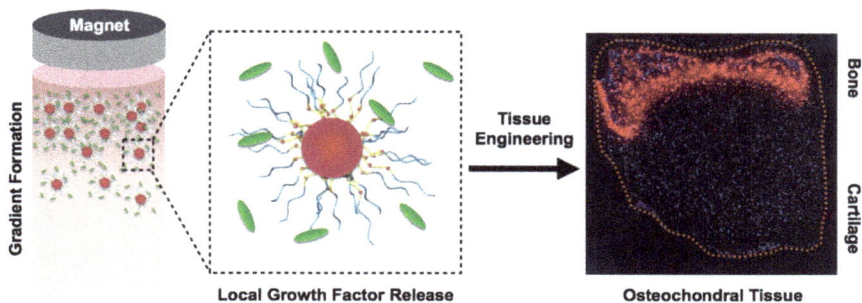

Figure 3. Application of SPIONs to create a gradient of growth factors for tissue engineering (adapted from Biomaterials 176 (2018) 24-33 [144])

Studies have reported the use of SPIONs to distribute growth factors over the constructs developed in tissue engineering. Glycosylated SPIONs, conjugated with Bone Morphogenic Protein 2 (BMP-2) in agarose hydrogel, were applied to create a gradient for osteogenesis. The applied magnetic field distributed the SPIONs conjugated BMP-2 molecules through the hydrogel and provided varying concentrations of growth factors [144]. In another study, the biomimetic method for mineralization in bone tissue engineering has used a combined composite of chitosan, hyaluronic acid, and bovine serum albumin with calcium phosphate associated with MNPs [145]. The magnetic scaffolds synthesized by the composite reportedly reduced the rate of enzymatic degradation of the biopolymers by restricting the access of enzymes to composite. A reduced rate of degradation resulted in the slower release of calcium phosphate and the regulation of mineralization of bone tissue. Studies have reported the applications of SPIONs in tissue engineering and regeneration via influencing the cellular signaling. A particular study on human bone-derived mesenchymal stem cells concluded the exact mechanisms for the role of SPIONs in tissue engineering [146]. The study reported that the SPIONs induced activation of the Mitogen-Activated Protein Kinase (MAPK) signal pathway and thus regulated the downstream molecules for osteogenesis. Application of the magnetic field to SPIONs results in the agglomeration of SPIONs, and thus varied physicochemical properties change the resultant biological properties [147]. Change in size and zeta potential of SPIONs decreases their cellular uptake. Variance in the surface charge also influences the protein corona formation which plays a significant role in therapeutics [148].

SPIONs have been used to influence various biological processes and cell signaling. A study has reported the use of SPIONs linked synthetic peptide 'UM206' to stimulate the Wnt receptor for the regulation of Wnt signaling [149]. After the administration of NPs, magnetic stimulations induced the SPIONs-UM206 and initiated matrix formation in the hMSCs during osteogenesis. Similar results were observed for SPIONs-UM206 injected in chick femurs. The particles enhanced mineralization and in synergy with BMP2 assist the bone formation. Therefore, facilitating the control over the signal transduction, MNPs have proved their potential in tissue development and regulation.

Utilizing a different approach, the *in vivo* synthesis of SPIONs within the target cells through iron storage proteins like ferritin can be applied for inducing cellular signals. Fe^{2+} ions bind to transferrin receptors, and the mineralization of iron occurs via endocytosis. The ferritin gene can be modified to increase the iron-carrying capacity through which it gives naturally occurring SPIONs [150-152]. A study on the heat-inducible expression of the transgene has used SPIONs to develop a control system to regulate the expression in human hepatoblastoma HepG2 cells. The control is carried out by SPIONs with the help of ferritin, an iron storage protein [153]. The heat generated by the exposure of an AMF induced the heat shock proteins to stimulate the transgene expression in both monolayer and multilayer cell sheets without raising the temperature of the whole system. Remote activation of the transgene with a controlled manner widened the scope of application of MNPs in cell-based therapy and tissue engineering.

SPIONs coated with 2, 3-dimercaptosuccinic acid (DMSA) showed cardioprotective activity when used on the Guinea pig Langendroff heart (test model at tissue level) and neonatal cardiomyocyte of rats [154]. Fe_2O_3-DMSA prevents intracellular free radical activities and the influx of calcium ions and thus prevents peroxidation injury to membrane lipids of cells. The cardioprotective activity of Fe_2O_3 was observed to be dependant on the size of the nanoparticle core and not on the coating. The study showed better results compared to commercially available drugs like Verapamil (calcium channel blockers) and *Salvia miltiorrhiza* extract (antioxidants). A study proposed that magnetic materials promote re-endothelization in injured blood vessels and facilitate cell coverage of high density for the implant during the first 24 h [155]. Bacterial cellulose fibrils synthesized by *in situ*-precipitation technique and linked to dextran-coated Fe_3O_4 NPs were applied to target cells *in vitro* to increase cell homecoming. The NPs with ten emu/g of magnetization employed within hydrogel retained an improved number of smooth muscle cells under pulsatile flow. The study confirmed the role of magnetic NPs in the protection against inflammation restenosis after implantation.

However, fewer studies have also reported certain disadvantages of the use of SPIONs in tissue engineering. Exposure of SPIONs to Static Magnetic Field (SMF) showed the adverse effects on the hepatocytes and resulted in reduced cell viability, aberrations of the cell cycle, and apoptosis. SMF caused the formation of SPIONs aggregates, which were subsequently proven to be cytotoxic. However, this disadvantage can be overcome by suitable surface modifications that may inhibit or reduce aggregate formations [156].

Tissue engineering has excellent potential to reproduce functional body tissues and organs. However, with the inability to provide exact *in vivo* conditions during the development of tissue constructs, the desired results seem challenging to obtain. The concentration of SGFs at the site governs the characteristics of the constructed tissue. Magnetic materials can spread the growth factors in the gradient as per the desired concentrations. SPIONs have reported their biocompatibility as well as their ability to function with conjugated molecules. Different SGFs can be linked to SPIONs and spread over the tissue with an applied magnetic field to create a gradient. Additionally, SPIONs used in the construction of scaffolds reduces the rate of disintegration and can affect cellular signaling in a controllable manner. Tuning of the disintegration rate can support the extensive growth of the tissue and thus helps to achieve the success in Tissue Engineering.

2.3   Medical imaging

Magnetic Resonance Imaging (MRI) is a non-invasive imaging technique offering an anatomical and functional assessment of body tissues based on the contrast in the relaxations of proton spins within the tissue under the applied magnetic fields. The controlled magnetic field (1.5 to 3T) applies high-frequency non-ionizing electromagnetic radiations to obtain high-quality images from cross-sectional images of the body [157]. MRI technique is used to differentiate normal and pathological tissues and to analyze the affected area. MRI offers advantages such as non-invasiveness, reliability, and avoiding exposure to irradiation.

A strong magnetic field excites the protons of water molecules, and the energy is released when protons get back to their ground state. The release of energy is associated with longitudinal relaxation (T_1) and the transversal relaxation (T_2) of protons. Relaxation depends upon the surrounding environment of the protons, mainly on the viscosity of the medium. For example, water and several fats show different relaxivities due to different viscosities; thus, different tissue types give different values of T_1 and T_2. The released energy is measured to produce resonance images. The intensity of signals depends on the relaxation time and proton density [158]. Contrast agents alter the intensities of the relaxations and thus enhance the contrast in images. T_1 contrast agents enhance the longitudinal relaxation and yield brighter image, while T_2 contrast agents shorten transverse relaxation to produce a darker image [159]. Contrast can be changed using a scanner console in accordance to get distinct areas based on intensities.

T_1 or spin-lattice relaxation is the time taken by magnetization to return to its 63% of actual value [160, 161]. Gd^{3+} and Mn^{2+} increase the T_1 relaxation and hence are used as T_1 contrast agents [162-166]. T_2 or spin-spin relaxation is the time in which the transverse component of magnetization loss its 63% of energy [160, 161]. SPIONs are used as T_2 relaxation agents as they reduce the T_2 relaxation time by dephasing protons when exposed to an external magnetic field; dephasing of the proton results in the reduction of T_2 time and intensity.

The contrast agents used in the MRI reduce the relaxation time and thus increase the signal intensity. Cancer cells usually take up the contrast agents resulting in enhanced signals to show distinct characteristics of the area. Gd-based contrast agents (T_1 contrast agents) are in use for a long time and offer advantages of a larger magnetic moment with its +7 unpaired electrons [167-169]. However, dissociation of Gd from its chelated complex results into the formation of Gd^{3+} ions and leads to accumulation within the various body tissues [170-172]. Gd^{3+} is reportedly a toxic lanthanide heavy metal to human health, and accumulation of Gd in tissues can cause severe renal impairments such as nephrogenic systemic fibrosis in long term effects [173, 174]. Constituting a similar size, the competitive replacement of Ca^{2+} by Gd^{3+} and other undesirable interactions result in various complications of biological systems [175-180].

The particle size of the NP is the foremost parameter to be used in MRI imaging for a particular tissue. For instance, a particle with a size ≥ 50 nm is easily captured by the reticuloendothelial system (RES) when injected into the body. RES are majorly present in the liver, spleen, and lungs, thus making the particles suitable for imaging of these organs. However, NPs with sizes ≤ 50 nm stay in the blood for a longer time and can leak out to intercellular spaces [181]. Longer half time of NP makes the particles suitable for better analysis and imaging. Leaked out NPs may be captured by macrophages, and the resulting accumulations of such particles in the lymph nodes are suitable for imaging of the lymphatic system. Macrophages present in healthy tissues engulf SPIONs. However, the pathological tissues such as cysts, tumors, and fibrosis tissues do not take away SPIONs through phagocytosis. The contrast is developed due to such different accumulation of the SPIONs and can be used to differentiate and detect lesions in different tissues [182, 183].

Retention and distribution of contrast agents in the blood or tissues play a significant role in MRI and depend upon their surface properties and concentrations [184]. When administered in the body, contrast agents must show tissue-specific biodistribution to avoid further doses of higher concentrations. Tissue specificity increases the local accumulation of contrast agents and enhances the sensitivity for lesion detections. Conjugation of NPs with tissue-specific antibody, peptide, ligand, or targeting via an external magnetic field can help to achieve specificity [185-187]. With the important aspect of biodistribution, clearance of the injected materials is also necessary. After injection, contrast agents get distributed over extracellular spaces except for the prevention of natural barriers and then are slowly cleared by glomerular filtration and the biliary system. Repetitive administrations resulted in the slow accumulation of contrast agents within tissues. The contrast agents administered may not be stable during the time taken for clearance and may result in dissociation products to react with other biological processes [188, 189]. Gd-based contrast agents have reported the dissociation Gd^{3+} from its chelated complex and induce toxicity over time [190].

SPIONs are a better alternative to Gd-based contrast agents with their superparamagnetism, biocompatibility, and high relaxivity. SPIONs reduce T_2 time to increase the contrast in very low concentrations and hence referred to as a T_2 contrast agent. Features like enhanced sensitivity, ease of surface modifications, and potentiality of multimodal imaging agents have made SPIONs popular for bioimaging applications [191, 192]. SPIONs based contrast agents such as *Endorem, Sinerem, Feridex, Resovist, Ferumoxytol* are commercially available [182, 193].

SPIONs synthesized by co-precipitation are used in commercially available contrast agents but are not monodispersed due to irregular morphology and variable size. Thermal decomposition methods give monodispersed SPIONs, but water dispersibility is low. Surface functionalization is used as an alternative to overcome such a problem. Dextran, silica, PEG, PVP coatings over SPIONs show improved water dispersibility. In a study, dextran sulfate (DS) coated, and poly(glycerol methacrylate) linked SPIONs prepared by co-precipitation were used as a contrast agent in MRI of atherosclerosis [194]. Macrophages mediated cellular uptake studies were performed using dextran-coated (Dex) SPIONs as a control. T_2-weighted MRI of cells reported a lower signal intensity of DS-SPIONs compared to Dex-SPIONs, proving DS-SPIONs as a potential contrast agent for MRI of atherosclerosis.

Contrast agents currently in use show contrast enhancement of only one relaxation time, either T_1 or T_2. However, the requirement of improved image visualization and reconstructions has become critical nowadays for a more accurate diagnosis of the diseased area. Thus, novel contrast agents with the ability to create contrast enhancement by simultaneous altering both relaxivities are necessary to be used [195-197]. Researchers have reported studies on the synthesis of such materials having dual-mode contrast enhancement ability. A study reported the combination of T_1 and T_2 agents, i.e., Gd and SPIONs, to achieve control over both the relaxivities [198]. Here, Fe_3O_4 coated with silica (21 nm) were linked to paramagnetic Gd complex. Using RGD peptide for the target specificity, the resultant molecule Fe_3O_4-SiO_2-(Gd-DTPA)-RGD NPs showed the T_1 relaxivity value of 4.2 mM/s and T_2 relaxivity value of 17.4 mM/s when tested *in vivo* and *in vitro* on brain cells. Results were obtained for Gd: Fe ratio of 0.3:1 and particle size of 27 nm, thus confirming the ability of NPs to act as dual-mode contrasting agents.

A recent attempt to use SPIONs as both positive and negative contrast agents (dual-mode) is reported by Alipour *et al.* [199]. In the study, SPIONs of cubic and spherical shape were examined for their ability to act as a contrast agent. MRI scan was performed on the rat model with 1 mg/kg of contrast agent. Among the tested cubic and spherical shaped SPIONs, 11 nm cubic SPIONs offered optimal relaxivities with enhanced contrast values by affecting the T_1 and T_2 times simultaneously. Experimentally, 11 nm cubic SPIONs showed 64% and 48% of contrast enhancement for T_1 and T_2 weighted images, respectively, and demonstrated their potential use as a dual-mode contrast agent. Fig. 4 graphically represents the application of SPIONs in dual-mode MRI.

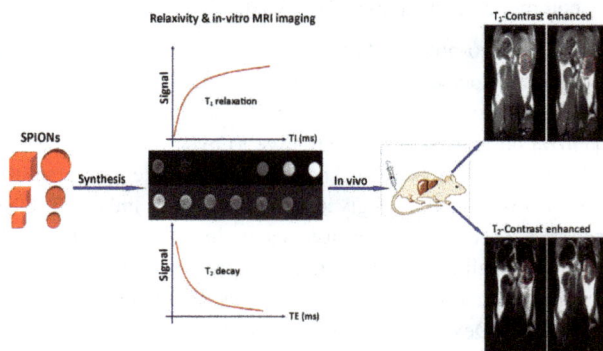

Figure 4. Application of SPIONs in dual mode Medical Resonance Imaging (adapted from Magnetic Resonance Imaging 49 (2018) 16–24 [199])

Medical imaging offers an analysis of anatomical structure within the body effectively. Relaxation of protons in the water molecules is responsible for the contrast generated via the application of the magnetic field. Contrast agents enhance the intensities obtained from relaxation and give high-quality images. SPIONs are used as a T_2 contrast agent and offer better stability over Gd-based T_1 contrast agents. Conjugation of SPIONs with tissue-specific molecules increases the efficiency of the doses provided and improve the analysis quality. Contrast agents working for both the relaxivities are beneficial to obtain quality images and reduced doses. SPIONs with God-based dual-mode agents are reported to give better results over conventional contrast agents.

2.4   Drug delivery

The success of disease treatment depends on drug specificity and efficacy. Adequate drug concentrations determine the efficacy and the time required for cure. Pharmacokinetics of the drug determines the fate of the drug within the body after administered in the body. Surface and chemical properties of drug molecules regulate adsorption, distribution, metabolism, and excretion of the drug [200, 201]. For efficacy, a particular amount of drug with its active structure should reach the target site and is dependent on the stability of molecules. However, variable physiological conditions within the body may cause structural deformation and lead to the loss function of the drug. A considerable portion of the intravenously injected drug is retained in the bloodstream and eliminated through the kidney. Orally administered drugs lose their significance due to a high acidic pH in the stomach and eliminated via excretion [202, 203]. Moreover, nonspecific adsorption of the drug molecules also reduces the amount required for the desired actions and may cause other side effects with accumulation within the body. These problems can be tackled by targeting drug molecules to the specific region by tuning surface properties of drug molecules or encapsulating them within the carrying molecules and directing them to target sites.

Targeted drug delivery is an efficient alternative to deliver the required amount of drug molecules to a specific site and thus reduces the required number of doses and side effects. Drug delivery is used in cancer therapy, neural system diseases, sensorineural hearing loss, and many more [204-206]. Drug molecules are linked to nano-sized carriers via the linkers or encapsulation and can be targeted to the desired site by providing target specificity. Some specific triggering action controls the release of the drug. The triggering action may include a change in temperature or pH locally or some enzymatic action resulting in the dissociation of drug molecules from the carrier [63, 67, 207]. The release of the drug can be categorized into three types: linker breaking, structural disassembly, and membrane breaking. Controlled breaking of linking polymer such as

PEG, dextran, or peptides releases the drug molecules from the carrier [207]. In structural disassembly, drug molecules assembled with nanostructure like micelles are released by destabilizing the structure [208]. Membrane breakdown involves opening/leakage of the encapsulating membrane of drug holding molecules to release the drug [209]. The amount of attached drug molecules depends on the carrying capacity or loading capacity of the carrier molecules.

There are many possible ways to deliver a drug to target sites, which include magnetically driven drug delivery, aptamers functionalized drug delivery, antibody-mediated drug delivery, and protein-based drug delivery [210-212]. Emphasizing on the magnetically driven drug delivery, the basic idea is to encapsulate or attach the therapeutic drug molecules to magnetic nanoparticles/nanocomposite. The molecules are then driven to the desired target site by applying an external magnetic field for controlled release. The nature of the magnetic field defines the scope of application for the SPIONs drug delivery. In the static magnetic field, a single magnet or assembly of magnets is kept over the surface near the target site. Due to magnetic attraction, the injected magnetic molecules get accumulated in the surrounding region of the target site. In some cases, multiple static fields are applied to direct the drug-carrying materials towards the target site. The varying magnetic field is used to create the gradient of the magnetic field in order to control the path of the molecules in a stepwise manner. However, the application of the external magnetic field is effective for the surface tumor. The efficiency of the magnetic field is considerably reduced when applied for deep tissue tumors. Therefore, the location of the tumor and strength of the applied field must be considered while planning drug delivery treatment.

SPIONs, possessing excellent biocompatibility and ease of functionalization, are a better option to use for drug delivery applications. SPIONs offer an exclusive advantage of heat generation in AMF, which can trigger the release of the drug. Some surface modifications are required to attach the drug molecules to SPIONS. Polymer like PEG, poly(vinyl alcohol); polysaccharides like dextran, starch; inorganic materials like silica; cysteine were used to modify the SPIONs surface in drug delivery applications [213-215]. Section 1.3 describes the requirement of surface modification of SPIONs.

A SPIONs based drug delivery system comprising a chemotherapeutic agent and photosensitizer molecule was studied for cancer treatment. COOH functionalized SPIONs were modified to have amine groups to which aptamers and dsDNA hairpin was conjugated (Apt-gc3-SPION). Chemotherapeutic drug DNM and photosensitizer TMPyP were attached to the functionalized SPIONs to give final drug delivery system molecules. Functionalized SPIONs with an average size of 15-20 nm were added to A549 cells and observed with a 4T magnet at the desired target site in a petri plate for 24h. Results showed that region with 4T magnetic field significant amount of cells appeared floating and in round shape compared to the control region confirming the cell death in the target region. The study confirmed the effect of the external magnetic field on the local concentration of SPIONs, ultimately drug molecules of DNM.

In the other study, thermally cross-linked SPIONs were functionalized with branched polyethyleneimine via EDC/NHS and used to deliver anticancer drugs, namely cyclodextrin conjugate and paclitaxel conjugate *in vivo* and *in vitro* [216]. Different cell lines such as mouse colon cancer cells CT26, human cervical carcinoma cells (HeLa) and human breast carcinoma cells (MCF-7) were used for *in vitro* and kept in a magnetic field for 4h. When injected intravenously in mouse models with colon cancer, the external magnetic field was used to direct the drug molecules to the tumor site for five days. Results showed higher cytotoxicity for *in vitro* study with the cells samples exposed to the magnetic field compared to other samples. In *in vivo* experiments, mice treated with drug conjugated SPIONs in the absence of a magnetic field showed growth of the tumor. In contrast, mice models with exposed magnetic field showed reduced tumor growth and also reported a decrease in body weights. The study confirmed that the magnetic field targeted has improved the drug concentrations in the tumor site to enhance the effect. Fig. 5 shows a magnetically driven delivery of anticancer drugs.

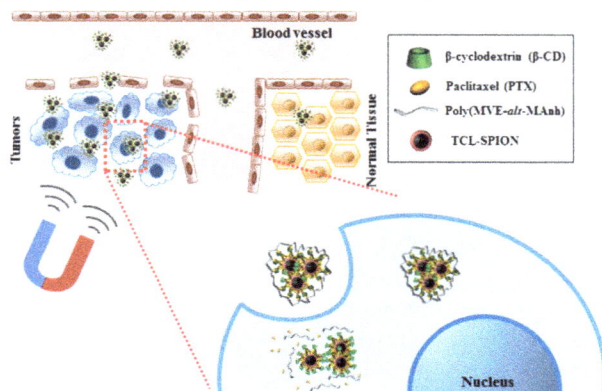

Figure. 5. Schematic illustration of pPTX/CD-SPION based drug delivery (adapted from Journal of Controlled Release 231 (2016) 68–76 [216])

Site-specific drug delivery improves the effectiveness of treatment and reduces the side effects due to the nonspecific interaction of drugs. SPIONs linked drug molecules are targeted by the externally applied magnetic field and thus results in the accumulation of a considerable amount of drug molecules within the target site. The magnetic targeting of a drug is easy and very much effective for the target tissues near to the surface. However, deep body tissues are difficult to target via an externally applied magnetic field. Therefore, the dynamic magnetic field with high field strength can be used to target the desired area. Surface modified SPIONs can be linked to tissue-specific ligands and drug molecules. Such SPIONs can be directed more efficiently to the desired target site and increase the success rate of treatment via Drug Delivery.

Conclusions and future perspectives

Magnetic properties of the nanomaterials offer distinct characteristics that can be used in various aspects of biomedical applications. Controllable and hence directable MNPs can be applied to obtain desired outcomes at specific sites. SPIONs are popularly used MNPs in the field of cancer therapy, MRI, drug delivery, and tissue engineering. SPIONs offer excellent biocompatibility, ease of surface functionalization, controllable size, and most importantly superparamagnetism. Superparamagnetic properties result in zero M_R with the removal of AMF and thus offer better control over the characteristics of nanoparticles. SPIONs generate the heat due to Neel and Brownian relaxations within the medium when exposed to AMF. This released heat is used in hyperthermia to kill cancer cells and in drug delivery to release the drug molecules at specific target sites. SPIONs used in tissue engineering provide enhanced scaffold stability and gradient distribution of SGFs within the scaffolds. While in MRI, SPIONs alter transversal relaxivities and hence used as a T_2 contrast agent. Surface modifications of SPIONs improve the biocompatibility, target specificity, and loading capacity, therefore, enhances the performances in various applications.

Still, there is a vast scope for improvement in the MNPs based biomedical applications. Firstly, the limited efficiency of the magnetic field within deep tissues is the main restriction in the application of SPIONs for hyperthermia and drug delivery. Ferrites based MNPs showing higher magnetic moments can be used over such limitations. Also, alloyed MNPs of iron with other materials may improve the magnetic moments and can be used for deep tissue target sites. Secondly, the low resolution of the MRI images is the major problem in the clinical diagnostics. Administration of the dual-mode contrast agents can enhance the contrast between both the relaxivities. Tuning of size, magnetic moment, and elimination of such dual-mode contrast agents are primary demanding goals for researchers. Third, achieving better biocompatibility with a higher loading capacity of the SPIONs is one of the major challenges to deal with. Surface functionalization of MNPs can improve their biocompatibility and loading capacity, but biodistribution and elimination of these MNPs are needed to be tuned accordingly. Understanding of biodistribution, metabolism and elimination of SPIONs can help in the application of MNPs without any undesired side effects.

References

[1] N.G. Shetake, A. Kumar, B.N. Pandey, Iron-oxide nanoparticles target intracellular HSP90 to induce tumor radio-sensitization, Biochim. Biophys. Acta Gen. Sub. 1863 (2019) 857-869. https://doi.org/10.1016/j.bbagen.2019.02.010

[2] C.F. Chee, B.F. Leo, C.W. Lai, Superparamagnetic iron oxide nanoparticles for drug delivery, Applications of Nanocomposite Materials in Drug Delivery, Elsevier 2018, pp. 861-903. https://doi.org/10.1016/B978-0-12-813741-3.00038-8

[3] L. Mohammed, H.G. Gomaa, D. Ragab, J. Zhu, Magnetic nanoparticles for environmental and biomedical applications: A review, Particuology 30 (2017) 1-14. https://doi.org/10.1016/j.partic.2016.06.001

[4] K. Qiao, W. Tian, J. Bai, L. Wang, J. Zhao, Z. Du, X. Gong, Application of magnetic adsorbents based on iron oxide nanoparticles for oil spill remediation: A review, J. Taiwan Institute Chem. Eng. 97 (2019) 227-236. https://doi.org/10.1016/j.jtice.2019.01.029

[5] H. Shokrollahi, Structure, synthetic methods, magnetic properties and biomedical applications of ferrofluids, Mater. Sci. Eng. C 33 (2013) 2476-2487. https://doi.org/10.1016/j.msec.2013.03.028

[6] T. Maurer, F. Ott, G. Chaboussant, Y. Soumare, J.-Y. Piquemal, G. Viau, Magnetic nanowires as permanent magnet materials, Appl. Phys. Lett. 91 (2007) 172501. https://doi.org/10.1063/1.2800786

[7] O. Tegus, E. Brück, K. Buschow, F. De Boer, Transition-metal-based magnetic refrigerants for room-temperature applications, Nature 415 (2002) 150-152. https://doi.org/10.1038/415150a

[8] P. Tartaj, M. Morales, T. Gonzalez-Carreno, S. Veintemillas-Verdaguer, C. Serna, Advances in magnetic nanoparticles for biotechnology applications, J. Magn. Magn. Mater. 290 (2005) 28-34. https://doi.org/10.1016/j.jmmm.2004.11.155

[9] S. Amiri, H. Shokrollahi, The role of cobalt ferrite magnetic nanoparticles in medical science, Mater. Sci. Eng. C 33 (2013) 1-8. https://doi.org/10.1016/j.msec.2012.09.003

[10] E. Mazario, N. Menéndez, P. Herrasti, M. Cañete, V. Connord, J. Carrey, Magnetic hyperthermia properties of electrosynthesized cobalt ferrite nanoparticles, J. Phys. Chem. C 117 (2013) 11405-11411. https://doi.org/10.1021/jp4023025

[11] A. Doaga, A. Cojocariu, W. Amin, F. Heib, P. Bender, R. Hempelmann, O. Caltun, Synthesis and characterizations of manganese ferrites for hyperthermia applications, Mater. Chem. Phys. 143 (2013) 305-310. https://doi.org/10.1016/j.matchemphys.2013.08.066

[12] B.D. Cullity, C.D. Graham, Introduction to magnetic materials, John Wiley & Sons 2011. https://doi.org/10.1002/9780470386323

[13] M. Willard, L. Kurihara, E. Carpenter, S. Calvin, V. Harris, Chemically prepared magnetic nanoparticles, Int. Mater. Rev. 49 (2004) 125-170. https://doi.org/10.1179/095066004225021882

[14] I. Khan, S. Park, J. Hong, Temperature dependent magnetic properties of Dy-doped $Fe_{16}N_2$: Potential rare-earth-lean permanent magnet, Intermetallics 108 (2019) 25-31. https://doi.org/10.1016/j.intermet.2019.02.001

[15] M. Srivastava, S. Alla, S.S. Meena, N. Gupta, R. Mandal, N. Prasad, Magnetic field regulated controlled hyperthermia with $Li_xFe_{3-x}O4$ ($0.06 \leq x \leq 0.3$) nanoparticles, Ceram. Int. 45 (2019) 12028-12034. https://doi.org/10.1016/j.ceramint.2019.03.097

[16] S. Pandey, A. Quetz, A. Aryal, I. Dubenko, D. Mazumdar, S. Stadler, N. Ali, Thermosensitive Ni-based magnetic particles for self-controlled hyperthermia applications, J. Magn. Magn. Mater. 427 (2017) 200-205. https://doi.org/10.1016/j.jmmm.2016.11.049

[17] C. Gomez-Polo, V. Recarte, L. Cervera, J. Beato-Lopez, J. Lopez-Garcia, J. Rodríguez-Velamazán, M. Ugarte, E. Mendonça, J. Duque, Tailoring the structural and magnetic properties of Co-Zn nanosized ferrites for hyperthermia applications, J. Magn. Magn. Mater. 465 (2018) 211-219. https://doi.org/10.1016/j.jmmm.2018.05.051

[18] A.S. Teja, P.-Y. Koh, Synthesis, properties, and applications of magnetic iron oxide nanoparticles, Prog. Cryst. Growth Charact. Mater. 55 (2009) 22-45. https://doi.org/10.1016/j.pcrysgrow.2008.08.003

[19] T.K. Jain, M.K. Reddy, M.A. Morales, D.L. Leslie-Pelecky, V. Labhasetwar, Biodistribution, clearance, and biocompatibility of iron oxide magnetic nanoparticles in rats, Mol. Pharm. 5 (2008) 316-327. https://doi.org/10.1021/mp7001285

[20] M.M. Can, M. Coşkun, T. Fırat, A comparative study of nanosized iron oxide particles; magnetite (Fe_3O_4), maghemite (γ-Fe_2O_3) and hematite (α-Fe_2O_3), using ferromagnetic resonance, J. Alloys Compd. 542 (2012) 241-247. https://doi.org/10.1016/j.jallcom.2012.07.091

[21] M. Manjunatha, R. Kumar, A.V. Anupama, V.B. Khopkar, R. Damle, K.P. Ramesh, B. Sahoo, XRD, internal field-NMR and Mössbauer spectroscopy study of

composition, structure and magnetic properties of iron oxide phases in iron ores, J. Mater. Res. Technol. 8 (2019) 2192-2200. https://doi.org/10.1016/j.jmrt.2019.01.022

[22] R. Zboril, L. Machala, M. Mashlan, V. Sharma, Iron (III) oxide nanoparticles in the thermally induced oxidative decomposition of Prussian blue, $Fe_4[Fe(CN)_6]_3$, Cryst. Growth Design 4 (2004) 1317-1325. https://doi.org/10.1021/cg049748+

[23] R.P. Feynman, R.B. Leighton, M. Sands, The feynman lectures on physics; vol. i, Am. J. Phys. 33 (1965) 750-752. https://doi.org/10.1119/1.1972241

[24] M. Catti, G. Valerio, R. Dovesi, Theoretical study of electronic, magnetic, and structural properties of α-Fe_2O_3 (hematite), Phys. Rev.B 51 (1995) 7441-7450. https://doi.org/10.1103/PhysRevB.51.7441

[25] R.M. Cornell, U. Schwertmann, The iron oxides: structure, properties, reactions, occurrences and uses, John Wiley & Sons 2003. https://doi.org/10.1002/3527602097

[26] W. Wei, W. Zhaohui, Y. Taekyung, J. Changzhong, K. Woo-Sik, Recent progress on magnetic iron oxide nanoparticles: synthesis, surface functional strategies and biomedical applications, Sci. Technol. Adv. Mater. 16(2) (2015) 023501. https://doi:10.1088/1468-6996/16/2/023501

[27] Z. Shaterabadi, G. Nabiyouni, M. Soleymani, Optimal size for heating efficiency of superparamagnetic dextran-coated magnetite nanoparticles for application in magnetic fluid hyperthermia, Physica C: Superconductivity Appl. 549 (2018) 84-87. https://doi.org/10.1016/j.physc.2018.02.060

[28] V. Ganesan, B. Lahiri, C. Louis, J. Philip, S.P. Damodaran, Size-controlled synthesis of superparamagnetic magnetite nanoclusters for heat generation in an alternating magnetic field, J. Mol. Liq. 281 (2019) 315-323. https://doi.org/10.1016/j.molliq.2019.02.095

[29] D. Jiles, Introduction to magnetism and magnetic materials, CRC press 2015. https://doi.org/10.1201/b18948

[30] C. Hasirci, O. Karaagac, H. Köçkar, Superparamagnetic zinc ferrite: A correlation between high magnetizations and nanoparticle sizes as a function of reaction time via hydrothermal process, J. Magn. Magn. Mater. 474 (2019) 282-286. https://doi.org/10.1016/j.jmmm.2018.11.037

[31] Y.C. Park, J.B. Smith, T. Pham, R.D. Whitaker, C.A. Sucato, J.A. Hamilton, E. Bartolak-Suki, J.Y. Wong, Effect of PEG molecular weight on stability, T_2 contrast, cytotoxicity, and cellular uptake of superparamagnetic iron oxide nanoparticles (SPIONs), Colloids Surf. B. Biointerfaces 119 (2014) 106-114. https://doi.org/10.1016/j.colsurfb.2014.04.027

[32] A. Luchini, Y. Gerelli, G. Fragneto, T. Nylander, G.K. Pálsson, M.S. Appavou, L. Paduano, Neutron Reflectometry reveals the interaction between functionalized SPIONs and the surface of lipid bilayers, Colloids Surf. B. Biointerfaces 151 (2017) 76-87. https://doi.org/10.1016/j.colsurfb.2016.12.005

[33] P. Lemal, S. Balog, C. Geers, P. Taladriz-Blanco, A. Palumbo, A.M. Hirt, B. Rothen-Rutishauser, A. Petri-Fink, Heating behavior of magnetic iron oxide nanoparticles at clinically relevant concentration, J. Magn. Magn. Mater. 474 (2019) 637-642. https://doi.org/10.1016/j.jmmm.2018.10.009

[34] K. Woo, J. Hong, S. Choi, H.W. Lee, J.-P. Ahn, C.S. Kim, S.W. Lee, Easy synthesis and magnetic properties of iron oxide nanoparticles, Chem. Mater. 16 (2004) 2814-2818. https://doi.org/10.1021/cm049552x

[35] T. Lastovina, A. Bugaev, S. Kubrin, E. Kudryavtsev, A. Soldatov, Structural studies of magnetic nanoparticles doped with rare-earth elements, J. Struct. Chem. 57 (2016) 1444-1449. https://doi.org/10.1134/S0022476616070209

[36] K.T. Nguyen, Y. Zhao, Engineered hybrid nanoparticles for on-demand diagnostics and therapeutics, Acc. Chem. Res. 48 (2015) 3016-3025. https://doi.org/10.1021/acs.accounts.5b00316

[37] Y. Wei, B. Han, X. Hu, Y. Lin, X. Wang, X. Deng, Synthesis of Fe_3O_4 nanoparticles and their magnetic properties, Procedia Engineering 27 (2012) 632-637. https://doi.org/10.1016/j.proeng.2011.12.498

[38] L. Gholami, R.K. Oskuee, M. Tafaghodi, A.R. Farkhani, M. Darroudi, Green facile synthesis of low-toxic superparamagnetic iron oxide nanoparticles (SPIONs) and their cytotoxicity effects toward Neuro2A and HUVEC cell lines, Ceram. Int. 44 (2018) 9263-9268. https://doi.org/10.1016/j.ceramint.2018.02.137

[39] K. Kalantari, M.B. Ahmad, K. Shameli, R. Khandanlou, Synthesis of talc/Fe_3O_4 magnetic nanocomposites using chemical co-precipitation method, Int. J. Nanomedicine 8 (2013) 1817-1823. https://doi.org/10.2147/IJN.S43693

[40] A. Gholoobi, Z. Meshkat, K. Abnous, M. Ghayour-Mobarhan, M. Ramezani, F.H. Shandiz, K. Verma, M. Darroudi, Biopolymer-mediated synthesis of Fe3O4 nanoparticles and investigation of their in vitro cytotoxicity effects, J. Mol. Struct. 1141 (2017) 594-599. https://doi.org/10.1016/j.molstruc.2017.04.024

[41] M. Azhdarzadeh, F. Atyabi, A.A. Saei, B.S. Varnamkhasti, Y. Omidi, M. Fateh, M. Ghavami, S. Shanehsazzadeh, R. Dinarvand, Theranostic MUC-1 aptamer targeted gold coated superparamagnetic iron oxide nanoparticles for magnetic resonance

imaging and photothermal therapy of colon cancer, Colloids Surf. B. Biointerfaces 143 (2016) 224-232. https://doi.org/10.1016/j.colsurfb.2016.02.058

[42] F. Arriagada, K. Osseo-Asare, Synthesis of nanosize silica in a nonionic water-in-oil microemulsion: effects of the water/surfactant molar ratio and ammonia concentration, J. Colloid Interface Sci. 211 (1999) 210-220. https://doi.org/10.1006/jcis.1998.5985

[43] A.E. Silva, G. Barratt, M. Chéron, E.S.T. Egito, Development of oil-in-water microemulsions for the oral delivery of amphotericin B, Int. J. Pharm. 454 (2013) 641-648. https://doi.org/10.1016/j.ijpharm.2013.05.044

[44] L. Cai, Z.-Z. Chen, M.-Y. Chen, H.-W. Tang, D.-W. Pang, MUC-1 aptamer-conjugated dye-doped silica nanoparticles for MCF-7 cells detection, Biomaterials 34 (2013) 371-381. https://doi.org/10.1016/j.biomaterials.2012.09.084

[45] M.A. Malik, M.Y. Wani, M.A. Hashim, Microemulsion method: A novel route to synthesize organic and inorganic nanomaterials: 1st Nano Update, Arabian J. Chem. 5 (2012) 397-417. https://doi.org/10.1016/j.arabjc.2010.09.027

[46] T. Aubert, F. Grasset, S. Mornet, E. Duguet, O. Cador, S. Cordier, Y. Molard, V. Demange, M. Mortier, H. Haneda, Functional silica nanoparticles synthesized by water-in-oil microemulsion processes, J. Colloid Interface Sci. 341 (2010) 201-208. https://doi.org/10.1016/j.jcis.2009.09.064

[47] M. Pileni, Control of the size and shape of inorganic nanocrystals at various scales from nano to macrodomains, J. Phys. Chem. C 111 (2007) 9019-9038. https://doi.org/10.1021/jp070646e

[48] M.R. Housaindokht, A.N. Pour, Precipitation of hematite nanoparticles via reverse microemulsion process, J. Natural Gas Chem. 20 (2011) 687-692. https://doi.org/10.1016/S1003-9953(10)60234-4

[49] J. Livage, M. Henry, C. Sanchez, Sol-gel chemistry of transition metal oxides, Prog. Solid State Chem. 18 (1988) 259-341. https://doi.org/10.1016/0079-6786(88)90005-2

[50] L.L. Hench, J.K. West, The sol-gel process, Chem. Rev. 90 (1990) 33-72. https://doi.org/10.1021/cr00099a003

[51] H. Schmidt, Chemistry of material preparation by the sol-gel process, J. Non-Cryst. Solids 100 (1988) 51-64. https://doi.org/10.1016/0022-3093(88)90006-3

[52] M. Tadić, V. Kusigerski, D. Marković, M. Panjan, I. Milošević, V. Spasojević, Highly crystalline superparamagnetic iron oxide nanoparticles (SPION) in a silica matrix, J. Alloys Compd. 525 (2012) 28-33. https://doi.org/10.1016/j.jallcom.2012.02.056

[53] X. Liu, S. Tao, Y. Shen, Preparation and characterization of nanocrystalline α-Fe_2O_3 by a sol-gel process, Sensors Actuators B: Chem. 40 (1997) 161-165. https://doi.org/10.1016/S0925-4005(97)80256-0

[54] Z. Haijun, J. Xiaolin, Y. Yongjie, L. Zhanjie, Y. Daoyuan, L. Zhenzhen, The effect of the concentration of citric acid and pH values on the preparation of $MgAl_2O_4$ ultrafine powder by citrate sol–gel process, Mater. Res. Bull. 39 (2004) 839-850. https://doi.org/10.1016/j.materresbull.2004.01.006

[55] P. Vaqueiro, M. Lopez-Quintela, Influence of complexing agents and pH on yttrium– Iron garnet synthesized by the sol–gel method, Chem. Mater. 9 (1997) 2836-2841. https://doi.org/10.1021/cm970165f

[56] S.H. Vajargah, H.M. Hosseini, Z. Nemati, Synthesis of nanocrystalline yttrium iron garnets by sol–gel combustion process: The influence of pH of precursor solution, Mater. Sci. Eng. B 129 (2006) 211-215. https://doi.org/10.1016/j.mseb.2006.01.014

[57] A. Mali, A. Ataie, Influence of the metal nitrates to citric acid molar ratio on the combustion process and phase constitution of barium hexaferrite particles prepared by sol–gel combustion method, Ceram. Int. 30 (2004) 1979-1983. https://doi.org/10.1016/j.ceramint.2003.12.178

[58] A. Akbar, H. Yousaf, S. Riaz, S. Naseem, Role of precursor to solvent ratio in tuning the magnetization of iron oxide thin films–A sol-gel approach, J. Magn. Magn. Mater. 471 (2019) 14-24. https://doi.org/10.1016/j.jmmm.2018.09.008

[59] A.K. Gupta, S. Wells, Surface-modified superparamagnetic nanoparticles for drug delivery: preparation, characterization, and cytotoxicity studies, IEEE Trans. NanoBiosci. 3 (2004) 66-73. https://doi.org/10.1109/TNB.2003.820277

[60] C. Janko, J. Zaloga, M. Pöttler, S. Dürr, D. Eberbeck, R. Tietze, S. Lyer, C. Alexiou, Strategies to optimize the biocompatibility of iron oxide nanoparticles–"SPIONs safe by design", J. Magn. Magn. Mater. 431 (2017) 281-284. https://doi.org/10.1016/j.jmmm.2016.09.034

[61] M. Singh, A. Savchenko, I. Shetinin, A. Majouga, An original route to target delivery via core-shell modification of SPIONs, Mater. Today Proceedings 3 (2016) 2652-2661. https://doi.org/10.1016/j.matpr.2016.06.009

[62] N. Singh, J. Nayak, S.K. Sahoo, R. Kumar, Glutathione conjugated superparamagnetic Fe_3O_4-Au core shell nanoparticles for pH controlled release of DOX, Mater. Sci. Eng. C 100 (2019) 453-465. https://doi.org/10.1016/j.msec.2019.03.031

[63] S. Ullah, K. Seidel, S. Türkkan, D.P. Warwas, T. Dubich, M. Rohde, H. Hauser, P. Behrens, A. Kirschning, M. Köster, Macrophage entrapped silica coated superparamagnetic iron oxide particles for controlled drug release in a 3D cancer model, J. Controlled Release 294 (2019) 327-336. https://doi.org/10.1016/j.jconrel.2018.12.040

[64] S.A.C. Lima, A. Gaspar, S. Reis, L. Durães, Multifunctional nanospheres for co-delivery of methotrexate and mild hyperthermia to colon cancer cells, Mater. Sci. Eng. C 75 (2017) 1420-1426. https://doi.org/10.1016/j.msec.2017.03.049

[65] S. Haracz, M. Hilgendorff, J. Rybka, M. Giersig, Effect of surfactant for magnetic properties of iron oxide nanoparticles, Nuclear Instruments and Methods in Physics Research Section B: Beam Interactions with Materials and Atoms 364 (2015) 120-126. https://doi.org/10.1016/j.nimb.2015.08.035

[66] S.E. Favela-Camacho, E.J. Samaniego-Benítez, A. Godínez-García, L.M. Avilés-Arellano, J.F. Pérez-Robles, How to decrease the agglomeration of magnetite nanoparticles and increase their stability using surface properties, Colloids Surf. Physicochem. Eng. Aspects 574 (2019) 29-35. https://doi.org/10.1016/j.colsurfa.2019.04.016

[67] B. Shaghaghi, S. Khoee, S. Bonakdar, Preparation of multifunctional Janus nanoparticles on the basis of SPIONs as targeted drug delivery system, Int. J. Pharm. 559 (2019) 1-12. https://doi.org/10.1016/j.ijpharm.2019.01.020

[68] M. Galli, B. Rossotti, P. Arosio, A.M. Ferretti, M. Panigati, E. Ranucci, P. Ferruti, A. Salvati, D. Maggioni, A new catechol-functionalized polyamidoamine as an effective SPION stabilizer, Colloids Surf. B. Biointerfaces 174 (2019) 260-269. https://doi.org/10.1016/j.colsurfb.2018.11.007

[69] D.K. Chatterjee, P. Diagaradjane, S. Krishnan, Nanoparticle-mediated hyperthermia in cancer therapy, Ther. Deliv. 2 (2011) 1001-1014. https://doi.org/10.4155/tde.11.72

[70] D. Ortega, Q.A. Pankhurst, Magnetic hyperthermia, Nanosci.1 (2013) 60-88. https://doi.org/10.1039/9781849734844-00060

[71] M. Falk, R. Issels, Hyperthermia in oncology, Int. J. Hyperthermia 17 (2001) 1-18. https://doi.org/10.1080/02656730150201552

[72] H. Mamiya, Y. Takeda, T. Naka, N. Kawazoe, G. Chen, B. Jeyadevan, Practical solution for effective whole-body magnetic fluid hyperthermia treatment, J. Nanomater. 2017 (2017) 1047697. https://doi.org/10.1155/2017/1047697

[73] S. Toraya-Brown, M.R. Sheen, P. Zhang, L. Chen, J.R. Baird, E. Demidenko, M.J. Turk, P.J. Hoopes, J.R. Conejo-Garcia, S. Fiering, Local hyperthermia treatment of

Materials Research Forum LLC
https://doi.org/10.21741/9781644900611-9

tumors induces CD8+ T cell-mediated resistance against distal and secondary tumors, Handbook of immunological properties of engineered nanomaterials: Volume 3: engineered nanomaterials and the immune cell function, World Scientific2016, pp. 309-347. https://doi.org/10.1142/9789813140479_0012

[74] S. Gao, M. Zheng, X. Ren, Y. Tang, X. Liang, Local hyperthermia in head and neck cancer: mechanism, application and advance, Oncotarget 7 (2016) 57367-57378. https://doi.org/10.18632/oncotarget.10350

[75] Z. Behrouzkia, Z. Joveini, B. Keshavarzi, N. Eyvazzadeh, R.Z. Aghdam, Hyperthermia: how can it be used?, Oman Med. J. 31 (2016) 89-97. https://doi.org/10.5001/omj.2016.19

[76] H. Kok, P. Wust, P.R. Stauffer, F. Bardati, G. Van Rhoon, J. Crezee, Current state of the art of regional hyperthermia treatment planning: a review, Radiation Oncology 10 (2015) 196. https://doi.org/10.1186/s13014-015-0503-8

[77] J. Crezee, C. van Leeuwen, A. Oei, L. van Heerden, A. Bel, L. Stalpers, P. Ghadjar, N. Franken, H. Kok, Biological modelling of the radiation dose escalation effect of regional hyperthermia in cervical cancer, Radiation Oncology 11 (2016) 14. https://doi.org/10.1186/s13014-016-0592-z

[78] M. Lévy, C. Wilhelm, J.M. Siaugue, O. Horner, J.C. Bacri, F. Gazeau, Magnetically induced hyperthermia: size-dependent heating power of γ-Fe_2O_3 nanoparticles, J. Phys.: Condens. Matter 20 (2008) 204133. https://doi.org/10.1088/0953-8984/20/20/204133

[79] M. Suto, Y. Hirota, H. Mamiya, A. Fujita, R. Kasuya, K. Tohji, B. Jeyadevan, Heat dissipation mechanism of magnetite nanoparticles in magnetic fluid hyperthermia, J. Magn. Magn. Mater. 321 (2009) 1493-1496. https://doi.org/10.1016/j.jmmm.2009.02.070

[80] E. Lima, E. De Biasi, R.D. Zysler, M.V. Mansilla, M.L. Mojica-Pisciotti, T.E. Torres, M.P. Calatayud, C. Marquina, M.R. Ibarra, G.F. Goya, Relaxation time diagram for identifying heat generation mechanisms in magnetic fluid hyperthermia, J. Nanopart. Res. 16 (2014) 2791. https://doi.org/10.1007/s11051-014-2791-6

[81] M. Coïsson, G. Barrera, F. Celegato, L. Martino, S.N. Kane, S. Raghuvanshi, F. Vinai, P. Tiberto, Hysteresis losses and specific absorption rate measurements in magnetic nanoparticles for hyperthermia applications, Biochim. Biophys. Acta Gen. Sub. 1861 (2017) 1545-1558. https://doi.org/10.1016/j.bbagen.2016.12.006

[82] C.W. Kartikowati, Q. Li, S. Horie, T. Ogi, T. Iwaki, K. Okuyama, Aligned Fe_3O_4 magnetic nanoparticle films by magneto-electrospray method, RSC Adv. 7 (2017) 40124-40130. https://doi.org/10.1039/C7RA07944C

[83] L. Maldonado-Camargo, I. Torres-Díaz, A. Chiu-Lam, M. Hernández, C. Rinaldi, Estimating the contribution of Brownian and Néel relaxation in a magnetic fluid through dynamic magnetic susceptibility measurements, J. Magn. Magn. Mater. 412 (2016) 223-233. https://doi.org/10.1016/j.jmmm.2016.03.087

[84] M. Brollo, J. M. Orozco-Henao, R. López-Ruiz, D. Muraca, C. S. B. Dias, K. R. Pirota, M. Knobel, Magnetic hyperthermia in brick-like Ag@ Fe3O4 core–shell nanoparticles, J. Magn. Magn. Mater. 397 (2016) 20-27. https://doi.org/10.1016/j.jmmm.2015.08.081

[85] A. Hervault and N.T.K. Thanh, Magnetic nanoparticle-based therapeutic agents for thermo-chemotherapy treatment of cancer. Nanoscale 6(20) (2014) 11553-11573. https://doi.org/10.1039/c4nr03482a

[86] Z. Shaterabadi, G. Nabiyouni, and M. Soleymani, Physics responsible for heating efficiency and self-controlled temperature rise of magnetic nanoparticles in magnetic hyperthermia therapy. Prog. Biophys. Mol. Biol. 133 (2018) 9-19. https://doi.org/10.1016/j.pbiomolbio.2017.10.001

[87] R.E. Rosensweig, Heating magnetic fluid with alternating magnetic field, J. Magn. Magn. Mater. 252 (2002) 370-374. https://doi.org/10.1016/S0304-8853(02)00706-0

[88] X. Wang, H. Gu, Z. Yang, The heating effect of magnetic fluids in an alternating magnetic field, J. Magn. Magn. Mater. 293 (2005) 334-340. https://doi.org/10.1016/j.jmmm.2005.02.028

[89] E.C. Abenojar, S. Wickramasinghe, J. Bas-Concepcion, A.C.S. Samia, Structural effects on the magnetic hyperthermia properties of iron oxide nanoparticles, Progress Natural Sci. Mater. Int. 26 (2016) 440-448. https://doi.org/10.1016/j.pnsc.2016.09.004

[90] J.-P. Fortin, F. Gazeau, C. Wilhelm, Intracellular heating of living cells through Néel relaxation of magnetic nanoparticles, Eur. Biophys. J. 37 (2008) 223-228. https://doi.org/10.1007/s00249-007-0197-4

[91] R. Ramanujan, L. Lao, Magnetic particles for hyperthermia treatment of cancer, Proc. First Intl. Bioengg. Conf, 2004, pp. 69-72.

[92] M. Mallory, E. Gogineni, G. C. Jones, L. Greer, C. B. Simone, Therapeutic hyperthermia: The old, the new, and the upcoming, Crit. Rev. Oncol. Hematol 97 (2016) 258-542. https://doi.org/10.1016/j.critrevonc.2015.08.003

[93] B. Frey, E.-M. Weiss, Y. Rubner, R. Wunderlich, O.J. Ott, R. Sauer, R. Fietkau, U.S. Gaipl, Old and new facts about hyperthermia-induced modulations of the immune system, Int. J. Hyperthermia 28 (2012) 528-542. https://doi.org/10.3109/02656736.2012.677933

[94] N. Datta, S.G. Ordóñez, U. Gaipl, M. Paulides, H. Crezee, J. Gellermann, D. Marder, E. Puric, S. Bodis, Local hyperthermia combined with radiotherapy and-/or chemotherapy: Recent advances and promises for the future, Cancer Treat. Rev. 41 (2015) 742-753. https://doi.org/10.1016/j.ctrv.2015.05.009

[95] T. Chen, J. Guo, C. Han, M. Yang, X. Cao, Heat shock protein 70, released from heat-stressed tumor cells, initiates antitumor immunity by inducing tumor cell chemokine production and activating dendritic cells via TLR4 pathway, J. Immunology 182 (2009) 1449-1459. https://doi.org/10.4049/jimmunol.182.3.1449

[96] P. Schildkopf, B. Frey, O.J. Ott, Y. Rubner, G. Multhoff, R. Sauer, R. Fietkau, U.S. Gaipl, Radiation combined with hyperthermia induces HSP70-dependent maturation of dendritic cells and release of pro-inflammatory cytokines by dendritic cells and macrophages, Radiother. Oncol. 101 (2011) 109-115. https://doi.org/10.1016/j.radonc.2011.05.056

[97] A. Jordan, R. Scholz, P. Wust, H. Fähling, R. Felix, Magnetic fluid hyperthermia (MFH): Cancer treatment with AC magnetic field induced excitation of biocompatible superparamagnetic nanoparticles, J. Magn. Magn. Mater. 201 (1999) 413-419. https://doi.org/10.1016/S0304-8853(99)00088-8

[98] R.R. Shah, T.P. Davis, A.L. Glover, D.E. Nikles, C.S. Brazel, Impact of magnetic field parameters and iron oxide nanoparticle properties on heat generation for use in magnetic hyperthermia, J. Magn. Magn. Mater. 387 (2015) 96-106. https://doi.org/10.1016/j.jmmm.2015.03.085

[99] I. Apostolova, J. Wesselinowa, Possible low-TC nanoparticles for use in magnetic hyperthermia treatments, Solid State Commun. 149 (2009) 986-990. https://doi.org/10.1016/j.ssc.2009.04.015

[100] K. McNerny, Y. Kim, D.E. Laughlin, M.E. McHenry, Chemical synthesis of monodisperse γ-Fe–Ni magnetic nanoparticles with tunable Curie temperatures for self-regulated hyperthermia, J. Appl. Phys. 107 (2010) 09A312. https://doi.org/10.1063/1.3348738

[101] S. Laurent, S. Dutz, U.O. Häfeli, M. Mahmoudi, Magnetic fluid hyperthermia: focus on superparamagnetic iron oxide nanoparticles, Adv. Colloid Interface Sci. 166 (2011) 8-23. https://doi.org/10.1016/j.cis.2011.04.003

[102] R. Di Corato, A. Espinosa, L. Lartigue, M. Tharaud, S. Chat, T. Pellegrino, C. Ménager, F. Gazeau, C. Wilhelm, Magnetic hyperthermia efficiency in the cellular environment for different nanoparticle designs, Biomaterials 35 (2014) 6400-6411. https://doi.org/10.1016/j.biomaterials.2014.04.036

[103] B.E. Kashevsky, S.B. Kashevsky, T.I. Terpinskaya, V.S. Ulashchik, Magnetic hyperthermia with hard-magnetic nanoparticles: In vivo feasibility of clinically relevant chemically enhanced tumor ablation, J. Magn. Magn. Mater. 475 (2019) 216-222. https://doi.org/10.1016/j.jmmm.2018.11.083

[104] S. Jadhav, P. Shewale, B. Shin, M. Patil, G. Kim, A. Rokade, S. Park, R. Bohara, Y. Yu, Study of structural and magnetic properties and heat induction of gadolinium-substituted manganese zinc ferrite nanoparticles for in vitro magnetic fluid hyperthermia, J. Colloid Interface Sci. 541 (2019) 192-203. https://doi.org/10.1016/j.jcis.2019.01.063

[105] J. Matuszak, J. Zaloga, R.P. Friedrich, S. Lyer, J. Nowak, S. Odenbach, C. Alexiou, I. Cicha, Endothelial biocompatibility and accumulation of SPION under flow conditions, J. Magn. Magn. Mater. 380 (2015) 20-26. https://doi.org/10.1016/j.jmmm.2014.09.005

[106] B. Ankamwar, T.-C. Lai, J.-H. Huang, R.-S. Liu, M. Hsiao, C.-H. Chen, Y. Hwu, Biocompatibility of Fe_3O_4 nanoparticles evaluated by in vitro cytotoxicity assays using normal, glia and breast cancer cells, Nanotechnology 21 (2010) 075102. https://doi.org/10.1088/0957-4484/21/7/075102

[107] N. Singh, G.J. Jenkins, R. Asadi, S.H. Doak, Potential toxicity of superparamagnetic iron oxide nanoparticles (SPION), Nano Rev. 1 (2010) 5358. https://doi.org/10.3402/nano.v1i0.5358

[108] A.R. Yasemian, M.A. Kashi, A. Ramazani, Surfactant-free synthesis and magnetic hyperthermia investigation of iron oxide (Fe_3O_4) nanoparticles at different reaction temperatures, Mater. Chem. Phys. 230 (2019) 9-16. https://doi.org/10.1016/j.matchemphys.2019.03.032

[109] C. Ispas, D. Andreescu, A. Patel, D.V. Goia, S. Andreescu, K.N. Wallace, Toxicity and developmental defects of different sizes and shape nickel nanoparticles in zebrafish, Environ. Sci. Technol. 43 (2009) 6349-6356. https://doi.org/10.1021/es9010543

[110] S. Chattopadhyay, S.K. Dash, S. Tripathy, B. Das, D. Mandal, P. Pramanik, S. Roy, Toxicity of cobalt oxide nanoparticles to normal cells; an in vitro and in vivo study, Chem. Biol. Interact. 226 (2015) 58-71. https://doi.org/10.1016/j.cbi.2014.11.016

[111] S.M. Hussain, A.K. Javorina, A.M. Schrand, H.M. Duhart, S.F. Ali, J.J. Schlager, The interaction of manganese nanoparticles with PC-12 cells induces dopamine depletion, Toxicol. Sci. 92 (2006) 456-463. https://doi.org/10.1093/toxsci/kfl020

[112] A. Hütten, D. Sudfeld, I. Ennen, G. Reiss, W. Hachmann, U. Heinzmann, K. Wojczykowski, P. Jutzi, W. Saikaly, G. Thomas, New magnetic nanoparticles for biotechnology, J. Biotechnol. 112 (2004) 47-63. https://doi.org/10.1016/j.jbiotec.2004.04.019

[113] S. Araújo-Barbosa, M. Morales, Nanoparticles of Ni1− xCux alloys for enhanced heating in magnetic hyperthermia, J. Alloys Compd. 787 (2019) 935-943. https://doi.org/10.1016/j.jallcom.2019.02.148

[114] D. Bharathi, R. Ranjithkumar, S. Vasantharaj, B. Chandarshekar, V. Bhuvaneshwari, Synthesis and characterization of chitosan/iron oxide nanocomposite for biomedical applications, Int. J. Biol. Macromol. 132 (2019) 880-887. https://doi.org/10.1016/j.ijbiomac.2019.03.233

[115] S.W. Lee, S. Bae, Y. Takemura, I.B. Shim, T.M. Kim, J. Kim, H.J. Lee, S. Zurn, C.S. Kim, Self-heating characteristics of cobalt ferrite nanoparticles for hyperthermia application, J. Magn. Magn. Mater. 310 (2007) 2868-2870. https://doi.org/10.1016/j.jmmm.2006.11.080

[116] E.L. Verde, G.T. Landi, J.D.A. Gomes, M.H. Sousa, A.F. Bakuzis, Magnetic hyperthermia investigation of cobalt ferrite nanoparticles: Comparison between experiment, linear response theory, and dynamic hysteresis simulations, J. Appl. Phys. 111 (2012) 123902. https://doi.org/10.1063/1.4729271

[117] R. Valenzuela, Novel applications of ferrites, Physics Research International 2012 (2012) 591839. https://doi.org/10.1155/2012/591839

[118] P. Pradhan, J. Giri, G. Samanta, H.D. Sarma, K.P. Mishra, J. Bellare, R. Banerjee, D. Bahadur, Comparative evaluation of heating ability and biocompatibility of different ferrite-based magnetic fluids for hyperthermia application, J. Biomed. Mater. Res. Part B: Applied Biomater. 81 (2007) 12-22. https://doi.org/10.1002/jbm.b.30630

[119] E.L. Verde, G.T. Landi, M. Carrião, A.L. Drummond, J.d.A. Gomes, E.D. Vieira, M.H. Sousa, A.F. Bakuzis, Field dependent transition to the non-linear regime in magnetic hyperthermia experiments: Comparison between maghemite, copper, zinc, nickel and cobalt ferrite nanoparticles of similar sizes, AIP Adv. 2 (2012) 032120. https://doi.org/10.1063/1.4739533

[120] Y. Sheng, S. Li, Z. Duan, R. Zhang, J. Xue, Fluorescent magnetic nanoparticles as minimally-invasive multi-functional theranostic platform for fluorescence imaging,

312

MRI and magnetic hyperthermia, Mater. Chem. Phys. 204 (2018) 388-396. https://doi.org/10.1016/j.matchemphys.2017.10.076

[121] V. Khot, A. Salunkhe, N. Thorat, R. Ningthoujam, S. Pawar, Induction heating studies of dextran coated $MgFe_2O_4$ nanoparticles for magnetic hyperthermia, Dalton Transactions 42 (2013) 1249-1258. https://doi.org/10.1039/C2DT31114C

[122] M.A. Kerroum, A. Essyed, C. Iacovita, W. Baaziz, D. Ihiawakrim, O. Mounkachi, M. Hamedoun, A. Benyoussef, M. Benaissa, O. Ersen, The effect of basic pH on the elaboration of $ZnFe_2O_4$ nanoparticles by co-precipitation method: Structural, magnetic and hyperthermia characterization, J. Magn. Magn. Mater. 478 (2019) 239-246. https://doi.org/10.1016/j.jmmm.2019.01.081

[123] V. Kusigerski, E. Illes, J. Blanusa, S. Gyergyek, M. Boskovic, M. Perovic, V. Spasojevic, Magnetic properties and heating efficacy of magnesium doped magnetite nanoparticles obtained by co-precipitation method, J. Magn. Magn. Mater. 475 (2019) 470-478. https://doi.org/10.1016/j.jmmm.2018.11.127

[124] C.R. De Silva, S. Smith, I. Shim, J. Pyun, T. Gutu, J. Jiao, Z. Zheng, Lanthanide (III)-doped magnetite nanoparticles, J. Am. Chem. Soc. 131 (2009) 6336-6337. https://doi.org/10.1021/ja9014277

[125] S. Nag, A. Roychowdhury, D. Das, S. Das, S. Mukherjee, Structural and magnetic properties of erbium (Er^{3+}) doped nickel zinc ferrite prepared by sol-gel auto-combustion method, J. Magn. Magn. Mater. 466 (2018) 172-179. https://doi.org/10.1016/j.jmmm.2018.06.084

[126] M. Hashim, M. Raghasudha, S.S. Meena, J. Shah, S.E. Shirsath, S. Kumar, D. Ravinder, P. Bhatt, R. Kumar, R. Kotnala, Influence of rare earth ion doping (Ce and Dy) on electrical and magnetic properties of cobalt ferrites, J. Magn. Magn. Mater. 449 (2018) 319-327. https://doi.org/10.1016/j.jmmm.2017.10.023

[127] T.S. Atabaev, N.H. Hong, Enhanced optical properties of ZrO_2: Eu^{3+} powders codoped with gadolinium ions, J. Sol-Gel Sci. Technol. 82 (2017) 15-19. https://doi.org/10.1007/s10971-017-4347-6

[128] X. Zhou, Y. Zhou, L. Zhou, J. Wei, J. Wu, D. Yao, Effect of Gd and La doping on the structure, optical and magnetic properties of NiZnCo ferrites, Ceram. Int. 45 (2019) 6236-6242. https://doi.org/10.1016/j.ceramint.2018.12.102

[129] S. Deka, V. Saxena, A. Hasan, P. Chandra, L.M. Pandey, Synthesis, characterization and in vitro analysis of α-Fe_2O_3-$GdFeO_3$ biphasic materials as therapeutic agent for magnetic hyperthermia applications, Materials Science and Engineering: C 92 (2018) 932-941. https://doi.org/10.1016/j.msec.2018.07.042

[130] K. Arda, S. Akay, C. Erisken, Effect of gadolinium concentration on temperature change under magnetic field, PLoS One 14 (2019) e0214910. https://doi.org/10.1371/journal.pone.0214910

[131] I. Hilger, W.A. Kaiser, Iron oxide-based nanostructures for MRI and magnetic hyperthermia, Nanomedicine 7 (2012) 1443-1459. https://doi.org/10.2217/nnm.12.112

[132] A. Rękorajska, G. Cichowicz, M.K. Cyranski, M. Pękała, P. Krysinski, Synthesis and characterization of Gd^{3+}-and Tb^{3+}-doped iron oxide nanoparticles for possible endoradiotherapy and hyperthermia, J. Magn. Magn. Mater. 479 (2019) 50-58. https://doi.org/10.1016/j.jmmm.2019.01.102

[133] V. Daboin, S. Briceño, J. Suárez, L. Carrizales-Silva, O. Alcalá, P. Silva, G. Gonzalez, Magnetic SiO_2-$Mn_{1-x}Co_xFe_2O_4$ nanocomposites decorated with $Au@Fe_3O_4$ nanoparticles for hyperthermia, J. Magn. Magn. Mater. 479 (2019) 91-98. https://doi.org/10.1016/j.jmmm.2019.02.002

[134] M. Ognjanović, D.M. Stanković, Y. Ming, H. Zhang, B. Jančar, B. Dojčinović, Ž. Prijović, B. Antić, Bifunctional (Zn, Fe)$_3O_4$ nanoparticles: Tuning their efficiency for potential application in reagentless glucose biosensors and magnetic hyperthermia, J. Alloys Compd. 777 (2019) 454-462. https://doi.org/10.1016/j.jallcom.2018.10.369

[135] L. Li, N. Yi, X. Wang, X. Lin, T. Zeng, T. Qiu, Novel triazolium-based ionic liquids as effective catalysts for transesterification of palm oil to biodiesel, J. Mol. Liq. 249 (2018) 732-738. https://doi.org/10.1016/j.molliq.2017.11.097

[136] S. Shaw, A. Biswas, A. Gangwar, P. Maiti, C. Prajapat, S.S. Meena, N. Prasad, Synthesis of Exchange Coupled Nanoflowers for Efficient Magnetic Hyperthermia, J. Magn. Magn. Mater. 484 (2019) 437-444. https://doi.org/10.1016/j.jmmm.2019.04.056

[137] I. Andreu, E. Natividad, L. Solozábal, O. Roubeau, Same magnetic nanoparticles, different heating behavior: Influence of the arrangement and dispersive medium, J. Magn. Magn. Mater. 380 (2015) 341-346. https://doi.org/10.1016/j.jmmm.2014.10.114

[138] S. El-Dek, M.A. Ali, S.M. El-Zanaty, S.E. Ahmed, Comparative investigations on ferrite nanocomposites for magnetic hyperthermia applications, J. Magn. Magn. Mater. 458 (2018) 147-155. https://doi.org/10.1016/j.jmmm.2018.02.052

[139] M. Sadat, R. Patel, J. Sookoor, S.L. Bud'ko, R.C. Ewing, J. Zhang, H. Xu, Y. Wang, G.M. Pauletti, D.B. Mast, Effect of spatial confinement on magnetic hyperthermia via dipolar interactions in Fe_3O_4 nanoparticles for biomedical applications, Mater. Sci. Eng. C 42 (2014) 52-63. https://doi.org/10.1016/j.msec.2014.04.064

[140] P. Shete, R. Patil, N. Thorat, A. Prasad, R. Ningthoujam, S. Ghosh, S. Pawar, Magnetic chitosan nanocomposite for hyperthermia therapy application: preparation, characterization and in vitro experiments, Appl. Surf. Sci. 288 (2014) 149-157. https://doi.org/10.1016/j.apsusc.2013.09.169

[141] A. Gangwar, S. Varghese, S.S. Meena, C. Prajapat, N. Gupta, N. Prasad, Fe_3C nanoparticles for magnetic hyperthermia application, J. Magn. Magn. Mater. 481 (2019) 251-256. https://doi.org/10.1016/j.jmmm.2019.03.028

[142] S.E. Minaei, S. Khoei, S. Khoee, F. Vafashoar, V.P. Mahabadi, In vitro anti-cancer efficacy of multi-functionalized magnetite nanoparticles combining alternating magnetic hyperthermia in glioblastoma cancer cells, Mater. Sci. Eng. C 101 (2019) 575-587. https://doi.org/10.1016/j.msec.2019.04.007

[143] Y. Xia, J. Sun, L. Zhao, F. Zhang, X.-J. Liang, Y. Guo, M.D. Weir, M.A. Reynolds, N. Gu, H.H. Xu, Magnetic field and nano-scaffolds with stem cells to enhance bone regeneration, Biomaterials 183 (2018) 151-170. https://doi.org/10.1016/j.biomaterials.2018.08.040

[144] C. Li, J.P. Armstrong, I.J. Pence, W. Kit-Anan, J.L. Puetzer, S.C. Carreira, A.C. Moore, M.M. Stevens, Glycosylated superparamagnetic nanoparticle gradients for osteochondral tissue engineering, Biomaterials 176 (2018) 24-33. https://doi.org/10.1016/j.biomaterials.2018.05.029

[145] F.D. Cojocaru, V. Balan, M.I. Popa, A. Lobiuc, A. Antoniac, I.V. Antoniac, L. Verestiuc, Biopolymers–Calcium phosphates composites with inclusions of magnetic nanoparticles for bone tissue engineering, Int. J. Biol. Macromol. 125 (2019) 612-620. https://doi.org/10.1016/j.ijbiomac.2018.12.083

[146] Q. Wang, B. Chen, M. Cao, J. Sun, H. Wu, P. Zhao, J. Xing, Y. Yang, X. Zhang, M. Ji, Response of MAPK pathway to iron oxide nanoparticles in vitro treatment promotes osteogenic differentiation of hBMSCs, Biomaterials 86 (2016) 11-20. https://doi.org/10.1016/j.biomaterials.2016.02.004

[147] K. Tschulik, R.G. Compton, Nanoparticle impacts reveal magnetic field induced agglomeration and reduced dissolution rates, PCCP 16 (2014) 13909-13913. https://doi.org/10.1039/C4CP01618A

[148] A.l. Jedlovszky-Hajdú, F.B. Bombelli, M.P. Monopoli, E. Tombácz, K.A. Dawson, Surface coatings shape the protein corona of SPIONs with relevance to their application in vivo, Langmuir 28 (2012) 14983-14991. https://doi.org/10.1021/la302446h

[149] M. Rotherham, J.R. Henstock, O. Qutachi, A.J. El Haj, Remote regulation of magnetic particle targeted Wnt signaling for bone tissue engineering, Nanomed. Nanotechnol. Biol. Med. 14 (2018) 173-184. https://doi.org/10.1016/j.nano.2017.09.008

[150] T. Kim, D. Moore, M. Fussenegger, Genetically programmed superparamagnetic behavior of mammalian cells, J. Biotechnol. 162 (2012) 237-245. https://doi.org/10.1016/j.jbiotec.2012.09.019

[151] A. Treffry, Z. Zhao, M.A. Quail, J.R. Guest, P.M. Harrison, Dinuclear center of ferritin: studies of iron binding and oxidation show differences in the two iron sites, Biochemistry 36 (1997) 432-441. https://doi.org/10.1021/bi9618301

[152] M. Uchida, M.L. Flenniken, M. Allen, D.A. Willits, B.E. Crowley, S. Brumfield, A.F. Willis, L. Jackiw, M. Jutila, M.J. Young, Targeting of cancer cells with ferrimagnetic ferritin cage nanoparticles, J. Am. Chem. Soc. 128 (2006) 16626-16633. https://doi.org/10.1021/ja0655690

[153] A. Ito, R. Teranishi, K. Kamei, M. Yamaguchi, A. Ono, S. Masumoto, Y. Sonoda, M. Horie, Y. Kawabe, M. Kamihira, Magnetically triggered transgene expression in mammalian cells by localized cellular heating of magnetic nanoparticles, J. Biosci. Bioeng. 128(3) (2019) 355-364. https://doi.org/10.1016/j.jbiosc.2019.03.008

[154] F. Xiong, H. Wang, Y. Feng, Y. Li, X. Hua, X. Pang, S. Zhang, L. Song, Y. Zhang, N. Gu, Cardioprotective activity of iron oxide nanoparticles, Sci. Rep. 5 (2015) 8579. https://doi.org/10.1038/srep08579

[155] S.L. Arias, A. Shetty, J. Devorkin, J.-P. Allain, Magnetic targeting of smooth muscle cells in vitro using a magnetic bacterial cellulose to improve cell retention in tissue-engineering vascular grafts, Acta Biomater. 77 (2018) 172-181. https://doi.org/10.1016/j.actbio.2018.07.013

[156] J.E. Bae, M.I. Huh, B.-K. Ryu, J.-Y. Do, S.U. Jin, M.J. Moon, J.C. Jung, Y. Chang, E. Kim, S.G. Chi, The effect of static magnetic fields on the aggregation and cytotoxicity of magnetic nanoparticles, Biomaterials 32 (2011) 9401-9414. https://doi.org/10.1016/j.biomaterials.2011.08.075

[157] G. Katti, S.A. Ara, A. Shireen, Magnetic resonance imaging (MRI)–A review, Int. J. Dental clinics 3 (2011) 65-70.

[158] U.A. van der Heide, M. Frantzen-Steneker, E. Astreinidou, M.E. Nowee, P.J. van Houdt, MRI basics for radiation oncologists, Clinical and Translational Radiation Oncology 18 (2019) 74-79. https://doi.org/10.1016/j.ctro.2019.04.008

[159] B.R. Vahid, M. Haghighi, Biodiesel production from sunflower oil over $MgO/MgAl_2O_4$ nanocatalyst: Effect of fuel type on catalyst nanostructure and performance, Energy Convers. Manage. 134 (2017) 290-300. https://doi.org/10.1016/j.enconman.2016.12.048

[160] M. Bautista, O. Bomati-Miguel, X. Zhao, M. Morales, T. Gonzalez-Carreno, R.P. de Alejo, J. Ruiz-Cabello, S. Veintemillas-Verdaguer, Comparative study of ferrofluids based on dextran-coated iron oxide and metal nanoparticles for contrast agents in magnetic resonance imaging, Nanotechnology 15 (2004) S154. https://doi.org/10.1088/0957-4484/15/4/008

[161] D. Zhu, F. Liu, L. Ma, D. Liu, Z. Wang, Nanoparticle-based systems for T_1-weighted magnetic resonance imaging contrast agents, International Journal of Molecular Sciences 14 (2013) 10591-10607. https://doi.org/10.3390/ijms140510591

[162] E. Chabanova, V. Logager, J.M. Moller, H. Dekker, J. Barentsz, H.S. Thomsen, Imaging liver metastases with a new oral manganese-based contrast agent, Acad. Radiol. 13 (2006) 827-832. https://doi.org/10.1016/j.acra.2006.03.013

[163] M.L. Belyanin, E.V. Stepanova, R.R. Valiev, V.D. Filimonov, V.Y. Usov, O.Y. Borodin, H. Ågren, Design, synthesis and evaluation of a new Mn–Contrast agent for MR imaging of myocardium based on the DTPA-phenylpentadecanoic acid complex, Chem. Phys. Lett. 665 (2016) 111-116. https://doi.org/10.1016/j.cplett.2016.10.058

[164] S.M. Kim, G.H. Im, D.-G. Lee, J.H. Lee, W.J. Lee, I.S. Lee, Mn2+-doped silica nanoparticles for hepatocyte-targeted detection of liver cancer in T_1-weighted MRI, Biomaterials 34 (2013) 8941-8948. https://doi.org/10.1016/j.biomaterials.2013.08.009

[165] J.R. Young, I. Orosz, M. Franke, H. Kim, D. Woodworth, B. Ellingson, N. Salamon, W. Pope, Gadolinium deposition in the paediatric brain: T_1-weighted hyperintensity within the dentate nucleus following repeated gadolinium-based contrast agent administration, Clin. Radiol. 73 (2018) 290-295. https://doi.org/10.1016/j.crad.2017.11.005

[166] D.R. Roberts, K.R. Holden, Progressive increase of T_1 signal intensity in the dentate nucleus and globus pallidus on unenhanced T_1-weighted MR images in the pediatric brain exposed to multiple doses of gadolinium contrast, Brain Dev. 38 (2016) 331-336. https://doi.org/10.1016/j.braindev.2015.08.009

[167] E. Kanal, Gadolinium based contrast agents (GBCA): Safety overview after 3 decades of clinical experience, Magn. Reson. Imaging 34 (2016) 1341-1345. https://doi.org/10.1016/j.mri.2016.08.017

[168] G. Grechnev, A. Logosha, A. Panfilov, I. Zhuravleva, Effects of pressure on magnetic properties of gadolinium, Physica B: Condensed Matter 407 (2012) 4143-4147. https://doi.org/10.1016/j.physb.2012.06.038

[169] B. Want, F. Ahmad, P. Kotru, Magnetic moment measurements of gadolinium, holmium and ytterbium tartrate trihydrate crystals, J. Alloys Compd. 448 (2008) L5-L6. https://doi.org/10.1016/j.jallcom.2007.01.003

[170] T. Kanda, Y. Nakai, H. Oba, K. Toyoda, K. Kitajima, S. Furui, Gadolinium deposition in the brain, Magn. Reson. Imaging 34 (2016) 1346-1350. https://doi.org/10.1016/j.mri.2016.08.024

[171] J. Ramalho, M. Ramalho, M. Jay, L.M. Burke, R.C. Semelka, Gadolinium toxicity and treatment, Magn. Reson. Imaging 34 (2016) 1394-1398. https://doi.org/10.1016/j.mri.2016.09.005

[172] N. Murata, K. Murata, L.F. Gonzalez-Cuyar, K.R. Maravilla, Gadolinium tissue deposition in brain and bone, Magn. Reson. Imaging 34 (2016) 1359-1365. https://doi.org/10.1016/j.mri.2016.08.025

[173] M. Rogosnitzky, S. Branch, Gadolinium-based contrast agent toxicity: a review of known and proposed mechanisms, BioMetals 29 (2016) 365-376. https://doi.org/10.1007/s10534-016-9931-7

[174] T. Grobner, F. Prischl, Gadolinium and nephrogenic systemic fibrosis, Kidney Int. 72 (2007) 260-264. https://doi.org/10.1007/s10534-016-9931-7

[175] L.M. Canzoniero, M. Taglialatela, G. Di Renzo, L. Annunziato, Gadolinium and neomycin block voltage-sensitive Ca^{2+} channels without interfering with the Na+ □ Ca2+ antiporter in brain nerve endings, European Journal of Pharmacology: Molecular Pharmacology 245 (1993) 97-103. https://doi.org/10.1016/0922-4106(93)90116-Q

[176] A. Lacampagne, F. Gannier, J. Argibay, D. Garnier, J.-Y. Le Guennec, The stretch-activated ion channel blocker gadolinium also blocks L-type calcium channels in isolated ventricular myocytes of the guinea-pig, Biochimica et Biophysica Acta (BBA)-Biomembranes 1191 (1994) 205-208. https://doi.org/10.1016/0005-2736(94)90250-X

[177] T. Kimitsuki, T. Nakagawa, K. Hisashi, S. Komune, S. Komiyama, Gadolinium blocks mechano-electric transducer current in chick cochlear hair cells, Hearing Res. 101 (1996) 75-80. https://doi.org/10.1016/S0378-5955(96)00134-7

[178] L.M. Boland, T.A. Brown, R. Dingledine, Gadolinium block of calcium channels: influence of bicarbonate, Brain Res. 563 (1991) 142-150. https://doi.org/10.1016/0006-8993(91)91527-8

[179] F. Knoepp, J. Bettmer, M. Fronius, Gadolinium released by the linear gadolinium-based contrast-agent Gd-DTPA decreases the activity of human epithelial Na+ channels (ENaCs), Biochimica et Biophysica Acta (BBA)-Biomembranes 1859 (2017) 1040-1048. https://doi.org/10.1016/j.bbamem.2017.02.019

[180] J. Zhao, Z.Q. Zhou, J.C. Jin, L. Yuan, H. He, F.L. Jiang, X.G. Yang, J. Dai, Y. Liu, Mitochondrial dysfunction induced by different concentrations of gadolinium ion, Chemosphere 100 (2014) 194-199. https://doi.org/10.1016/j.chemosphere.2013.11.031

[181] R. Weissleder, M. Nahrendorf, M.J. Pittet, Imaging macrophages with nanoparticles, Nature materials 13 (2014) 125-138. https://doi.org/10.1038/nmat3780

[182] R. Jin, B. Lin, D. Li, H. Ai, Superparamagnetic iron oxide nanoparticles for MR imaging and therapy: design considerations and clinical applications, Curr. Opin. Pharmacol. 18 (2014) 18-27. https://doi.org/10.1016/j.coph.2014.08.002

[183] B. Bahrami, M. Hojjat-Farsangi, H. Mohammadi, E. Anvari, G. Ghalamfarsa, M. Yousefi, F. Jadidi-Niaragh, Nanoparticles and targeted drug delivery in cancer therapy, Immunol. Lett. 190 (2017) 64-83. https://doi.org/10.1016/j.imlet.2017.07.015

[184] D. Hao, T. Ai, F. Goerner, X. Hu, V.M. Runge, M. Tweedle, MRI contrast agents: basic chemistry and safety, J. Magn. Reson. Imaging 36 (2012) 1060-1071. https://doi.org/10.1002/jmri.23725

[185] Z. Zhou, M. Qutaish, Z. Han, R.M. Schur, Y. Liu, D.L. Wilson, Z.-R. Lu, MRI detection of breast cancer micrometastases with a fibronectin-targeting contrast agent, Nature communications 6 (2015) 7984. https://doi.org/10.1038/ncomms8984

[186] C.Y. Yeh, J.K. Hsiao, Y.P. Wang, C.H. Lan, H.C. Wu, Peptide-conjugated nanoparticles for targeted imaging and therapy of prostate cancer, Biomaterials 99 (2016) 1-15. https://doi.org/10.1016/j.biomaterials.2016.05.015

[187] D.A. Richards, A. Maruani, V. Chudasama, Antibody fragments as nanoparticle targeting ligands: a step in the right direction, Chemical science 8 (2017) 63-77. https://doi.org/10.1039/C6SC02403C

[188] R.M. Patil, N.D. Thorat, P.B. Shete, P.A. Bedge, S. Gavde, M.G. Joshi, S.A. Tofail, R.A. Bohara, Comprehensive cytotoxicity studies of superparamagnetic iron oxide nanoparticles, Biochemistry and Biophysics Reports 13 (2018) 63-72. https://doi.org/10.1016/j.bbrep.2017.12.002

[189] C. Briguori, D. Tavano, A. Colombo, Contrast agent-associated nephrotoxicity, Prog. Cardiovasc. Dis. 45 (2003) 493-503. https://doi.org/10.1053/pcad.2003.YPCAD16

[190] J. Ramalho, R. Semelka, M. Ramalho, R. Nunes, M. AlObaidy, M. Castillo, Gadolinium-based contrast agent accumulation and toxicity: An update, Am. J. Neuroradiol. 37 (2016) 1192-1198. https://doi.org/10.3174/ajnr.A4615

[191] D.-E. Lee, H. Koo, I.-C. Sun, J.H. Ryu, K. Kim, I.C. Kwon, Multifunctional nanoparticles for multimodal imaging and theragnosis, Chem. Soc. Rev. 41 (2012) 2656-2672. https://doi.org/10.1039/C2CS15261D

[192] C. Janko, T. Ratschker, K. Nguyen, L. Zschiesche, R. Tietze, S. Lyer, C. Alexiou, Functionalized superparamagnetic iron oxide nanoparticles (SPIONs) as platform for the targeted multimodal tumor therapy, Front. Oncol. 9 (2019). https://doi.org/10.3389/fonc.2019.00059

[193] Y.-X.J. Wang, Current status of superparamagnetic iron oxide contrast agents for liver magnetic resonance imaging, World J. Gastroenterol. 21 (2015) 13400-13402. https://doi.org/10.3748/wjg.v21.i47.13400

[194] D.G. You, G. Saravanakumar, S. Son, H.S. Han, R. Heo, K. Kim, I.C. Kwon, J.Y. Lee, J.H. Park, Dextran sulfate-coated superparamagnetic iron oxide nanoparticles as a contrast agent for atherosclerosis imaging, Carbohydr. Polym. 101 (2014) 1225-1233. https://doi.org/10.1016/j.carbpol.2013.10.068

[195] H. Jung, B. Park, C. Lee, J. Cho, J. Suh, J. Park, Y. Kim, J. Kim, G. Cho, H. Cho, Dual MRI T_1 and $T_2^{(*)}$ contrast with size-controlled iron oxide nanoparticles, Nanomed. Nanotechnol. Biol. Med. 10 (2014) 1679-1689. https://doi.org/10.1016/j.nano.2014.05.003

[196] N. Xiao, W. Gu, H. Wang, Y. Deng, X. Shi, L. Ye, T_1–T_2 dual-modal MRI of brain gliomas using PEGylated Gd-doped iron oxide nanoparticles, J. Colloid Interface Sci. 417 (2014) 159-165. https://doi.org/10.1016/j.jcis.2013.11.020

[197] G.H. Im, S.M. Kim, D.G. Lee, W.J. Lee, J.H. Lee, I.S. Lee, Fe_3O_4/MnO hybrid nanocrystals as a dual contrast agent for both T_1-and T_2-weighted liver MRI, Biomaterials 34 (2013) 2069-2076. https://doi.org/10.1016/j.biomaterials.2012.11.054

[198] H. Yang, Y. Zhuang, Y. Sun, A. Dai, X. Shi, D. Wu, F. Li, H. Hu, S. Yang, Targeted dual-contrast T_1-and T_2-weighted magnetic resonance imaging of tumors using multifunctional gadolinium-labeled superparamagnetic iron oxide nanoparticles, Biomaterials 32 (2011) 4584-4593. https://doi.org/10.1016/j.biomaterials.2011.03.018

[199] A. Alipour, Z. Soran-Erdem, M. Utkur, V.K. Sharma, O. Algin, E.U. Saritas, H.V. Demir, A new class of cubic SPIONs as a dual-mode T_1 and T_2 contrast agent for MRI, Magn. Reson. Imaging 49 (2018) 16-24. https://doi.org/10.1016/j.mri.2017.09.013

[200] A. Alalaiwe, The clinical pharmacokinetics impact of medical nanometals on drug delivery system, Nanomed. Nanotechnol. Biol. Med. 17 (2019) 47-61. https://doi.org/10.1016/j.nano.2019.01.004

[201] L.Z. Benet, D. Kroetz, L. Sheiner, J. Hardman, L. Limbird, Pharmacokinetics: the dynamics of drug absorption, distribution, metabolism, and elimination, Goodman and Gilman's the pharmacological basis of therapeutics (1996) 3-27.

[202] S. Chillistone, J.G. Hardman, Modes of drug elimination and bioactive metabolites, Anaesthesia & Intensive Care Medicine 18 (2017) 458-461. https://doi.org/10.1016/j.mpaic.2017.06.005

[203] B.J. Pleuvry, Modes of drug elimination, Anaesthesia & Intensive Care Medicine 6 (2005) 277-279. https://doi.org/10.1383/anes.2005.6.8.277

[204] C.-Y. Wu, Y.-C. Chen, Riboflavin immobilized Fe_3O_4 magnetic nanoparticles carried with n-butylidenephthalide as targeting-based anticancer agents, Artifi. cells Nanomedicine Biotechnol. 47 (2019) 210-220. https://doi.org/10.1080/21691401.2018.1548473

[205] J.C. Chen, L.M. Li, J.Q. Gao, Biomaterials for local drug delivery in central nervous system, Int. J. Pharm. 560 (2019) 92-100. https://doi.org/10.1016/j.ijpharm.2019.01.071

[206] J.Y. Yoon, K.J. Yang, K.Y. Lee, S.N. Park, D.K. Kim, J.D. Kim, Intratympanic delivery of oligoarginine-conjugated nanoparticles as a gene (or drug) carrier to the inner ear, Biomaterials 73 (2015) 243-253. https://doi.org/10.1016/j.biomaterials.2015.09.025

[207] D. Dheer, J. Nicolas, R. Shankar, Cathepsin-sensitive nanoscale drug delivery systems for cancer therapy and other diseases, Adv. Drug Del. Rev. 151-152 (2019) 130-151. https://doi.org/10.1016/j.addr.2019.01.010

[208] Z. Karami, S. Sadighian, K. Rostamizadeh, S.H. Hosseini, S. Rezaee, M. Hamidi, Magnetic brain targeting of naproxen-loaded polymeric micelles: pharmacokinetics and biodistribution study, Mater. Sci. Eng. C 100 (2019) 771-780. https://doi.org/10.1016/j.msec.2019.03.004

[209] N. Mauro, C. Scialabba, R. Puleio, P. Varvarà, M. Licciardi, G. Cavallaro, G. Giammona, SPIONs embedded in polyamino acid nanogels to synergistically treat tumor microenvironment and breast cancer cells, Int. J. Pharm. 555 (2019) 207-219. https://doi.org/10.1016/j.ijpharm.2018.11.046

[210] K. Min, H. Jo, K. Song, M. Cho, Y.-S. Chun, S. Jon, W.J. Kim, C. Ban, Dual-aptamer-based delivery vehicle of doxorubicin to both PSMA (+) and PSMA (−)

prostate cancers, Biomaterials 32 (2011) 2124-2132.
https://doi.org/10.1016/j.biomaterials.2010.11.035

[211] M. Wang, W. Liu, Y. Zhang, M. Dang, Y. Zhang, J. Tao, K. Chen, X. Peng, Z. Teng, Intercellular adhesion molecule 1 antibody-mediated mesoporous drug delivery system for targeted treatment of triple-negative breast cancer, J. Colloid Interface Sci. 538 (2019) 630-637. https://doi.org/10.1016/j.jcis.2018.12.032

[212] Y. Krishnan, H.A. Rees, C.P. Rossitto, S.-E. Kim, H.-H.K. Hung, E.H. Frank, B.D. Olsen, D.R. Liu, P.T. Hammond, A.J. Grodzinsky, Green fluorescent proteins engineered for cartilage-targeted drug delivery: Insights for transport into highly charged avascular tissues, Biomaterials 183 (2018) 218-233.
https://doi.org/10.1016/j.biomaterials.2018.08.050

[213] M. Licciardi, C. Scialabba, R. Puleio, G. Cassata, L. Cicero, G. Cavallaro, G. Giammona, Smart copolymer coated SPIONs for colon cancer chemotherapy, Int. J. Pharm. 556 (2019) 57-67. https://doi.org/10.1016/j.ijpharm.2018.11.069

[214] S. Mohapatra, M. Asfer, M. Anwar, S. Ahmed, F.J. Ahmad, A.A. Siddiqui, Carboxymethyl Assam Bora rice starch coated SPIONs: Synthesis, characterization and in vitro localization in a micro capillary for simulating a targeted drug delivery system, Int. J. Biol. Macromol. 115 (2018) 920-932.
https://doi.org/10.1016/j.ijbiomac.2018.04.152

[215] T.N. Britos, C.E. Castro, B.M. Bertassoli, G. Petri, F.L. Fonseca, F.F. Ferreira, P.S. Haddad, In vivo evaluation of thiol-functionalized superparamagnetic iron oxide nanoparticles, Mater. Sci. Eng. C 99 (2019) 171-179.
https://doi.org/10.1016/j.msec.2019.01.118

[216] H. Jeon, J. Kim, Y.M. Lee, J. Kim, H.W. Choi, J. Lee, H. Park, Y. Kang, I.-S. Kim, B.-H. Lee, Poly-paclitaxel/cyclodextrin-SPION nano-assembly for magnetically guided drug delivery system, J. Controlled Release 231 (2016) 68-76.
https://doi.org/10.1016/j.jconrel.2016.01.006

Materials Research Forum LLC
https://doi.org/10.21741/9781644900611-10

Chapter 10

Magnetic Nanomaterials for Spintronics

Prasun Banerjee [1*], Adolfo Franco Jr[2], D. Baba Basha[3] and K. Chandra Babu Naidu[1]

[1]Department of Physics, Gandhi Institute of Technology and Management (GITAM) University, Bangalore-561203, India

[2]Instituto de Física, Universidade Federal de Goiás, Goiânia, Brazil

[3]College of Computer and Information Sciences, Majmaah University Al'Majmaah, K.S.A-11952

*prasun.banerjee@gitam.edu

Abstract

Magnetic nanomaterials are known to be prominent materials for development of next level of spintronic devices. The different magnetic nanomaterials can find application in spintronics devices especially some dilute magnetic semiconductors (DMSs) such as GaP: Mn, CdGeP$_2$: Mn, GaN: Mn, ZnO: Co, TiO$_2$: Co and CeO$_2$: Co are specially mentioned in this chapter. The 1D nanowire of GaP: Mn is having clear magnetic hysteresis loop at room temperature which indicates its ferromagnetic nature. Chalcopyrite CdGeP$_2$: Mn is another important DMS with a band gap value of 1.72 eV and Curie temperature situated at 293K. The strong exchange coupling between the spin polarization states in Mn-doped GaN DMS makes a very high value of Curie temperature at 940K. On the other hand due to the spin-split band and large Fermi level at higher magnetic fields ZnO: Co DMS shows both positive and negative magnetoresistance with high Curie temperature. TiO$_2$: Co NCs shows room temperature ferromagnetism properties due to the defect and domain structure. The room temperature magnetic hysteresis loop for the CeO$_2$: Co NPs shows ferromagnetic nature with giant magnetic properties up to 6.8 Bohr magnetron. Hence in this chapter, we specifically discussed the developments of the spintronics devices by using the DMS magnetic nanomaterials.

Keywords

Dilute Magnetic Semiconductor (DMS), Curie Temperature, Ferromagnetic, Magneto Resistance, Nanowire, Hysteresis Loop

Contents

1.   Introduction

Some of the biggest discoveries of 19th centuries such as a diode or transistors which apparently changed the overall electronic industries is just because of the control of the currents in these devices by mediating charge transport characteristic of the electrons [1]. But at that time it is also known to the scientific communities that electrons possess an angular momentum in addition to the charge transport properties. With the beginning of the new century and discovery of the new advanced technologies, scientists start exploring new possibilities by differentiating electrons into two different types by using their spins according to its projection axis i.e. ± ½ which start giving birth to new science known as spintronics [2- 7]. In spintronics devices spin has been used as logic bits to store information in nanomagnetic elements from one part of the functional devices to another part of it which increases efficiency, integration density as well as the speed of the devices [8]. The spintronics devices also having the advantage of easy manipulation of spins from up to down or vice versa just applying a magnetic field when the charge transport properties of the electrons which are having electrostatic screening effect with a applied static electric field for the devices like diodes and transistors [9]. There is also no need to increase or decrease the barrier potential in spintronics devices with respect to devices like diodes. Here the spin polarization state of electrons i.e. down or up state can be marked as logic bits of zero or one. The spin polarization can be switched 180° in the magnetic domains of the materials by applying magnetic field whereas for the switching in the charge information devices it is required to control the motion of the electrons around the potential barriers [10]. The correlation between the two polarization spin states is quite fundamental due to the scattering process which leads to the coupling between the two states. The spin current density (j) is directly proportional to the electrochemical potential gradient ($\delta\mu/\delta x$) in spintronics materials [11]. The spin injection has been achieved when the sample is connected with a magnetic electrode. For instance, in the Johnson-Silsbee spin injection system, the sample is non-magnetic in nature whereas the spin injector electrode is ferromagnetic material [12]. This difference in the properties between these two causes non-equilibrium situation at the interface when spin is injected. Hence the relaxation length (λ) and the relaxation time (τ) are the two most important factors in the spintronics devices because they decided the rate at which the spin accumulation at the interface can achieve the equilibrium state back again [13]. The spintronics can find interesting application in the field of magnetoresistance RAM [14], spin FET [15] and spin LED [16] etc.

It is a well-known fact that the spin of the electron generates magnetic moment hence it is the magnetic properties of the material which is linked to the spintronics. The obvious choice of magnetic material is metals because of their abundance. Due to the presence of the broken symmetry, the two spin polarization states in the metals are unequal in numbers [17]. If the resultant coupling between these two spin polarizations state by superexchange interactions results in negative value then they are known as antiferromagnetic while the positive coupling results in ferromagnetic nature. Hence spintronics subdivided into two fields here known as antiferromagnetic spintronics while the ferromagnetic materials result in sub fields known as magnetoelectronics. Prof Peter Gunberg and Prof Albert Fert won Nobel Prize in physics in the year 2007 for the discovery of GMR in the fields of magnetoelectronics [18]. The discovery of the carbon-based magnetic materials also led to the new subfields in spintronics known as graphene spintronics with single electron physics [19]. But the most relevant question in the field of spintronics is whether it is possible to control both the spin and charge of electrons within the material? The obvious answer is lying with the semiconductor materials when mixed with metal impurities [20]. This enables to control degrees of freedom in the material as the semiconductor material contains the highest amount of relaxation length and relaxation times with respect to the metals. But in the search of magnetic semiconductor materials, it has been observed that a great no of semiconductor materials become non-magnetic even after the addition of metal impurities within the limit of phase change [21]. Some of them although magnetic in nature but the Curie temperature is below the room temperature which led them unsuitable for the spintronics applications. But the extensive efforts of the scientist for the discovery of magnetic semiconductor spintronics materials with having high Curie temperature finally successful with the discovery of a wide variety of materials such as GaP:Mn [22], CdGeP:Mn [23], GaN:Mn [24], ZnO:Co [25], TiO_2: Co [26] and CeO_2:Co [27] with improved spintronics capabilities.

2. Different structural magnetic nanomaterials for spintronics applications

2.1. GaP:Mn nanowires

GaP:Mn is a well-known dilute magnetic semiconductor (DMS) for the spintronics application. The other known DMS in these series with 5% Mn-doped GaAs semiconductor is having Curie temperature well below the room temperature [28]. But only 1% doping of Mn in GaP DMS can have Curie temperature of the order of 300K [29]. Fig 1 shows the 1D nanowire structure of GaP:Mn nanowires prepared with thermal evaporation technique [29]. Fig. 2 a) show that the 1D nanowire contains a feasible amount of remanence, coercive field and saturation magnetization in it which indicate ferromagnetic nature. The Fig. 2(b) of ZFC and FC curve clearly establish that the ΔM value becomes non zero up to 300K which clearly indicates its high Curie temperature. Fig 3 (a) shows the establishment of ohmic contacts between the 1D nanowire structures of GaP:Mn DMS. The exceptionally high value of the magnetoresistance can be seen in Fig. 3 (b) for GaP:Mn due to the efficient alignments of the spins along the 1D structure axis of the DMS. This property establishes the importance of the 1D nanowire structures of GaP:Mn DMS for the spintronics applications.

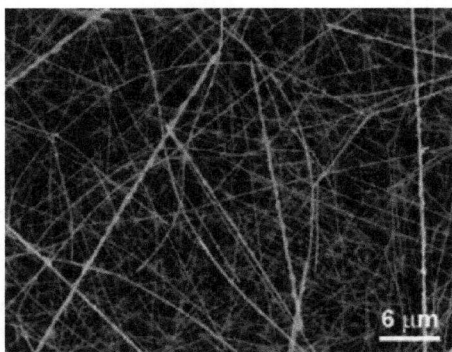

Figure 1. SEM image of 1D GaP:Mn nanowires. Reprint with the permission from Ref. [29]. Copyright 2005, American Chemical Society.

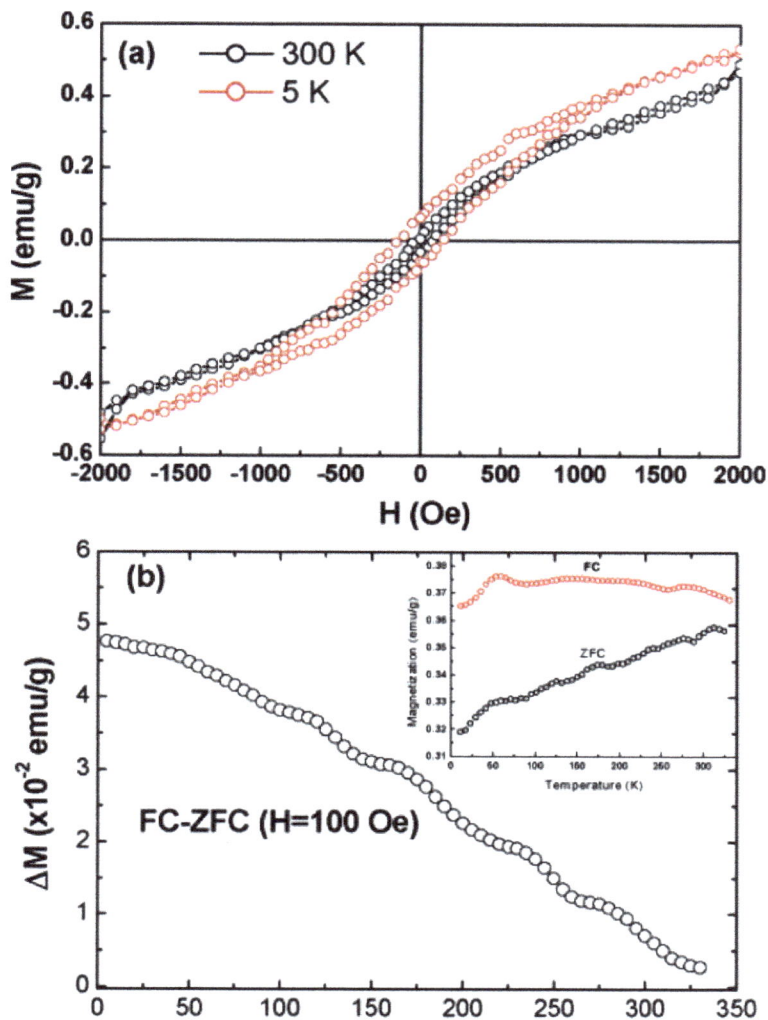

Figure 2. a) Hysteresis loop and b) ZFC and FC graph for 1D GaP:Mn nanowires.
Reprint with the permission from Ref. [29]. Copyright 2005, American Chemical Society.

Materials Research Forum LLC
https://doi.org/10.21741/9781644900611-10

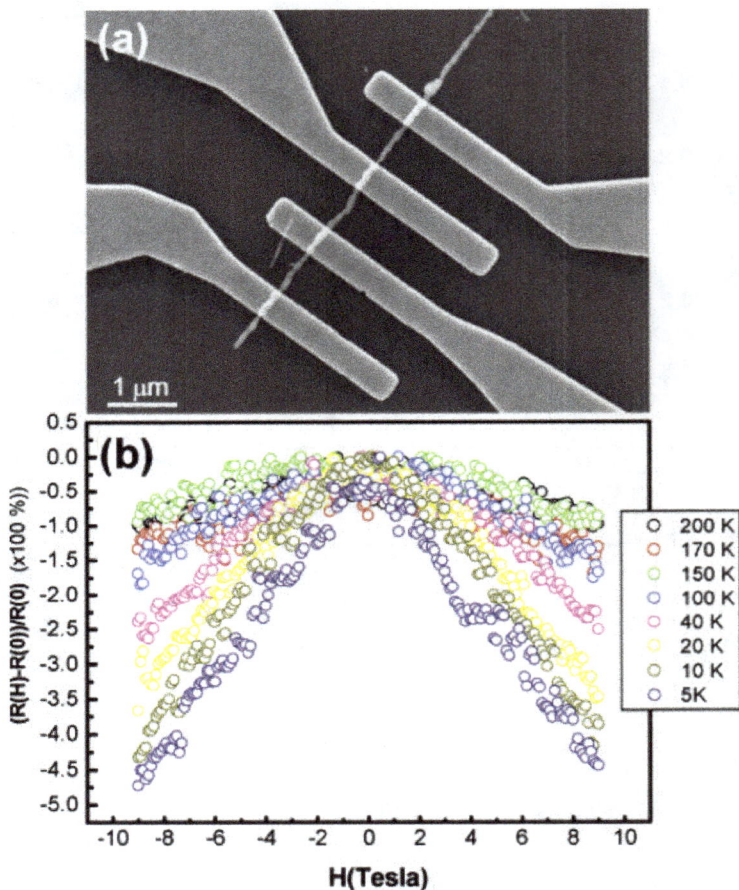

Figure 3. a) Ohmic contacts and b) Magnetoresistance measurements for 1D GaP:Mn nanowires. Reprint with the permission from Ref. [29]. Copyright 2005, American Chemical Society.

2.2. Chalcopyrite CdGeP$_2$:Mn

Chalcopyrite CdGeP2:Mn is another important DMS with II-VI-V$_2$ type with lattice parameters of c=10.775 Å and a=5.741 Å with the band gap value of 1.72 eV with high Curie temperature. The Mn atom can be doped inside the matrix of CdGeP$_2$ DMS using a solid-state reaction method without any phase change. The hysteresis loop of the chalcopyrite $CdGeP_2$:Mn nanopowder shows ferromagnetic nature up to 293K with clear remanence and coercivity shown in Fig. 4(a) [30]. The M-T graph shown in Fig. 4(b) shows that the chalcopyrite $CdGeP_2$:Mn nanopowder is having high Curie temperature which makes it unique for the spintronics applications.

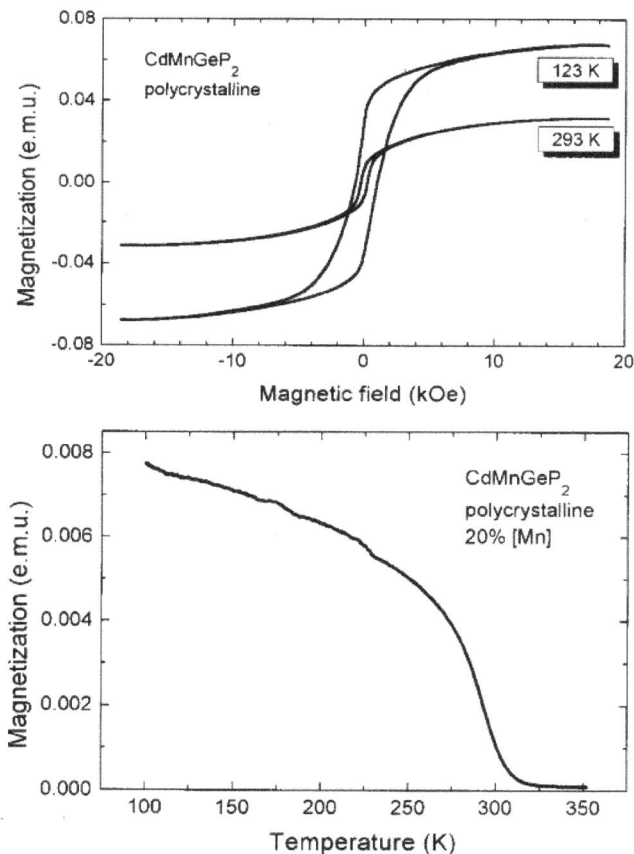

Figure 4. a) Hysteresis loop and b) M-T graph for chalcopyrite $CdGeP_2$:Mn nano powder. Reprint with the permission from Ref. [30]. Copyright 2001, AIP Publishing.

2.3.   Wurtzite GaN:Mn

Wurtzite Mn-doped GaN DMS is another very interesting group III-V based material which shows Curie temperature above room temperature. Fig. 5 shows the hysteresis loop with the ferromagnetic nature of MBE grown wurtzite Mn-doped GaN DMS [31]. The report suggested that within the noise limit the observed hysteresis loop for this material can go up to 400K temperature. The upper inset Fig. 5(a) shows the variation of magnetization up to 750K. The experimental graph fits well with the theoretical molecular field approximation graph which further shows that the magnetization becomes zero at 940K indicating very high Curie temperature. The lower inset in Fig. 5(b) shows the saturation magnetization at room temperature up to 7 T. The strong exchange coupling between the spin polarization states makes the high Courie temperature value possible for the spintronics application for the wurtzite Mn-doped GaN DMS.

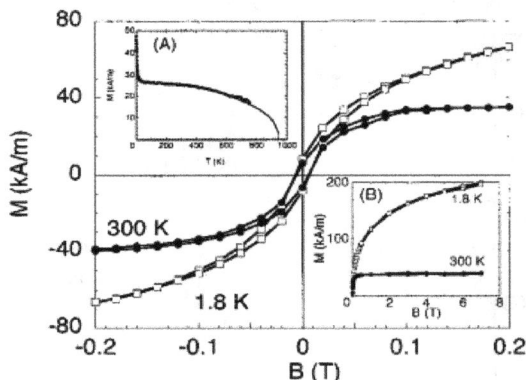

Figure 5. Hysteresis loop for wurtzite Mn doped GaN DMS. wurtzite Mn doped GaN DMS. wurtzite Mn doped GaN DMS. The inset a) shows the temperature dependence of magnetization and inset b) shows the M-B graph. Reprint with the permission from Ref. [31]. Copyright 2002, AIP Publishing.

Magnetochemistry - Materials and Applications   Materials Research Forum LLC

Materials Research Foundations **66** (2020) 323-341   https://doi.org/10.21741/9781644900611-10

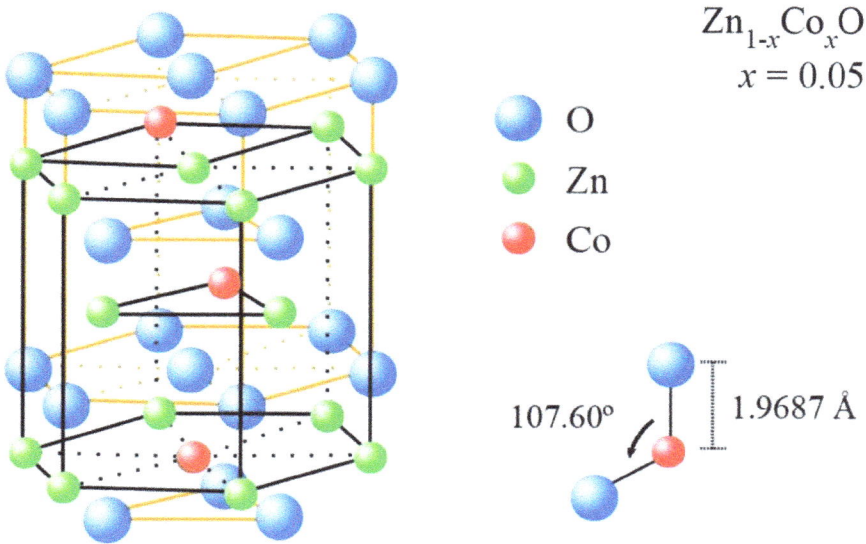

Figure. 6. 3-D representation of unit cell of ZnO:Co.

2.4. Cobalt-doped Zinc oxide nanowires

Another ideal material for spintronics application is zinc oxide DMS. Similar to GaN material the theoretical prediction for ZnO material also suggested a high Curie temperature. ZnO is having wurtzite structure and Fig. 6 shows the resultant structure when cobalt doped in the matrix of ZnO. XRD image is shown in Fig. 7 shows that the ZnO:Co is having with P63mc space group. In general, the pure ZnO contains s electrons in conduction band due to the doping with the Co it results in the spin-split band which results in positive magnetoresistance in the existence of the magnetic field at small temperature. But at relatively high-temperature magnetoresistance become negative due to large Fermi level at higher magnetic fields. Hence ZnO:Co shows both negative and positive magnetoresistance shown in Fig. 8. The SEM images of the ZnO:Co nanowires with silver contacts can be seen in Fig. 9. These unique properties help to use ZnO:Co nanowires to construct practical spintronics devices.

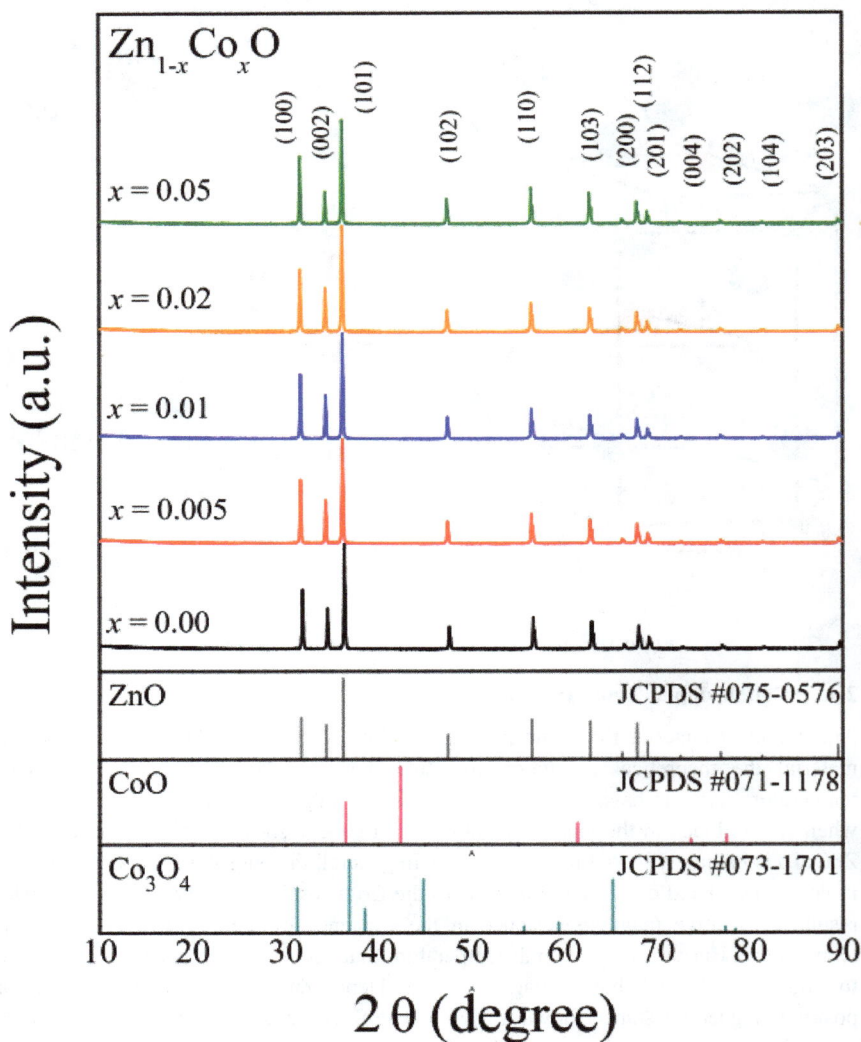

Figure. 7. XRD images for ZnO:Co material.

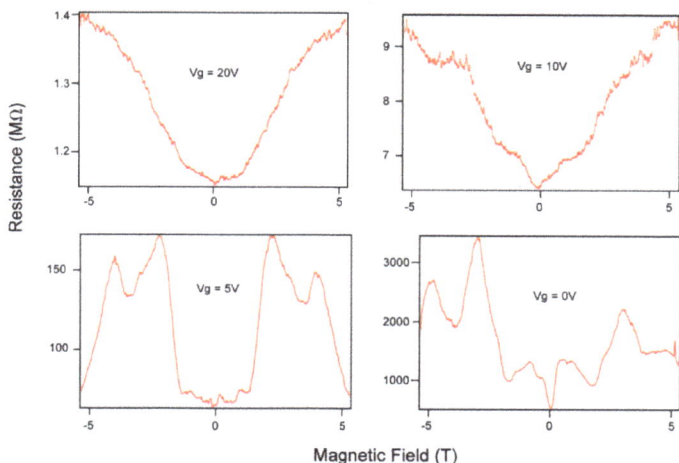

Figure 8. Magnetoresistance measurements results for ZnO:Co. Reprint with the permission from Ref. [32]. Copyright 2009, American Chemical Society.

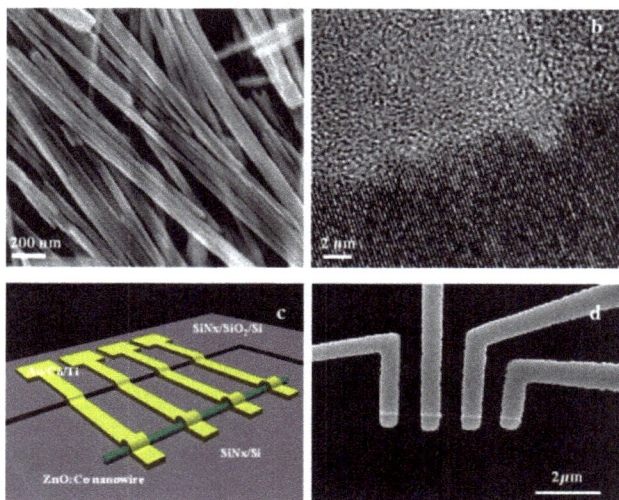

Figure 9. a) SEM, b) TEM, c) schematic and d) real contacts between ZnO:Co nanowires. Reprint with the permission from Ref. [32]. Copyright 2009, American Chemical Society.

2.5.   Cobalt-doped Titaniumoxide nanocrystals

TiO$_2$:Co is another high Curie temperature DMS material which increases speed and decreases power consumption in Spintronics devices. Fig. 10 demonstrates the TEM and SEM of spin-coated TiO$_2$:Co NCs [33]. The clear hysteresis loop is shown in Fig. 11(a) for TiO$_2$:Co NCs at 300K temperature proves the ferromagnetic properties for the DMS. Defects and domain structure due to Co doping are responsible for the room temperature ferromagnetism properties of TiO$_2$ NCs. To establish the intrinsic nature of this ferromagnetism property Fourier transform of the extended fine structure has been shown in Fig. 11. b).

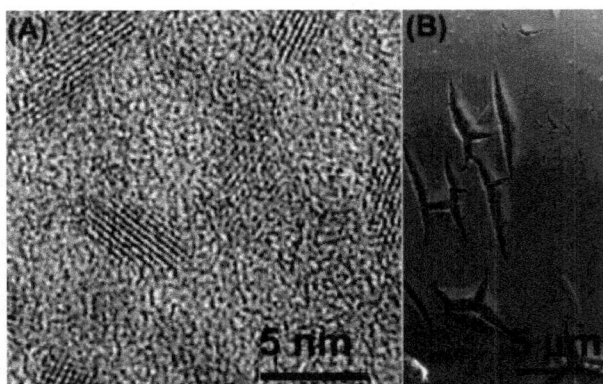

Figure 10. a) TEM and b) SEM images of spin coated TiO$_2$:Co NCs. Reprint with the permission from Ref. [33]. Copyright 2004, American Chemical Society.

2.6.   Cobalt-doped Cerium oxide nanoparticles

CeO$_2$ is a well-known DMS with cubic crystal symmetry which is having a high Curie temperature. Fig. 12 shows the thin film structure is phased pure for the CeO$_2$:Co NPs has grown with pulsed laser deposition techniques [34]. The room temperature magnetization curve for the CeO$_2$:Co NPs shows ferromagnetic nature with giant magnetic properties up to 6.8 Bohr magnetron, shown in Fig. 13. The magnetization vs temperature graph shown in Fig. 14. proves that the Curie temperature for the CeO$_2$:Co NPs is of the order of 725K. Hence CeO$_2$:Co NPs can be magnetically polarizable which is very important for the injection and detection at the interface of the spintronics devices.

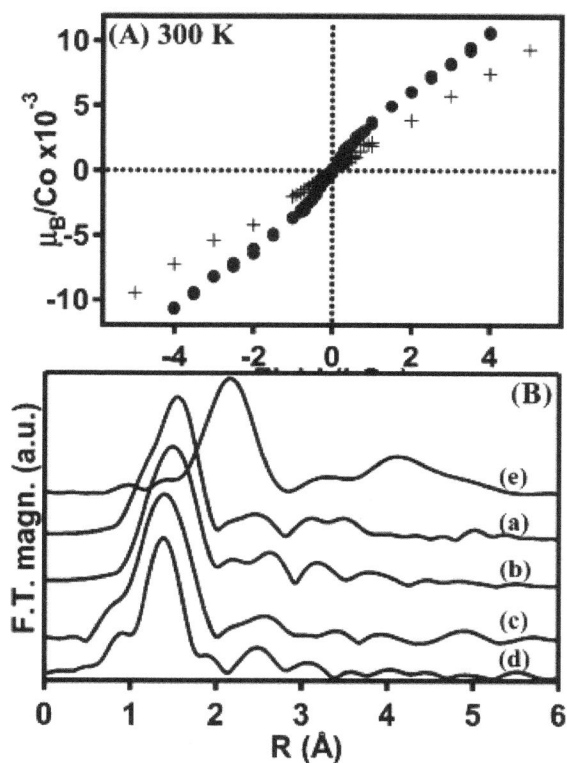

Figure 11. a) Magnetization data for TiO₂:Co NCs at room temperature. b) Magnitude of Fourier transform data EXAFS study. Reprint with the permission from Ref. [33]. Copyright 2004, American Chemical Society.

Materials Research Forum LLC

https://doi.org/10.21741/9781644900611-10

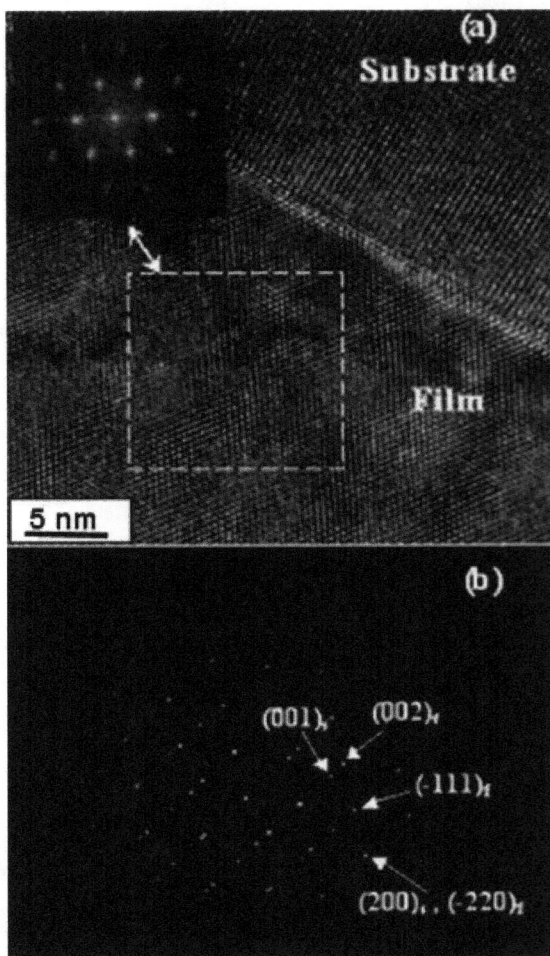

Figure 12. Thin film structure of CeO_2:Co NPs grown with pulsed laser deposition techniques. Reprint with the permission from Ref. [34]. Copyright 2006, AIP Publishing.

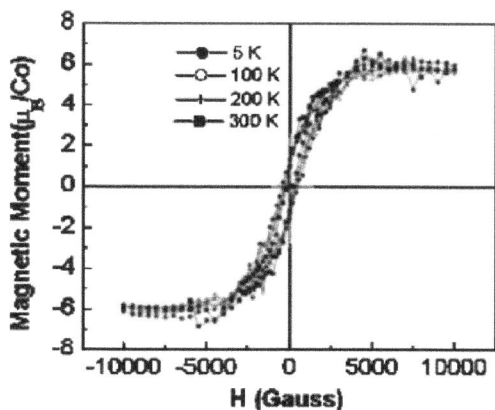

Figure 13. Magnetization curve for CeO₂:Co NPs upto 300K . Reprint with the permission from Ref. [34]. Copyright 2006, AIP Publishing.

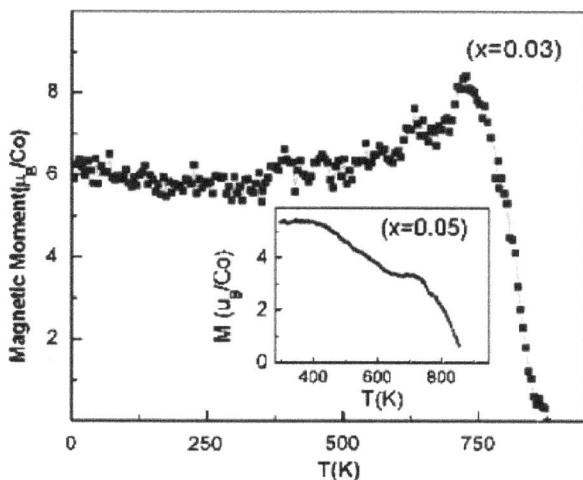

Figure 14. Magnetization vs temperature curve for CeO₂:Co NPs. Reprint with the permission from Ref. [34]. Copyright 2006, AIP Publishing.

Conclusions

In summary, magnetic nanomaterials are one of the prominent candidates for the developments of next generation spintronics devices. We specifically focus here on developments of spintronic devices using magnetic nanomaterials because they increase efficiency, integration density as well as the speed of the devices. The different magnetic nanomaterials can find application in spintronics devices such as GaP:Mn , $CdGeP_2$:Mn, GaN:Mn , ZnO:Co , TiO_2: Co and CeO_2:Co etc. due to their improved performances. GaP:Mn comes with Curie temperature of the order of 300K. The 1D the nanowire of GaP:Mn is having a feasible amount of remanence, coercive field and saturation magnetization at room temperature which indicates its ferromagnetic nature. Chalcopyrite $CdGeP_2$:Mn is another important DMS with II-VI-V_2 type with lattice parameters of c=10.775 Å and a=5.741 Å with the band gap value of 1.72 eV with Curie temperature at 293K. The strong exchange coupling between the spin polarization state makes the Courie temperature value possible at 940K for the wurtzite Mn-doped GaN DMS which makes it very ideal material for the spintronics application. Similar to GaN and GaP materials the theoretical prediction for ZnO material also suggested a high Curie temperature. Due to the spin-split band and large Fermi level at higher magnetic fields ZnO:Co shows both positive and negative magnetoresistance. These unique property helps to use ZnO:Co nanowires to construct practical spintronics devices. TiO_2:Co NCs shows room temperature ferromagnetism properties due to the defect and domain structure. The room temperature magnetic hysteresis loop for the CeO_2:Co NPs shows ferromagnetic nature with giant magnetic properties up to 6.8 Bohr magnetron. Hence the magnetic nanomaterials show higher stability with improved performance in the design of spintronics device applications.

Acknowledgments

The author would like to thank UGC, New Delhi for start-up grant with no. F.30-457/2018 (BSR). One of us (D. Baba Basha) was also thankful to Deanship of Scientific Research at Majmaah University, K.S.A. for supporting this work. We also acknowledge the support provided to A. Franco Jr. by CNPq, Brazil with grant No.307557/2015-4.

References

[1] CW. Deisch, Simple switching control method changes power converter into a current source, In 1978 IEEE Power Electronics Specialists Conference, (1978) pp. 300-306. https://doi.org/10.1109/PESC.1978.7072368

[2] J.F. Gregg, Spintronics: a growing science, Nat. Mater. 6 (2007) 798. https://doi.org/10.1038/nmat2049

[3] G.L. Rikken, A new twist on spintronics, Science 331 (2011) 864-865.
https://doi.org/10.1126/science.1201663

[4] J Sinova, Ž. Igor, New moves of the spintronics tango, Nat. Mater. 11 (2012) 368.
https://doi.org/10.1038/nmat3304

[5] S.A. Wolf, Y.C. Almadena, M.T. Daryl, Spintronics—A retrospective and
perspective, IBM J. Res. Development 50 (2006) 101-110.
https://doi.org/10.1147/rd.501.0101

[6] A. Fert, G.J. Marie, J. Henri, M. Richard, S. Pierre, The new era of spintronics,
Europhysics news 34 (2003) 227-229. https://doi.org/10.1051/epn:2003609

[7] S.D. Sarma, Spintronics: A new class of device based on electron spin, rather than on
charge, may yield the next generation of microelectronics, American Scientist 89
(2001) 516-523. https://doi.org/10.1511/2001.40.747

[8] S.A. Wolf, D.D. Awschalom, R.A. Buhrman, J.M. Daughton, S.V. Molnar, M.L.
Roukes, A.Y. Chtchelkanova, D.M. Treger, Spintronics: a spin-based electronics
vision for the future, Science 294 (2001) 1488-1495.
https://doi.org/10.1126/science.1065389

[9] A. McLaughlin, W.K. Eng, G. Vaio, T. Wilson, S.M. Laughlin. Dimethonium, a
divalent cation that exerts only a screening effect on the electrostatic potential
adjacent to negatively charged phospholipid bilayer membranes, J. Membrane
Biology 76 (1983) 183-193. https://doi.org/10.1007/BF02000618

[10]   N.A. Benedek, J.F Craig. Hybrid improper ferroelectricity: a mechanism for
controllable polarization-magnetization coupling, Phys. Rev. Lett.106 (2011) 107204.
https://doi.org/10.1103/PhysRevLett.106.107204

[11]   G. Schmidt, D. Ferrand, L.W. Molenkamp, A.T. Filip, B.J.V. Wees, Fundamental
obstacle for electrical spin injection from a ferromagnetic metal into a diffusive
semiconductor, Phys. Rev. B 62 (2000) R4790.
https://doi.org/10.1103/PhysRevB.62.R4790

[12]   M. Johnson, R.H. Silsbee, Spin-injection experiment, Phys. Rev. B 37 (1988)
5326. https://doi.org/10.1103/PhysRevB.37.5326

[13]   M. Tran, H. Jaffrès, C. Deranlot, J.M. George, A. Fert, A. Miard, A. Lemaître,
Enhancement of the spin accumulation at the interface between a spin-polarized
tunnel junction and a semiconductor, Phys. Rev. Lett. 102 (2009) 036601.
https://doi.org/10.1103/PhysRevLett.102.036601

[14]   G Jeong, W Cho, S Ahn, H Jeong, G Koh, Y Hwang, and K Kim. "A 0.24-/spl mu/m 2.0-V 1T1MTJ 16-kb nonvolatile magnetoresistance RAM with self-reference sensing scheme." IEEE Journal of solid-state circuits 38, no. 11 (2003): 1906-1910. https://doi.org/10.1109/JSSC.2003.818145

[15]   J Schliemann, J. C Egues, D Loss, Nonballistic spin-field-effect transistor, Physical review letters 90, no. 14 (2003): 146801. https://doi.org/10.1103/PhysRevLett.90.146801

[16]   R. Fiederling, M. Keim, G Reuscher, W. Ossau, G. Schmidt, A. Waag, L.W. Molenkamp. Injection and detection of a spin-polarized current in a light-emitting diode, Nature 402 (1999) 787. https://doi.org/10.1038/45502

[17]   L Noodleman, E.R. Davidson, Ligand spin polarization and antiferromagnetic coupling in transition metal dimers, Chem. Phys. 109 (1986) 131-143. https://doi.org/10.1016/0301-0104(86)80192-6

[18]   P A Grünberg, Nobel Lecture: From spin waves to giant magnetoresistance and beyond, Rev. Modern Phys. 80 (2008) 1531. https://doi.org/10.1103/RevModPhys.80.1531

[19]   W. Han, R.K. Kawakami, M. Gmitra, J. Fabian, Graphene spintronics, Nature Nanotechnol. 9 (2014) 794. https://doi.org/10.1038/nnano.2014.214

[20]   K Sato, H.K. Yoshida, Material design of GaN-based ferromagnetic diluted magnetic semiconductors, Japanese J. Appl. Phys. 40 (2001) L485. https://doi.org/10.1143/JJAP.40.L485

[21]   J. Alaria, M. Bouloudenine, G. Schmerber, S. Colis, A. Dinia, P. Turek, M. Bernard. Pure paramagnetic behavior in Mn-doped ZnO semiconductors, J. Appl. Phys. 99 (2006) 08M118. https://doi.org/10.1063/1.2172887

[22]   A. Wolos, M. Palczewska, M. Zajac, J. Gosk, M. Kaminska, A. Twardowski, M. Bockowski, I. Grzegory, S. Porowski, Optical and magnetic properties of Mn in bulk GaN, Phys. Rev. B 69 (2004) 115210. https://doi.org/10.1103/PhysRevB.69.115210

[23]   G.A. Medvedkin, T. Ishibashi, T. Nishi, K. Sato, A new magnetic semiconductor $Cd_{1-x}Mn_xGeP_2$, Semiconductors 35 (2001) 291-294. https://doi.org/10.1134/1.1356149

[24]   A. Wolos, A. Wysmolek, M. Kaminska, A. Twardowski, M. Bockowski, I. Grzegory, S. Porowski, M. Potemski, Neutral Mn acceptor in bulk GaN in high magnetic fields, Phys. Rev. B 70 (2004) 245202.

https://doi.org/10.1103/PhysRevB.70.245202

[25]   H.V.S. Pessoni, P. Banerjee, A. Franco, Colossal dielectric permittivity in Co-doped ZnO ceramics prepared by a pressure-less sintering method, Phys. Chem. Chem. Phys. 20 (2018) 28712-28719. https://doi.org/10.1039/C8CP04215B

[26]   R.P. Galhenage, H Yan, S.A. Tenney, N. Park, G. Henkelman, P. Albrecht, D.R. Mullins, D.A. Chen, Understanding the nucleation and growth of metals on TiO_2: Co compared to Au, Ni, and Pt, J. Phys. Chem. C 117 (2013) 7191-7201. https://doi.org/10.1021/jp401283k

[27]   L.R. Shah, B. Ali, H. Zhu, W.G. Wang, Y.Q. Song, H.W. Zhang, S.I. Shah, J.Q. Xiao, Detailed study on the role of oxygen vacancies in structural, magnetic and transport behavior of magnetic insulator: Co–CeO_2, J. Phys. Condensed Matter 21 (2009) 486004. https://doi.org/10.1088/0953-8984/21/48/486004

[28]   H. Ohno, F. Matsukura, A Ferromagnetic III–V Semiconductor:(Ga, Mn)As, Solid State Commun.117(3) (2001) 179-186. https://doi.org/10.1016/S0038-1098(00)00436-1

[29]   D.S. Han, S.Y. Bae, H.W. Seo, K. Chang, Synthesis and magnetic properties of manganese-doped GaP nanowires, J. Phys. Chem. B, 109 (2005) 9311-9316. https://doi.org/10.1021/jp050655s

[30]   K. Sato, G.A. Medvedkin, T. Nishi, Y. Hasegawa, R. Misawa, K. Hirose, T. Ishibashi, Ferromagnetic phenomenon revealed in the chalcopyrite semiconductor $CdGeP_2$:Mn, J. Appl. Phys. 89 (2001) 7027-7029. https://doi.org/10.1063/1.1357842

[31]   T. Sasaki, S. Sonoda, Y. Yamamoto, K.I. Suga, S. Shimizu, K. Kindo, H. Hori, Magnetic and transport characteristics on high Curie temperature ferromagnet of Mn-doped GaN, J. Appl. Phys. 91 (2002) 7911-7913. https://doi.org/10.1063/1.1451879

[32]   W. Liang, B.D. Yuhas, P. Yang, Magnetotransport in Co-doped ZnO nanowires, Nano Lett. 9 (2009) 892-896. https://doi.org/10.1021/nl8038184

[33]   J.D. Bryan, S.M. Heald, S.A. Chambers, D.R. Gamelin, Strong room-temperature ferromagnetism in Co^{2+}-doped TiO_2 made from colloidal nanocrystals, J. Am. Chem. Soc.126 (2004) 11640-11647. https://doi.org/10.1021/ja047381r

[34]   A. Tiwari, V.M. Bhosle, S. Ramachandran, N. Sudhakar, J. Narayan, S. Budak, A. Gupta, Ferromagnetism in Co doped CeO_2: Observation of a giant magnetic moment with a high Curie temperature, Appl. Phys. Lett. 88 (2006) 142511. https://doi.org/10.1063/1.2193431

Keyword Index

About the Editors

Dr. Inamuddin is currently working as Assistant Professor in the Chemistry Department, Faculty of Science, King Abdulaziz University, Jeddah, Saudi Arabia. He is a permanent faculty member (Assistant Professor) at the Department of Applied Chemistry, Aligarh Muslim University, Aligarh, India. He obtained Master of Science degree in Organic Chemistry from Chaudhary Charan Singh (CCS) University, Meerut, India, in 2002. He received his Master of Philosophy and Doctor of Philosophy degrees in Applied Chemistry from Aligarh Muslim University (AMU), India, in 2004 and 2007, respectively. He has extensive research experience in multidisciplinary fields of Analytical Chemistry, Materials Chemistry, and Electrochemistry and, more specifically, Renewable Energy and Environment. He has worked on different research projects as project fellow and senior research fellow funded by University Grants Commission (UGC), Government of India, and Council of Scientific and Industrial Research (CSIR), Government of India. He has received Fast Track Young Scientist Award from the Department of Science and Technology, India, to work in the area of bending actuators and artificial muscles. He has completed four major research projects sanctioned by University Grant Commission, Department of Science and Technology, Council of Scientific and Industrial Research, and Council of Science and Technology, India. He has published 147 research articles in international journals of repute and eighteen book chapters in knowledge-based book editions published by renowned international publishers. He has published 60 edited books with Springer (U.K.), Elsevier, Nova Science Publishers, Inc. (U.S.A.), CRC Press Taylor & Francis Asia Pacific, Trans Tech Publications Ltd. (Switzerland), IntechOpen Limited (U.K.), and Materials Research Forum LLC (U.S.A). He is a member of various journals' editorial boards. He is also serving as Associate Editor for journals (Environmental Chemistry Letter, Applied Water Science and Euro-Mediterranean Journal for Environmental Integration, Springer-Nature), Frontiers Section Editor (Current Analytical Chemistry, Bentham Science Publishers), Editorial Board Member (Scientific Reports-Nature), Editor (Eurasian Journal of Analytical Chemistry), and Review Editor (Frontiers in Chemistry, Frontiers, U.K.) He is also guest-editing various special thematic special issues to the journals of Elsevier, Bentham Science Publishers, and John Wiley & Sons, Inc. He has attended as well as chaired sessions in various international and national conferences. He has worked as a Postdoctoral Fellow, leading a research team at the Creative Research Initiative Center for Bio-Artificial Muscle, Hanyang University, South Korea, in the field of renewable energy, especially biofuel cells. He has also worked as a Postdoctoral Fellow at the Center of Research Excellence in Renewable Energy, King Fahd University of Petroleum and Minerals, Saudi Arabia, in the field of polymer electrolyte membrane fuel cells and computational fluid dynamics of polymer electrolyte membrane fuel cells. He is a life member of the Journal of the Indian Chemical Society. His research interest includes ion exchange materials, a sensor for heavy metal ions, biofuel cells, supercapacitors and bending actuators.

Dr. Rajender Boddula is currently working with Chinese Academy of Sciences-President's International Fellowship Initiative (CAS-PIFI) at National Center for Nanoscience and Technology (NCNST, Beijing). His academic honors include University Grants Commission National Fellowship and many merit scholarships, study-abroad fellowships from Australian Endeavour Research fellowship and CAS-PIFI. He has published many scientific articles in international peer-reviewed journals and has authored six book chapters, and also serving as editorial board member and referee for reputed international peer-reviewed journals. He has published edited books with Springer, United Kingdom, Elsevier, CRC Press Taylor & Francis Asia Pacific and Materials Research Forum LLC, U.S.A. His specialized areas of energy conversion and storage, which include nanomaterials, graphene, polymer composites, heterogeneous catalysis, photoelectrocatalytic water splitting, biofuel cell, and supercapacitors.

Prof. Abdullah M. Asiri is the Head of the Chemistry Department at King Abdulaziz University since October 2009 and he is the founder and the Director of the Center of Excellence for Advanced Materials Research (CEAMR) since 2010 till date. He is the Professor of Organic Photochemistry. He graduated from King Abdulaziz University (KAU) with B.Sc. in Chemistry in 1990 and a Ph.D. from University of Wales, College of Cardiff, U.K. in 1995. His research interest covers color chemistry, synthesis of novel photochromic and thermochromic systems, synthesis of novel coloring matters and dyeing of textiles, materials chemistry, nanochemistry and nanotechnology, polymers and plastics. Prof. Asiri is the principal supervisors of more than 20 M.Sc. and six Ph.D. theses. He is the main author of ten books of different chemistry disciplines. Prof. Asiri is the Editor-in-Chief of King Abdulaziz University Journal of Science. A major achievement of Prof. Asiri is the research of tribochromic compounds, a new class of compounds which change from slightly or colorless to deep colored when subjected to small pressure or when grind. This discovery was introduced to the scientific community as a new terminology published by International Union of Pure and Applied Chemistry (IUPAC) in 2000. This discovery was awarded a patent from European Patent office and from UK patent. Prof. Asiri involved in many committees at the KAU level and on the national level. He took a major role in the advanced materials committee working for King Abdulaziz City for Science and Technology (KACST) to identify the national plan for science and technology in 2007. Prof. Asiri played a major role in advancing the chemistry education and research in KAU. He has been awarded the best researchers from KAU for the past five years. He also awarded the Young Scientist Award from the Saudi Chemical Society in 2009 and also the first prize for the distinction in science from the Saudi Chemical Society in 2012. He also received a recognition certificate from the American Chemical Society (Gulf region Chapter) for the advancement of chemical science in the Kingdome. He received a Scopus certificate for the most publishing scientist in Saudi Arabia in chemistry in 2008. He is also a member of the editorial board of various journals of international repute. He is the Vice- President of Saudi Chemical Society (Western Province Branch). He holds four USA patents, more than one thousand publications in international journals, several book chapters and edited books.

www.ingramcontent.com/pod-product-compliance
Lightning Source LLC
Chambersburg PA
CBHW071321210326
41597CB00015B/1304